NanoScience and Technology

NanoScience and Technology

Series Editors:
P. Avouris   B. Bhushan   D. Bimberg   K. von Klitzing   H. Sakaki   R. Wiesendanger

The series NanoScience and Technology is focused on the fascinating nano-world, mesoscopic physics, analysis with atomic resolution, nano and quantum-effect devices, nanomechanics and atomic-scale processes. All the basic aspects and technology-oriented developments in this emerging discipline are covered by comprehensive and timely books. The series constitutes a survey of the relevant special topics, which are presented by leading experts in the field. These books will appeal to researchers, engineers, and advanced students.

Scanning Probe Microscopy
Analytical Methods
Editor: R. Wiesendanger

Biological Micro- and Nanotribology
Nature's Solutions
By M. Scherge and S.N. Gorb

Semiconductor Quantum Dots
Physics, Spectroscopy and Applications
Editors: Y. Masumoto and T. Takagahara

**Semiconductor Spintronics
and Quantum Computation**
Editors: D.D. Awschalom, N. Samarth,
D. Loss

Nano-Optoelectronics
Concepts, Physics and Devices
Editor: M. Grundmann

Noncontact Atomic Force Microscopy
Editors: S. Morita, R. Wiesendanger,
E. Meyer

Nanoelectrodynamics
Electrons and Electromagnetic Fields
in Nanometer-Scale Structures
Editor: H. Nejo

Single Organic Nanoparticles
Editors: H. Masuhara, H. Nakanishi,
K. Sasaki

Epitaxy of Nanostructures
By V.A. Shchukin, N.N. Ledentsov and D. Bimberg

Applied Scanning Probe Methods I
Editors: B. Bhushan, H. Fuchs, S. Hosaka

Nanostructures
Theory and Modeling
By C. Delerue and M. Lannoo

**Nanoscale Characterisation
of Ferroelectric Materials**
Scanning Probe Microscopy Approach
Editors: M. Alexe and A. Gruverman

**Magnetic Microscopy
of Nanostructures**
Editors: H. Hopster and H.P. Oepen

Silicon Quantum Integrated Circuits
Silicon-Germanium Heterostructure Devices: Basics and Realisations
By E. Kasper, D.J. Paul

The Physics of Nanotubes
Fundamentals of Theory, Optics
and Transport Devices
Editors: S.V. Rotkin and S. Subramoney

Single Molecule Chemistry and Physics
An Introduction
By C. Wang, C. Bai

C. Wang   C. Bai

Single Molecule Chemistry and Physics

An Introduction

With 146 Figures

 Springer

Professor Chunli Bai
Professor Chen Wang
National Center for Nanoscience and Technology
Beijing 100080, China
E-mail: clbai@iccas.ac.cn, wangch@nanoctr.cn

Series Editors:

Professor Dr. Phaedon Avouris
IBM Research Division
Nanometer Scale Science & Technology
Thomas J. Watson Research Center
P.O. Box 218
Yorktown Heights, NY 10598, USA

Professor Dr. Bharat Bhushan
Ohio State University
Nanotribology Laboratory
for Information Storage
and MEMS/NEMS (NLIM)
Suite 255, Ackerman Road 650
Columbus, Ohio 43210, USA

Professor Dr. Dieter Bimberg
TU Berlin, Fakutät Mathematik/
Naturwissenschaften
Institut für Festkörperphyisk
Hardenbergstr. 36
10623 Berlin, Germany

Professor Dr., Dres. h. c. Klaus von Klitzing
Max-Planck-Institut
für Festkörperforschung
Heisenbergstr. 1
70569 Stuttgart, Germany

Professor Hiroyuki Sakaki
University of Tokyo
Institute of Industrial Science
4-6-1 Komaba, Meguro-ku
Tokyo 153-8505, Japan

Professor Dr. Roland Wiesendanger
Institut für Angewandte Physik
Universität Hamburg
Jungiusstr. 11
20355 Hamburg, Germany

ISSN 1434-4904
ISBN-10 3-540-25369-6 Springer Berlin Heidelberg New York
ISBN-13 978-3-540-25369-3 Springer Berlin Heidelberg New York

Library of Congress Control Number: 2006927051

This work is subject to copyright. All rights are reserved, whether the whole or part of the material is concerned, specifically the rights of translation, reprinting, reuse of illustrations, recitation, broadcasting, reproduction on microfilm or in any other way, and storage in data banks. Duplication of this publication or parts thereof is permitted only under the provisions of the German Copyright Law of September 9, 1965, in its current version, and permission for use must always be obtained from Springer. Violations are liable to prosecution under the German Copyright Law.
Springer is a part of Springer Science+Business Media.
springer.com
© Springer-Verlag Berlin Heidelberg 2006
Printed in Germany
The use of general descriptive names, registered names, trademarks, etc. in this publication does not imply, even in the absence of a specific statement, that such names are exempt from the relevant protective laws and regulations and therefore free for general use.

Cover background image: "Single-wall nanotube METFET inverter: organic molecules wrap around a metallic tube channel and modify it to a semiconductor while the second tube is used for gating". Concept: S.V. Rotkin, Physics Department, Lehigh University, image: B. Grosser, Imaging Technology Group, Beckman Institute, UIUC.

Typesetting and prodcution: PTP-Berlin, Protago-TEX-Production GmbH, Berlin (www.ptp-berlin.com)
Cover design: *design& production*, Heidelberg

Printed on acid-free paper   57/3141/YU - 5 4 3 2 1 0

Preface

Single-molecule studies constitute a distinguishable category of focused research in nanoscience and nanotechnology. This book is dedicated to the introduction of recent advances on single-molecule studies. It will be illustrated that studying single molecules is both intellectually and technologically challenging, and also offers vast potential in opening up new scientific frontiers. We wish to present the readers with several different techniques for studying single molecules, such as electron-tunneling methods, interaction-force measurement techniques, optical spectroscopy, plus a number of directions where further progress could be pursued. We hope the work may assist the readers, especially graduate students and those who wish to explore single molecules, to become familiarized with the pace of the progress in this field and the relevant primary techniques.

Due to limitation of space, we are not able to elaborate on the technical details of all of the experimental methods that are vital in single molecule studies, so introductions to only selected experimental methods are touched in the context. Since the technical details and theoretical analysis of these techniques have already been thoroughly covered in many literatures, we only provide introductions to the basic principles of the detection techniques here, and focus on their experimental achievements in the area of single-molecule studies. These techniques have proven to be highly effective when independently used. The combinationof those techniques could lead to further advances in the detection capabilities. Additional readings on the theoretical analysis of techniques are crucial for understanding the advantages, as well as limitations, of these detection techniques, and therefore are highly recommended.

This work has been endeavored as the result of the encouragements from many colleagues working with the authors. We have benefited greatly through communications with our colleagues throughout the preparation of this manuscript. Dr. Fang Xiaohong, Dr. Shang Guangyi and Dr. Qiu Xiaohui kindly reviewed the manuscript and provided valuable suggestions. The understanding and generous support from the family members are also most gratefully acknowledged. We also wish to thank the Springer editor of this book, Dr. C. Ascheron, for his patience and support throughout the preparation of the manuscript. We are also indebted greatly to Dagmar Rossow for enormous effort putting this manuscript into the final book format.

Beijing
May 2006

Chen Wang
Chunli Bai

Contents

1	Introduction to Single Molecule Chemistry and Physics		1
2	Basics of Electron Tunneling Processes and Scanning Tunneling Microscopy		5
	2.1	Principles of Tunneling Processes	5
		2.1.1 Elastic Tunneling Process	5
		2.1.2 Inelastic Tunneling Process	8
		2.1.3 Two-Step Tunneling Process	14
		2.1.4 Resonant Tunneling Effect	14
	2.2	Introduction to Scanning Tunneling Microscopy (STM)	15
		2.2.1 Introduction to STM	15
		2.2.2 STM Contrast Mechanisms	20
		2.2.3 Scanning Tunneling Spectroscopy (STS)	22
		2.2.4 Measurement of Apparent Tunneling Barrier Height	24
3	Single Molecule Structural Characterization		29
	3.1	Molecular Imaging Mechanisms of STM	30
		3.1.1 Molecular Orbital Model	31
		3.1.2 HOMO-Ionization Potential Model	33
		3.1.3 Work Function Model	34
	3.2	Single Diatomic Molecules on Metal Surfaces	36
		3.2.1 CO	37
		3.2.2 O_2 Molecules	40
	3.3	Aromatic Molecules and Macrocyclic Molecules	42
		3.3.1 Single Benzene Molecules Observed by STM	42
		3.3.2 Phthalocyanines (Pc)	46
		3.3.3 Porphyrin	48
		3.3.4 Heterocyclic Molecules	51
		3.3.5 Fullerene	53
		3.3.6 Other Molecules	54
	3.4	Single Hydrocarbon Molecules	55
	3.5	Single Molecules Immobilized by Molecular Matrix	57
		3.5.1 Hydrogen-Bonded Networks and Single Molecule Inclusions	57

		3.5.2	Molecular Networks Stabilized	
			by van der Waals Interaction	59
	3.6	Single Molecule Adsorption on Organic Substrates		60
		3.6.1	Simple Alkane Lamella	61
		3.6.2	Alkylated Amino Acid Molecular Templates........	63
		3.6.3	Tridodecyl Amine (TDA) Templates	64
	3.7	Electron-Spin Resonance Study		
		of Single Molecules..................................		66

4 Single Molecule Diffusion and Chemical Reactions 69

	4.1	Molecular Diffusion on Surfaces..........................		69
		4.1.1	Thermal-Activated Single Molecule Diffusion	69
		4.1.2	Laser-Activated Single Molecule Diffusion	71
		4.1.3	Field-Induced Diffusion of Single Atoms	71
	4.2	Single Atom and Molecule Manipulations		74
		4.2.1	Controlled Manipulation of Single Xe Atoms	74
		4.2.2	Si Atoms	75
		4.2.3	Gold Atoms....................................	76
		4.2.4	CO Molecules	76
		4.2.5	C_{60} Molecules	77
	4.3	Single Molecule Chemical Reactions on Metal Surfaces......		78
		4.3.1	Single Molecule Oxidative Reaction	
			on Metal Surfaces..............................	79
		4.3.2	Dissociative Adsorption of H_2	82
		4.3.3	Dissociative Adsorption of NO...................	83
		4.3.4	Dissociation of NH_3	84
		4.3.5	CO Oxidation	85
		4.3.6	Dehydrogenation of Single Molecules	86
		4.3.7	Tip-Induced Reactions of Single Iodobenzene	
			on Cu(111).....................................	88
		4.3.8	Formation of Metal Ligand Complexes.............	89
		4.3.9	Other Reaction Model Systems	91
	4.4	Single Molecules on Semiconductor Surfaces		93
		4.4.1	Single H_2 Molecules on Si(100)	93
		4.4.2	Single NH_3 Molecules on Si Surfaces	94
		4.4.3	Single O_2 on Ge(111), Si(100) and Si(111)	95
		4.4.4	Other Molecules on Si Surfaces	96
	4.5	Single Molecule Reactions on Metal Oxide Surfaces		97
		4.5.1	TiO_2	98
		4.5.2	CO on $RuO_2(110)$	102
		4.5.3	Fe Oxide Surfaces...............................	103
		4.5.4	Other Oxide Surfaces............................	105

5 Molecular Scale Analysis Using Scanning Force Microscopy 107
- 5.1 Basic Principles of Atomic Force Microscopy (AFM) 107
  - 5.1.1 Introduction of Instrumentation 107
  - 5.1.2 Cantilever 108
  - 5.1.3 Cantilever Deflection Detection 109
  - 5.1.4 Cantilever Calibration 110
- 5.2 AFM Operating in Contact Mode 112
  - 5.2.1 Contact Mode 112
  - 5.2.2 Friction Force Microscopy 117
- 5.3 AFM Operating in Oscillatory Modes 118
  - 5.3.1 Tapping Mode 118
  - 5.3.2 Phase Imaging 120
  - 5.3.3 Operations Under Liquids 120
  - 5.3.4 Non-Contact Mode 122
- 5.4 Magnetic Force Microscopy (MFM) 123
  - 5.4.1 Basic Imaging Mechanism 123
  - 5.4.2 Examples of MFM Studies of Molecular Structures .. 125
  - 5.4.3 Imaging Single Molecule Magnets 126
- 5.5 Force Spectrum and Surface Mapping 127
  - 5.5.1 Force Spectrum and Imaging 127
  - 5.5.2 Chemical Force Microscopy 128

6 Intermolecular and Intramolecular Interactions 131
- 6.1 Techniques for Studying Intermolecular and Intramolecular Interactions 131
  - 6.1.1 Biomembrane Force Probe (BFP) 131
  - 6.1.2 Optical Tweezers 132
- 6.2 Static Force Measurements of Single Molecules 134
  - 6.2.1 Single Bond Interaction 134
  - 6.2.2 Single Pair Ligand–Receptor Interactions 139
  - 6.2.3 Guest–Host Interactions 141
  - 6.2.4 Desorption of Single Molecules at Interfaces 142
- 6.3 Intramolecular Interactions of Single Molecules 144
  - 6.3.1 Elasticity of DNA Molecules 144
  - 6.3.2 Folding and Refolding of Single Protein Molecules... 148
  - 6.3.3 Stretching Other Biomolecules 149
  - 6.3.4 Polysaccharides 152
  - 6.3.5 Other Polymers 153
- 6.4 Dynamic Force Measurements of Single Molecules 154
  - 6.4.1 Pulling Rate Effect on Force Spectrum Measurements 154
  - 6.4.2 Pulling Rate Effect on Rupture Force Measurements 155

7 Electrical Conductivity of Single Molecules ... 159
- 6.4.3 Force Measurements Relevant to Movements of Biomolecules ... 158
- 7.1 Introduction ... 159
  - 7.1.1 One-Dimensional Molecular Conductance Structures ... 159
  - 7.1.2 Methods for Measuring Molecular Conductivity ... 163
- 7.2 Electrical Conductivity of Molecular Monolayers ... 165
  - 7.2.1 Linear Alkane Derivatives ... 165
  - 7.2.2 Conjugated Molecules ... 166
  - 7.2.3 Rectification Molecular Conductance ... 167
  - 7.2.4 Switching Behavior of Molecular Conductance ... 169
- 7.3 Single Molecule Conductance ... 170
  - 7.3.1 Molecule–Electrode Contact Effect ... 170
  - 7.3.2 Conductance of Single Organic Molecules ... 174
  - 7.3.3 Conductance of Single Nanotubes and Nanowires ... 176
  - 7.3.4 DNA Molecules ... 177
  - 7.3.5 Single Molecule Devices ... 179

8 Single Molecule Fluorescence Imaging and Spectroscopy: Far-Field Studies ... 183
- 8.1 Introduction ... 183
  - 8.1.1 Fluorescence of Molecules ... 183
  - 8.1.2 General Considerations for Experimental Setup ... 185
  - 8.1.3 Criteria of Single Molecule Identification ... 187
- 8.2 Single Molecule Imaging in Far-Field Configuration ... 188
  - 8.2.1 Imaging by Confocal Fluorescence Microscopy ... 188
  - 8.2.2 Wide-Field Imaging: Epi-Illumination Microscopy ... 188
- 8.3 Low-Temperature Studies of Single Molecules in Solid Matrices ... 189
  - 8.3.1 Observation of Single Molecules in Crystalline Matrix ... 189
  - 8.3.2 Pump–Probe Effects ... 193
  - 8.3.3 Magnetic Resonance of Single Fluorescence Molecules ... 195
- 8.4 Single Fluorescence Molecules in Liquid Conditions ... 196
  - 8.4.1 Experimental Considerations ... 196
  - 8.4.2 Examples of Fluorescence of Single Molecules in Solutions ... 197
  - 8.4.3 Single Molecule Diffusions in Living Cells ... 200
  - 8.4.4 Single-Pair FRET ... 202
- 8.5 Single Molecules in Other Support Media ... 207
  - 8.5.1 Single Molecules in Polymer Hosts ... 207
  - 8.5.2 Lateral Diffusion Behavior of Single Molecules ... 210

		8.5.3	Fluorescence from Single Atomic Clusters and Defects 212
	8.6	Tip-Induced Single Molecule Fluorescence 212
	8.7	Dynamics of Single Polymeric Molecules Studied by Fluorescence Microscopy and Related Techniques 213
		8.7.1	Dynamics of Single Macromolecules in Solutions 213
		8.7.2	Single Molecules Moving Through Channels 215
		8.7.3	Migration of DNA Molecules on Flat Surfaces 217
		8.7.4	Single Molecule Condensation of DNA 219

9 Single Molecule Fluorescence Imaging and Spectroscopy: Near-Field Studies 223

- 9.1 Near-Field Scanning Optical Microscopy 223
  - 9.1.1 Introduction of Near-Field Effect 223
  - 9.1.2 NSOM Probe Designs 226
  - 9.1.3 Approaching Modes 229
- 9.2 Near-Field Scanning Optical Microscopy and Spectroscopy .. 230
  - 9.2.1 Near-Field Optical Microscopy 230
  - 9.2.2 Near-Field Optical Spectroscopy 232
  - 9.2.3 Fluorescence Resonance Energy Transfer (FRET) Studied by NSOM 235
- 9.3 Other Near-Field Optical Microscopy 237
  - 9.3.1 Near-Field Optical Chemical Sensors 237
  - 9.3.2 Scanning Exciton Microscopy 238

10 Surface-Enhanced Raman Scattering (SERS) of Single Molecules 241

- 10.1 Introduction of SERS Effect 241
- 10.2 SERS of Single Molecules 244
  - 10.2.1 Single Particle SERS Effect 244
  - 10.2.2 SERS of Nanoparticle Aggregates 245
- 10.3 Tip-Induced SERS 253
- 10.4 Near-Field SERS 254
- 10.5 Raman Spectroscopy of Carbon Nanotubes 256

References 259

Index 299

1 Introduction to Single Molecule Chemistry and Physics

We have learned a great deal about molecules from different aspects, thanks to pioneering investigations by generations of researchers. In particular, with recent major technological breakthroughs in manipulating individual atoms and molecules using scanning probe microscopy (SPM) and optical spectroscopy methods, the interest in examining individual atoms and molecules directly, based upon structural characteristics and other properties, have become unprecedentedly strong and also practical.

Characterization of individual atoms and molecules has continued to be a scientifically attractive and challenging task for a long time. The behavior of single molecules is closely related to their immediate environment. We will compare the uniqueness of single molecules bound to surfaces studied by scanning probe microscopy, and the behavior of single molecules in solutions, using fluorescence microscopy. The chemical specificity, structural geometry and other characteristics of surfaces and interfaces provide rich environments (mostly two-dimensional) for single molecules. Solutions, on the other hand, represent a different category of experimental conditions, which provide a homogeneous microscopic environment that can be adjusted by solvents.

With the aid of SPM, we have made direct observations of arrays of atoms and molecules on surfaces, and of individual adsorbed monodispersed species as well. The capability of precision control of probes positioned over the species provides us with solid means to investigate various aspects of molecules laying on surfaces. It could be considered that assessing the structural characteristics of surface-bound molecules is essentially straightforward. However, it needs to be cautioned that rigorous data collection still requires substantial experimental as well as theoretical efforts.

It has been shown that the presence of minute amounts of adsorbed atoms, less than one monolayer coverage, can appreciably alter the local density of states and barrier height. The information collected to date is far from conclusive in many cases, and very often significant background signals, either from the substrate or in the form of noise from the scanning probe, lead to ambiguous experimental data. Many classical theories and experimental techniques have been dedicated to looking for the optimal approach for individual molecules, with continued encouraging results in many respects.

The signature of atoms and molecules, in either the free-moving or surface-bound state, has been recorded by various spectroscopic measurements involving electronic, mechanical, magnetic, and optical properties. Traditional methods based on the tunneling technique have proven successful in certain domains in this field, and magnetic and optical approaches, in conjunction with the SPM technique, also show very promising potential. This information is crucial in identifying individual components needed for comparison with well-documented data. Tunneling spectroscopy is a seemingly very promising path in realizing our goal of single molecule chemical specificity recognition. Recent reports on the stretching mode of C_2H_2 by scanning tunneling spectroscopy (STS) certainly promote this work. The concept of nuclear magnetic resonance, combined with magnetic force microscopy (MFM), has injected new hope in this conquest.

Deeper investigations using the SPM method require not only a presumably atomically sharp probe, but furthermore a more chemically and physically specific probe. From the practical point of view, there are a large number of candidates to be explored for this purpose. As examples, C_{60} and carbon nanotubes have been shown to function as superb probes once attached to a tip apex, molecular crystals could substantially increase the throughput of the near-field scanning optical microscopy (NSOM) fiber probe, and sharp edges of crystallites could also be used as special probes. This exploratory work already suggests that these functional specific atomic (molecular) probes have great potential in leading to more powerful atomic microscopy.

The capability of nanometer-scale manipulation and fabrication further promotes chemical reactivity studies at the molecular level. It is clear that the recognizing capability of SPM is advancing beyond the structural regime to a regime of spectroscopic characters. The above-mentioned progress is not yet at the stage of recognizing individual atoms or molecules randomly, but concerns rather the functional specificity associated with the individual components. Nevertheless, the achievements have already produced far-reaching impacts on research endeavors at single molecule scale. This effort has greatly enriched the scope of SPM, leading to the development of new microscopy techniques. This could be benefecial for fundamental research in physics and chemistry as well as in fields with important application potential.

As a major category of SPM techniques, atomic force microscopy (AFM) and related microscopies are based on the interacting force between probe and sample. Versatile experimental conditions (vacuum, ambient, liquid) have led to AFM and related microscopies beeing commonly used in surface analytical studies. The pursuit of nanometer-scale mapping of surface compositions using AFM-derived techniques has produced many illuminating results. With high spatial resolution capabilities, such efforts should provide insightful information complementary to other surface characterization techniques. In addition, force spectroscopy based on AFM techniques provides an important approach for studying the mechanical properties of single molecules.

Single molecules have been an inspiring subject for molecular device studies. Electron transportation through various molecular systems can be affected by internal molecular structures and environmental factors, and has become a test ground for many experimental and theoretical methods. Advances in this field could enrich our understanding of the physical and chemical properties of single molecules.

Recent progress in optical microscopy has allowed scientists to optically detect individual molecules and to conduct single-molecule spectroscopy assessments, which form another frontier of related research. These optical techniques provide high temporal resolution suitable for dynamical studies, but have less spatial resolution. Whereas near-field microscopy is capable of imaging single molecules with a spatial resolution beyond the diffraction limit, conventional optical microscopes and spectroscopies can be used to investigate single molecule behavior for very dilute samples. The most widely used approach is fluorescence detection, used for single molecule work at both cryogenic and ambient temperatures.

Last but not least, the enhancement of Raman signals at single molecule level is briefly discussed. Owing to the extremely small scattering cross section of Raman detection for molecules, the enhancement effect is crucial for studying single molecule vibrational characteristics. The results from nanoparticle-induced enhancement and preliminary work on tip-induced enhancement effects reflect the novel achievements on this front.

2 Basics of Electron Tunneling Processes and Scanning Tunneling Microscopy

2.1 Principles of Tunneling Processes

The term electron tunneling process has been coined to describe the penetration of electrons through a classically impenetrable energy barrier. The theoretical treatments of such processes have been well established in the framework of quantum mechanics and successfully applied to resolve many important experimental phenomena. A recent example can be seen in the development of scanning tunneling microscopy (STM). Great effort and attention have been apparent from the early stage of STM research, in the drive for directly visualizing single molecules using STM. This chapter is aimed at providing a brief introduction of the background of tunneling phenomena, focusing on aspects that are closely related to STM studies. The theoretical aspects of STM are subsequently introduced, followed by discussions on several specific technical aspects of the STM method.

2.1.1 Elastic Tunneling Process

As described in many standard textbooks, the only way an electron can get through a classically, "forbidden", energy barrier is through the tunneling process. For elastic processes, electron energy is unchanged before and after passing the barrier. Using planar electrode approximation, we can illustrate the theoretical approaches for tunneling processes. The static tunneling model begins with the electron density distribution function $f(E)$ at the Fermi level:

$$f(E) = \frac{1}{1 + \exp \frac{E - E_F}{k_B T}} \quad (2.1)$$

E is the electron energy, E_F the Fermi level energy, k_B the Boltzmann constant and T the temperature. The tunneling current from one side of an electrode to the opposite electrode (represented by left to right (J_{LR}) or right to left (J_{RL})) can be expressed in a generalized form for the electron tunneling effect between similar electrodes separated by a thin insulating film [2.1–2.3], as schematically illustrated in Fig. 2.1.

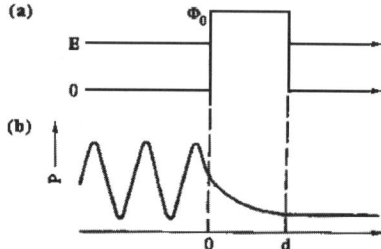

Fig. 2.1. Schematic of elastic tunneling process in which the electrons tunnel through the energy barrier without energy loss (illustrated in **a**). **b** The decay behavior of the electronic wave function across the energy barrier

$$J_{LR} = -2e \int \frac{d^3k}{8\pi^3} v_{x_1} f_L(E_L)[1 - f_R(E_R + eV)] D_{LR}(E)$$

$$J_{RL} = -2e \int \frac{d^3k}{8\pi^3} v_{x_2} f_R(E_R)[1 - f_L(E_L + eV)] D_{RL}(E) \qquad (2.2)$$

J is the current density, V the applied bias voltage across the tunneling junction, D the electron transmission probability, v_{x_1} and v_{x_2} the projection of speed perpendicular to the electrode surfaces (assigned as x direction). The net current is

$$J = \frac{4\pi m e}{h^3} \int_0^\infty \left[D(E_x) \int_0^\infty [f(E) - f(E + eV)] \, dE_t \right] dE_x \qquad (2.3)$$

where E_t is the energy component parallel to the electrode surface.

According to the Wentzel-Kramers-Brillouin (WKB) approximation, the probability (D) of the electrons tunneling through the energy barrier is:

$$\begin{aligned} D(E_x) &= \exp\left[\frac{-4\pi\sqrt{2m}}{h} \int_{\Delta s} (E_F + \phi(x) - E_x)^{1/2} dx \right] \\ &\approx \exp\left[-A(E_F + \overline{\phi} - E_x)^{1/2} \right] \end{aligned} \qquad (2.4)$$

Here the parameters are defined as following:

$$\overline{\phi} = \frac{1}{\Delta s} \int_{\Delta s} \phi(x) dx$$
$$A = 4\pi\beta\sqrt{2m}\Delta s/h$$
$$\beta = 1 - \frac{1}{8\overline{\phi}^2 \Delta s} \int_{\Delta s} [\phi(x) - \overline{\phi}]^2 dx$$

where $\overline{\phi}$ is the averaged barrier height. Under the approximation of $E = E_x + E_t$ and low temperature:

2.1 Principles of Tunneling Processes

$$J = \frac{e}{2\pi h}(\beta \Delta s)^{-2}\{\overline{\phi}\exp(-A\overline{\phi})^{1/2} - (\overline{\phi}+eV)\exp[-A(\overline{\phi}+eV)^{1/2}]\}$$
$$= J_0\{\overline{\phi}\exp(-A\overline{\phi})^{1/2} - (\overline{\phi}+eV)\exp[-A(\overline{\phi}+eV)^{1/2}]\} \quad (2.5)$$
$$J_0 = \frac{e}{2\pi h}(\beta \Delta s)^{-2}$$

The expression can be further simplified at different bias voltage ranges:

$$J \propto V: \text{ at small } V$$
$$J \propto J_L(V + \beta V^3): \text{ at large } V$$

Another widely used treatment of the tunneling process is the many-particle transition Hamiltonian method, which considers the electron tunneling as a transition between two electronic states separated by an energy barrier [2.4].

The electronic wave function at both sides of the energy barrier can be expressed in a general form [2.4]:

$$\phi_m = C p_x^{-1/2} \exp[i(p_y y + p_z z)] \sin(p_x x + \gamma) \quad \text{for } x < x_a$$
$$= \frac{1}{2} C \mid p_x \mid^{-1/2} \exp[i(p_y y + p_z z)] \exp\left(-\int_{x_a}^{x} \mid p_x \mid \mathrm{d}x\right)$$
$$\text{for } x_a < x < x_b \quad (2.6)$$

The tunneling junction is represented by the Hamiltonian equation:

$$H = H_1 + H_2 + H_3$$

The general solution to the Hamiltonian equation is in terms of ϕ_{mn} with energy W_{mn}:

$$\phi = a(t)\phi_0 e^{iW_0 t} + \sum_{\mathrm{mn}} b_{\mathrm{mn}}(t)\phi_{\mathrm{mn}} e^{-iW_{\mathrm{mn}} t} \quad (2.7)$$

By utilizing the Fermi rule, the transition probabilities between the two sides of the barrier under perturbation are:

$$L \to R \quad P_{\mathrm{LR}} = \frac{2\pi}{\hbar} \mid M_{\mathrm{LR}} \mid^2 \rho_{\mathrm{R}} f_{\mathrm{L}}(1 - f_{\mathrm{R}})$$
$$L \leftarrow R \quad P_{\mathrm{RL}} = \frac{2\pi}{\hbar} \mid M_{\mathrm{RL}} \mid^2 \rho_{\mathrm{L}} f_{\mathrm{R}}(1 - f_{\mathrm{L}})$$

M_{RL} is the transition matrix element. The net current J is thus:

$$J = \frac{4\pi e}{\hbar} \sum \mid M_{\mathrm{LR}} \mid^2 \rho_{\mathrm{L}} \rho_{\mathrm{R}} (f_{\mathrm{L}} - f_{\mathrm{R}}) \mathrm{d}E_t \quad (2.8)$$

$$M_{\mathrm{mn}} = \int [\phi_0^* H \phi_{\mathrm{mn}} - \phi_{\mathrm{mn}} H \phi_0^*] \mathrm{d}\tau \quad (2.9)$$

$$= -\frac{\hbar^2}{2m} \sum_i \int_i (\phi_0^* \nabla_i^2 \phi_{\mathrm{mn}} - \phi_{\mathrm{mn}} \nabla_i^2 \phi_0^*) S(x_i) \mathrm{d}\tau_i$$
$$= -i[J_{\mathrm{mn}}(x_1) - J_{\mathrm{mn}}(x_0)] \quad (2.10)$$

$$J = \frac{\hbar}{2mi}[\phi_0^* \frac{\partial \phi_{mn}}{\partial x} - \phi_{mn} \frac{\partial \phi_0^*}{\partial x}]$$

$J_{mn}(x_0) = 0$, according to the assumption of the model. Therefore,

$$M_{mn} = -iJ_{mn}(x_1) \quad (2.11)$$

Using the wave function definition, one can obtain the explicit expression of the transition matrix:

$$|M_{LR}|^2 = \left(\frac{\hbar^2}{2m}\right)^2 \frac{(k_x)_L}{L_1} \frac{(k_x)_R}{L_2} \exp\left[-2\int_0^s |k_x|\, dx\right]$$

$$|k_x| = \frac{\sqrt{2m}}{\hbar}(E_F + \phi(x) - E_x)^{1/2} \quad (2.12)$$

The equivalence of the above two treatments of the tunneling process can be illustrated under one-dimensional approximation [2.5]. Under the one-dimensional free electron approximation, the density of states can be written as:

$$\rho = \frac{L}{\pi}\left(\frac{dE}{dk_x}\right)^{-1} = \frac{mL}{\pi \hbar^2 k_x}$$

Therefore,

$$M_{LR} = \frac{1}{4\pi^2 \rho_L \rho_R} \exp\left[-2\int_0^s |k_x|\, dx\right]$$

$$J = \frac{e}{\pi \hbar} \sum_{k_t} \int dE_x (f_L - f_R) \exp\left[-2\int_0^s |k_x|\, dx\right] \quad (2.13)$$

This expression is identical to the result obtained from the static tunneling model. The equivalence is due to the fact that both models apply WKB approximation to obtain the electron tunneling probability. In addition, the momentum component parallel to the electrode is assumed as unchanged across the tunnel junction. The difference between the two models is that in principle the Hamiltonian method applies to small perturbations, whereas the static method could apply to general types of energy barriers.

In addition, a number of detailed studies have dealt with the elastic tunneling process. Green's function method can be used to deal with orthogonality and completeness of the wave functions [2.6]. The series expansion approach can be introduced to treat arbitrary-shaped energy barriers [2.7]. Feuchtwang [2.8] used perturbation theory to obtain general treatment of the tunneling process.

2.1.2 Inelastic Tunneling Process

During the inelastic tunneling process, the tunneling electron loses part of its energy before reaching the target electrode [2.9–2.15]. This phenomenon

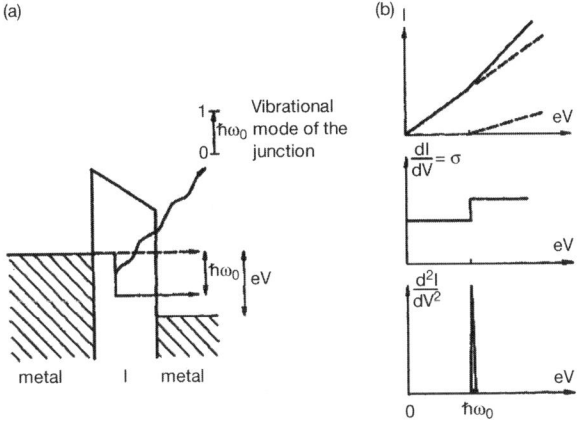

Fig. 2.2. Schematic illustration of **a** the energy diagram of the tunneling process and **b** the effect of inelastic interaction on measured tunneling spectra (extracted from [2.15])

was first discovered by Jaklevic and Lambe [2.9] in 1966. It was suggested that the energy dissipation is associated with the excitation of vibrational levels of the molecules embedded in the junction, represented as phonons. The excitation is also related to the electron density of states of the target electrode. Assuming the phonon frequency as ω, the energy required for such excitation is $eV > \hbar\omega$. Such excitations appear as discontinuous jumps in the tunneling current, and can be reflected in the first- and second-order derivatives, as illustrated in Fig. 2.2. These are the signature features of the inelastic effect in tunneling characteristics.

The consideration of the molecular adsorbate effect can commence from the interaction between the tunneling electron and the dipole moment of the molecule. The interaction potential can be expressed as the combination of the elastic tunneling potential, the Coulomb potential of the molecule and its mirror potential on the electrode [2.10]. The discussion can be simplified if one assumes the dimension of the junction is much larger than that of the molecule, and only the first-order approximation is considered. There are two types of inelastic tunneling spectra based on the vibrational characteristics, i.e., infrared and Raman. Infrared-type spectra originate from the interaction between polar molecules and the electric field, with the interaction energy of $(P \cdot E)$, where P is the molecular dipole moment, and E the electric field strength. Raman-type spectra are associated with the electric field-induced dipole moment, where the interaction termed αE^2 with polarizability α. The interaction potential of the surface-bound molecules and tunneling electrons is [2.11]:

$$U_{\text{int}} = \frac{2ep_x x}{(x^2 + r_\perp^2)^{3/2}} \tag{2.14}$$

The tunneling matrix is:

$$|M_{12}| \propto \exp\left\{-\int_0^l dx \left(\frac{2m}{\hbar^2}\right)^{1/2} \times [U(x) + U_{\text{int}}(x) - (E - E_\perp)]^{1/2}\right\} \tag{2.15}$$

Using the approximation of $U - (E - E_\perp) \approx \phi$ and expanding the expression in terms of $\frac{U_{\text{int}}}{\phi}$, one will have:

$$|M_{12}| \propto \left[\left(\frac{2m}{\phi}\right)^{1/2} \frac{ep_x}{\hbar \ell} g\left(\frac{r_\perp}{\ell}\right) + 1\right] \exp\left(-\left(\frac{2m\phi}{\hbar^2}\right)^{1/2} \ell\right)$$

$$g(y) = \frac{1}{y} - \frac{1}{(1+y^2)^{1/2}} \tag{2.16}$$

The first term in the above expression is the inelastic one, the second term the elastic one. The interaction between the molecule and the tunneling electron can be different for different vibration levels. It can be viewed as a reflection of vibrational mode of the molecules. The following discussion considers the interaction of molecular vibration-induced dipole moment with the tunneling electron.

Assuming a molecule with a vibration frequency ω_0, by transition law [2.11] the tunneling current can be expressed in the following form:

$$I_i(\omega_0, V) = \left(\frac{dj}{dV}\right)_0 |<1|p_x|0>|^2 \left[\frac{4\pi me}{\hbar^2 \phi}\right] \ln\left|\frac{\ell}{r_0}\right| \times$$

$$\int_{-\infty}^{\infty} dE f(E)[1 - f(E + eV - \hbar\omega_0)] N_1(E) N_2(E + eV - \hbar\omega_0)$$

$N_1(E_1)$, $N_2(E_2)$ are density of states of two pertinent vibrational levels, and $|<1|p_x|0>|^2$ the dipole moment of the vibrational transition. A practical treatment should include the summation of different vibration modes:

$$I_i(V) = N\left(\frac{dj}{dV}\right)_0 \left[\frac{4\pi me}{\hbar^2 \phi}\right] \ln\left|\frac{\ell}{r_0}\right| \sum_m |<m|P_z|0>|^2 \times$$

$$\int_{-\infty}^{\infty} dE f(E)[1 - f(E + eV - \hbar\omega_m)] N_1(E) N_2(E + eV - \hbar\omega_m)$$

$$\tag{2.17}$$

where N is the total number of vibration modes. Under low-temperature approximation, the derivative of the tunneling current is:

$$\frac{dI}{dV} = N\left(\frac{dj}{dV}\right)_0 \left(\frac{4\pi me^2}{\hbar^2 \phi}\right) \ln\left|\frac{\ell}{r_0}\right| \sum_m |<m|p_x|0>|^2 S_m(eV)$$

$$\frac{d^2 I}{dV^2} = N\left(\frac{dj}{dV}\right)_0 \left(\frac{4\pi me}{\hbar^2 \phi}\right) \ln\left|\frac{\ell}{r_0}\right| \sum_m |<m|p_x|0>|^2 \delta(eV - \hbar\omega_m)$$

$$\tag{2.18}$$

where $S_m(eV)$ is the unit step function at $eV = \hbar\omega_m$. The experimental realization of the measurements on vibrational characteristics is through I–V curves and the I versus dI/dV curves.

The tunneling electron can interact with the vibration-associated permanent dipole moment (infrared-type spectrum), as well as the field induced dipole (Raman-type spectrum):

$$P = \alpha E$$
$$U = -P \cdot E = \alpha E^2$$
$$U_{\text{int}}(\alpha) = -\frac{4e^2 \alpha x^2}{(x^2 + r_\perp^2)^3}$$

The transition matrix for the tunneling electron is [2.10]

$$|M| \propto \left[\left(\frac{2m}{\phi}\right)^{1/2} \frac{1}{\hbar} \frac{\alpha e^2}{4\ell^3} t\left(\frac{r_\perp}{\ell}\right) + 1\right] \exp\left(-\left(\frac{2m\phi}{\hbar^2}\right)^{1/2} \ell\right) \quad (2.19)$$

Here $t(y) = \frac{1}{y^2}[\frac{1-y^2}{(1+y^2)^2} + \frac{1}{y}\tan^{-1}(\frac{1}{y})]$. The inelastic tunneling current can be expressed as:

$$I(V) = N \left(\frac{dj}{dV}\right)_0 \left[\frac{4\pi m}{\hbar^2 \phi} \frac{e^3}{16\ell^6}\right] \left[\int_{r_0}^{\ell} t^2 r_\perp dr_\perp\right] \sum_m |<m|\alpha|0>|^2 \times$$
$$\int_{-\infty}^{\infty} dE f(E)[1 - f(E + eV - \hbar\omega_m)] N_1(E) N_2(E + eV - \hbar\omega_m)$$
$$(2.20)$$

where $<m|\alpha|0>$ is the matrix element of polarizability and is a scalor.

There are a number of theoretical models dealing with the inelastic tunneling process. Appelbaum and Brinkman [2.12] utilized the approach by Bardeen [2.4] for the elastic tunneling process to study the assisted tunneling process associated with the excited states residing in the energy barrier. Both elastic and inelastic effects were taken into consideration in this approach. Brailsford and Davis [2.13], and Davis [2.14] extended the static tunneling approach to the multi-electron method and obtained consistent results with experiments.

We now turn to the quantitative analysis of tunneling spectra based on the principles presented above. The position and width of spectrum peaks are two important factors. The width of the spectrum peak can be affected by temperature, the intrinsic width of energy level of dopant molecules, and the amplitude of the detection signal. Lambe and Jaklevic [2.11] analyzed the temperature-induced broadening of peak width. For an electrode in normal state, the tunneling current is:

$$I_i = c \int_{-\infty}^{\infty} \left(1 + \exp\left(\frac{E}{k_B T}\right)\right)^{-1} \left\{1 - \left[1 + \exp\left(E + \frac{e(V - V_0)}{k_B T}\right)\right]^{-1}\right\} dE$$
$$(2.21)$$

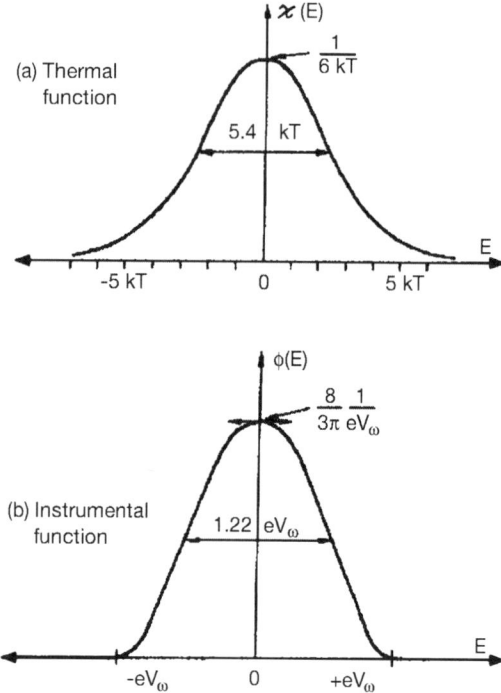

Fig. 2.3. The spectrum broadening of the tunneling feature due to **a** thermal and **b** detection signal amplitude (extracted from [2.15])

Here, the constant c includes all quantities independent of the energy E and temperature T. V_0 is the center position of the peak of interest. The integral will yield:

$$I_i = ce(V - V_0) \frac{\exp[e(V - V_0)/k_B T]}{\exp[e(V - V_0)/k_B T] - 1} \quad (2.22)$$

The characteristic line shape of the temperature-broadened spectrum is illustrated in Fig. 2.3. The width at half height is 5.4 $k_B T$. Further analysis for an electrode in superconducting state suggested that the half height width will be reduced to 2.9 $k_B T$ [2.11].

Another factor that affects the width is the amplitude of the detection signal. In a typical measurement, a small AC signal will be coupled to the tunneling bias voltage. The derivative curves are obtained by measuring the amplitude of the harmonics. Klein et al. [2.15] demonstrated the effect of the detection signal amplitude on peak width:

The tunneling current is (including the detection signal $V_\omega \cos \omega t$)

$$I = I(eV_0 + eV_\omega \cos \omega t)$$

Here V_ω and ω are the amplitude and frequency of the modulating signal, respectively, and V_0 the DC component. The second harmonic term is measured by the following:

$$I_{2\omega} = \frac{2}{\tau} \int_\tau I(eV_0 + eV_\omega \cos \omega t) \cos 2\omega t \, dt$$

Assuming $E = eV \cos(\omega t)$ and by partial integration, one can obtain:

$$I_{2\omega} = \frac{2}{3\pi} \int_{-eV_\omega}^{eV_\omega} I''(eV_0 + E) \frac{(e^2 V_\omega^2 - E^2)^{3/2}}{V_\omega^2} \, dE \tag{2.23}$$

The actual measured amplitude of the second harmonic is $4I_{2\omega}/V_\omega^2$, which is the convolution of the second order derivative of the current and the following function:

$$\phi(E) = \frac{8}{3\pi} \frac{(e^2 V_\omega^2 - E^2)^{3/2}}{(eV_\omega)^4}, \quad \text{for } |E| < eV_\omega$$
$$\phi(E) = 0, \quad \text{for } |E| > eV_\omega$$

The half height width of the $\phi(E)$ function is $1.22 eV_\omega$, as shown in Fig. 2.3. The experimentally measured width is determined by the above two factors, together with the intrinsic width of the electronic states. The broadening of the combination of these two factors is:

$$\Delta E = [(5.4 k_B T)^2 + (1.22 eV_\omega)^2]^{1/2} \tag{2.24}$$

Under low-temperature conditions, for example, $T = 4.2\,\text{K}$, and $V_\omega \sim 2\,\text{mV}$, the peak broadening is about 3 mV. Therefore, special care should be taken in studying closely spaced peak features.

In addition, strongly bonded molecules at electrode surfaces will lead to relatively strong interaction with tunneling electrons, resulting in possible shifts of the peak positions. Furthermore, the spatial position of the dopant molecule within the junction will also affect the mirror potential, and thus can significantly alter the peak position. Kirtley and Hansma [2.16] confirmed that the vibrational modes could be affected by the material composition of the electrode. The effect was attributed to the mirror potential of the molecular dipole moment. The strength of the molecular dipole in the mirror potential can be approximated as $\sim 1/d^3$, d being the distance between the molecule and electrode surface. The change of vibration frequency can be expressed as:

$$\Delta\omega_\perp = \frac{-q_1^2}{8 m \omega_0 n_1^2 d^3} \left\{ 1 + \frac{3}{2} \beta \frac{q_0}{q_1} s \left[1 - \left(1 + \frac{a}{2d}\right)^{-2} \right] \right\} \tag{2.25}$$

$$\Delta\omega_{//} = \frac{-q_1^2}{16 m \omega_0 n_1^2 d^3} \tag{2.26}$$

Here the static molecular dipole is $\rho \overline{P_0} = q_0 \overline{a}$, where $\overline{a}$ is bond length. The vibrational dipole moment is $\overline{P} = q_1 \overline{x}$, m is the effective mass of the molecule, n_1 the diffraction index of the oxide layer, $\beta^2 = (m\omega_0^2)/2E_D$, and E_D the bond dissociation energy.

2.1.3 Two-Step Tunneling Process

As discussed above, the molecules embedded within the tunneling junction can interact with tunneling electrons by dipole interactions. The molecular electronic energy state can also participate in the interaction with tunneling electrons as an intermediate state [2.17, 2.18]. Define D as the transmission probability for the process without intermediate state, whereby D_1 and D_2 represent the probability from one side of the electrode in the intermediate state. These quantities can be expressed as:

$$D = \exp\left[-2\int_0^t k(x)\,\mathrm{d}x\right] \quad \text{without intermediate state}$$

$$D_1 = \exp\left[-2\int_0^{x_1} k(x)\,\mathrm{d}x\right] \quad \text{for } x = 0 \longrightarrow x = x_1$$

$$D_2 = \exp\left[-2\int_{x_1}^t k(x)\,\mathrm{d}x\right] \quad \text{for } x = x_1 \longrightarrow x = t$$

Denote the corresponding electron tunneling probability as W_1, W_2:

$$\begin{aligned} x = 0 &\longrightarrow x = x_1 & W_1 &\cong N_t(1 - f_1)D_1 \\ x = x_1 &\longrightarrow x = t & W_2 &\cong N_t f_2 D_2 \end{aligned}$$

where N_t is the density of states of the intermediate electrode. For a static process, W_1 equals W_2, and therefore:

$$f_1 = \frac{D_1}{D_1 + D_2}$$

The two-step tunneling probability is:

$$W = W_1 = W_2 = \frac{N_t D_1 D_2}{D_1 + D_2} = \frac{N_t D}{D_1 + D_2} \tag{2.27}$$

W is maximum when D_1 equals D_2:

$$W = \frac{1}{2} N_T D^{1/2} \propto \exp\left[-\int_0^t k(x)\,\mathrm{d}x\right] \tag{2.28}$$

This result differs from that of the simple tunneling junction in that the tunneling probability is appreciably increased.

2.1.4 Resonant Tunneling Effect

Another effect that could enhance tunneling probability is associated with resonant tunneling processes. When the kinetic energy of the incoming electron is matched with the bound state of the energy barrier, the possible interference can lead to an enhanced tunneling probability [2.18, 2.19].

In addition to the adsorbed atoms and molecules, resonant tunneling effects have been observed in a range of tunneling experiments. For examples, the quantum dots and quantum wires were found to display resonant tunneling effects, which could benefit the study of quantum state structures [2.20, 2.21].

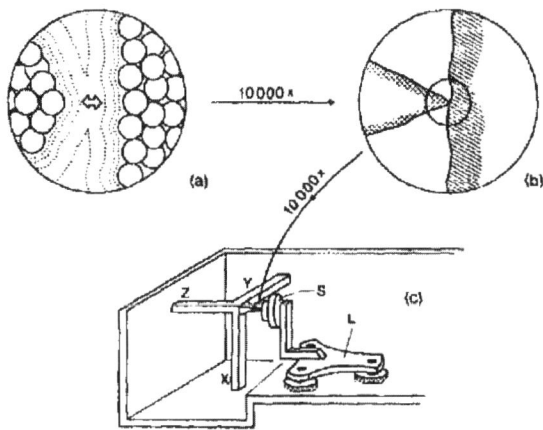

Fig. 2.4. a,b Illustration of the atomically sharp tunneling junction in the vicinity of the tip apex. **c** Schematic of an STM with tripod scanner (extracted from [2.22])

2.2 Introduction to Scanning Tunneling Microscopy (STM)

The application of the electron tunneling principle to the detection of single molecules may be best demonstrated in the invention and development of scanning tunneling microscopy (STM). Probably the main difference between STM and other microscopy techniques is that there is no need for lenses and special light or electron sources. Rather, the bound electrons already existing in the sample under investigation serve as the exclusive source of radiation, as shown schematically in Fig. 2.4 [2.22–2.25]. STM-based analysis techniques are capable of revealing the surface structure of conductors and semiconductors with high resolution in real time under different experimental conditions, such as under vacuum, in air and solutions, and at low temperatures. The information on the surface electronic structure can also be obtained by using scanning tunneling spectroscopy (STS).

2.2.1 Introduction to STM

There have been a number of designs for scanners, as shown schematically in Fig. 2.5. The tube scanner and tripod scanners are widely adopted in STM designs, and their mechanical stabilities have been systematically analyzed. Since tunneling processes involve electronic states at the Fermi level, which may themselves have a complex spatial structure, we must expect that the electronic structure of the surface and tip may contribute to the imaging mechanisms in a complex way. Based on the analogy with the one-dimensional tunneling problem described by the WKB approximation, the full elucidation

16   2 Basics of Electron Tunneling Processes

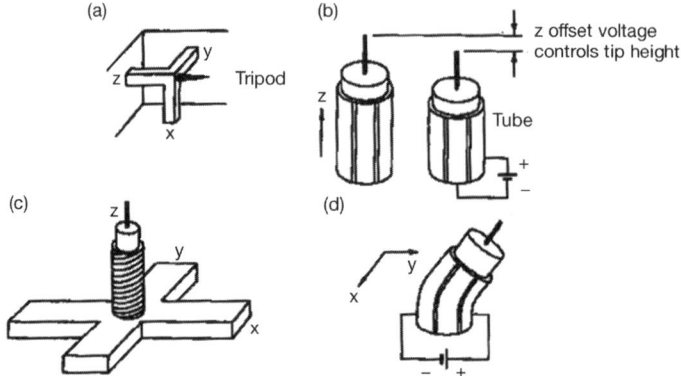

Fig. 2.5. Several typical designs of STM scanners. **a** Tripod design where the tip motion in three directions (x, y, z) is provided by three independent piezo arms. **b** A tube scanner that controls the tip motion in z direction by expanding/retracting the tube wall. **c** A combined piezo cross arm and tube scanner. **d** A tube scanner that controls the tip motion in x, y directions by bending the quadrant sections of tube walls

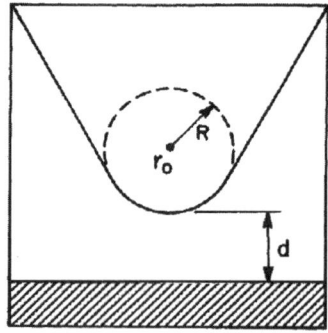

Fig. 2.6. Schematic of tip and sample geometry used for calculation of tunneling current in STM (extracted from [2.26])

of the STM imaging mechanism should be considered in three-dimensional models. These theoretical concepts have been carefully addressed by many groups [2.26, 2.27]. In addition to delineating the atomic topography of a surface, STM has made it technically possible to directly probe the electronic structures of materials at an atomic level by spatially resolved tunneling spectroscopy, which is also a vital part of STM studies. Conventional STM is based on the control of the tunneling current through the potential barrier between the sample surface and the probing metal tip. If a small bias voltage is applied between the sample surface and the tip (in the best case,

2.2 Introduction to Scanning Tunneling Microscopy (STM)

an atomically sharp tip), a tunneling current will flow between the tip and sample when the gap between them is reduced to a few angstrom. It takes advantage of the strong dependence of the tunneling probability of electrons on the electrode separation. It is generally considered that the atomic resolution of STM originates from the atomic scale tip. The analysis of the imaging mechanism is usually started from the ideal model of an atomic tip, with an extension to consider atomic cluster as a tip. In the pioneering work of Tersoff and Hamann [2.26], the tip is approximated by a spherical object of atomic dimension, and the sample is represented by a metallic surface (Fig. 2.6). Assuming $\varphi_\nu(\boldsymbol{r}_0)$ is the sample wave function at position $\boldsymbol{r}_0$, φ_μ is the wave function for the tip, and φ_ν–φ_μ are orthogonal eigenfunctions, the wave function for the sample surface is expressed in the general form [2.26]:

$$\varphi_\nu = \Omega_s^{-1/2} \sum_G a_G \exp[-(\kappa^2 + |-\boldsymbol{\kappa}_G|^2)^{1/2} z] \exp(i\boldsymbol{\kappa}_G \cdot \boldsymbol{X}) \qquad (2.29)$$

$$\kappa = \hbar^{-1}(2m\phi)^{1/2}, \qquad \boldsymbol{\kappa}_G = \boldsymbol{\kappa}_{//} + \boldsymbol{G}$$

where Ω_s is the sample volume, $\boldsymbol{G}$ the surface inverse wave vector, and $\boldsymbol{\kappa}_{//}$ the surface Bloch wave vector. The spherical tip can be expressed as:

$$\varphi_\mu = \Omega_t^{-1/2} c_t \kappa R \, e^{\kappa R} (\kappa \, |\boldsymbol{r} - \boldsymbol{r}_0|)^{-1} e^{-\kappa|\boldsymbol{r} - \boldsymbol{r}_0|} \qquad (2.30)$$

$$\kappa = \frac{\sqrt{2m\phi}}{\hbar}$$

where Ω_t is the tip volume. Following the transition Hamiltonian method [2.4], the tunneling current I is expressed as:

$$\begin{aligned} I &= \frac{2\pi e}{\hbar} \sum_{\mu,\nu} \{f(E_\mu)(1 - f(E_\nu + eV)) \, |M_{\mu\nu}|^2 \, \delta(E_\mu - E_\nu) \\ &\quad - f(E_\nu + eV)(1 - f(E_\mu)) \, |M_{\mu\nu}|^2 \, \delta(E_\mu - E_\nu - eV)\} \\ &= \frac{2\pi e}{\hbar} \sum_{\mu,\nu} |M_{\mu\nu}|^2 \, [f(E_\mu) - f(E_\nu + eV)] \delta(E_\mu - E_\nu) \qquad (2.31) \end{aligned}$$

$f(E)$ is the Fermi distribution function. It can be seen that the electronic states contributing to the tunneling current are in the vicinity of the Fermi level. Under the approximation of small voltage and low temperature, the expression can be simplied to the following:

$$I = \frac{2\pi}{\hbar} e^2 V \sum_{\mu,\nu} |M_{\mu\nu}|^2 \, \delta(E_\nu - E_\mathrm{F}) \delta(E_\mu - E_\mathrm{F})$$

$$M_{\mu\nu} = \frac{\hbar^2}{2m} \int d\boldsymbol{S} \cdot (\varphi_\mu^* \nabla \varphi_\nu - \varphi_\nu \nabla \varphi_\mu^*)$$

By expanding the probe wave function in terms of the surface wave function:

$$\frac{e^{-\kappa|r|}}{\kappa|r|} = \int d^2q\, b(q) \exp[-(\kappa^2 + q^2)^{1/2} |z|] \exp(iq \cdot x)$$

$$b(q) = (2\pi)^{-1}\kappa^{-2}(1 + \frac{q^2}{\kappa^2})^{-1/2}$$

The transition matrix can be expressed as:

$$M_{\mu\nu} = \frac{\hbar^2}{2m} 4\pi\kappa^{-1}\Omega_t^{-1/2} \kappa R e^{\kappa R} \varphi_n u(r_0)$$

$$I = 32\pi^3 \hbar^{-1} e^2 \phi^2 D_t(E_F) R^2 \kappa^{-4} e^{2\kappa R} \times \sum_\nu |\varphi_\nu(r_0)|^2 \delta(E_\nu - E_F)$$

$$\propto \sum_\nu |\varphi_\nu(r_0)|^2 \delta(E_\nu - E_F)$$

where D_t is the density of states of the probe. The local density of states is defined as:

$$\rho(r_0, E_F) \equiv \sum_\nu |\varphi_\nu(r_0)|^2 \delta(E_\nu - E_F)$$

Due to the fact that

$$|\varphi_\nu(r_0)|^2 \propto e^{-2\kappa(R+d)}$$

The tunneling current can be simplified to [2.26]:

$$I \propto e^{-2\kappa d} \tag{2.32}$$

where $2\kappa = 1.025\phi^{1/2}$, and ϕ is the average barrier height, i.e., the mean barrier height between the two electrodes. For typical metals ($\phi \approx 5$ eV), the predicted change in current by one order of magnitude for the change $\Delta d \approx 1$ Å has been verified. If the current is kept constant to within, e.g., 2%, then the gap d remains constant to within 0.01 Å. This fact represents the basis for interpreting the image as simply a contour of constant height above the surface.

It should be noted that the practical tip geometry may be far different from the idealized single atom model. As a matter of fact, very rugged tip geometry can often be observed in microscopic images of tip geometry. The exponential dependence of the tunneling current on the tip-sample separation stresses that the tunneling current is dominated by the outmost atom on the tip, rather than by the collective effect of all the atoms at the tip apex.

The STM can be operated in either constant-current or constant-height mode, as shown in Fig. 2.7 [2.29, 2.30]. In the basic constant-current mode of operation, the tip is raster scanned across the surface at pre-set a tunneling current, which is maintained at a pre-set value by continuously adjusting the vertical tip position with the feedback voltage V_z. In the case, of an electronically homogeneous surface, the topographic height of surface features

2.2 Introduction to Scanning Tunneling Microscopy (STM)

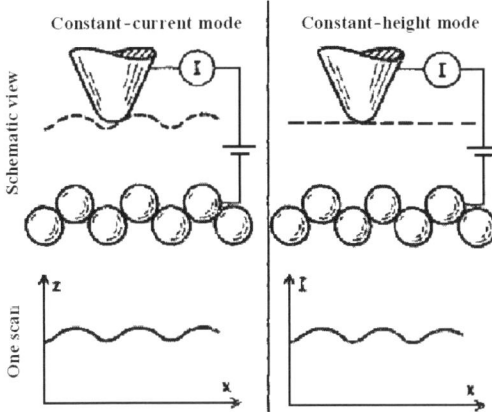

Fig. 2.7. Illustration of operation modes of constant current and constant height of STM (extracted from [2.30])

of a sample can be visualized. The height of the tip as a function of position is read and processed subsequently. Alternatively, in the constant-height mode a tip can be scanned rapidly across the surface at nearly constant height and constant voltage V_z while the tunneling current is monitored, as shown in Fig. 2.1b. In this case, the electronic feedback network is slowed down to keep the average tunneling current constant or even turned off completely. The rapid variations in the tunneling current due to the tip passing over surface features are recorded and plotted as a function of scan position.

Each mode has its own advantages. The basic constant-current mode was originally employed and can be used to track surfaces that are not atomically flat. The height of surface features can be derived from V_z and the sensitivity of the piezoelectric driver element. On the other hand, a disadvantage of this mode is that the finite response time of the feedback network and of the piezoelectric driver sets relatively low limits for the scan speed. The constant-height mode allows for much faster imaging of atomically flat surfaces, since the feedback loop and the piezoelectric driver do not have to re-pinned to the surface features passing under the tip. Fast imaging is important, since it may enable studies on dynamic processes on surfaces as well as reducing data collection time. Fast imaging also minimizes the image distortion due to piezoelectric creep, hysteresis and thermal drifts. In contrast to the constant-current mode, however, deriving topographic height information from recorded variations of the tunneling current in the constant-height mode is not easy because an independent determination of $\phi^{1/2}$ is required to calibrate z, as illustrated in (2.32). In both modes, the tunneling voltage and/or the z position can be modulated to obtain information about the local spectroscopy and/or spatially resolved local tunneling barrier height, respectively.

2.2.2 STM Contrast Mechanisms

The above discussion on STM imaging mechanisms involves the spherical wave function (s-wave) as the probe wave function. In realistic situations, one should also consider the effect of other types of wave function on STM imaging processes. The consideration is reflected in the analysis of Chen [2.27] and other studies. As will be seen below, it is convenient to express wave functions in terms of spherical harmonics using Green's function method [2.27].

The Schrodinger equation for states under vacuum conditions can be written as:

$$(\nabla^2 - \kappa^2)\varphi_\mu(\boldsymbol{r}) = 0 \tag{2.33}$$
$$\kappa = (2m\phi)^{1/2}/\hbar$$

The general solution can be expressed as the expansion of Bessel's functions

$$\varphi_{\ell m}(\boldsymbol{r}) = C_{\ell m} K_\ell(\kappa\rho) Y_{\ell m}(\theta, \varphi) \tag{2.34}$$

The Green function corresponding to the solution of Schrodinger's equation can be obtained from the following:

$$(\nabla^2 - \kappa^2) G(\boldsymbol{r} - \boldsymbol{r}_0) = \delta(\boldsymbol{r} - \boldsymbol{r}_0)$$
$$G(\boldsymbol{r} - \boldsymbol{r}_0) = \frac{\exp(-\kappa \mid \boldsymbol{r} - \boldsymbol{r}_0 \mid)}{4\pi \mid \boldsymbol{r} - \boldsymbol{r}_0 \mid} = \frac{\kappa}{4\pi} k_0(\kappa\rho) \tag{2.35}$$

Thus, the s-wave of the tip can be rewritten as:

$$\varphi_s(\boldsymbol{r}) = \frac{4\pi c}{\kappa} G(\boldsymbol{r} - \boldsymbol{r}_0) \tag{2.36}$$

Taking the derivative of z_0 and using the known equation,

$$k_0(u) - k_1(u) = 0$$

one can obtain

$$\frac{\partial}{\kappa \partial z_0} G(\boldsymbol{r} - \boldsymbol{r}_0) = \frac{\kappa}{4\pi} \cdot \frac{z - z_0}{\rho} k_1(\kappa\rho)$$

where $\frac{z-z_0}{\rho} = \cos\theta$. The above equation can be rewritten as:

$$\varphi_{p_z}(\boldsymbol{r}) = \frac{4\pi c}{\kappa} \frac{\partial}{\kappa \partial z_0} G(\boldsymbol{r} - \boldsymbol{r}_0) \tag{2.37}$$

2.2 Introduction to Scanning Tunneling Microscopy (STM)

Similar expressions can be obtained:

$$\varphi_{p_x}(\boldsymbol{r}) = \frac{4\pi c}{\kappa} \frac{\partial}{\kappa \partial x_0} G(\boldsymbol{r} - \boldsymbol{r}_0)$$

$$\varphi_{p_y}(\boldsymbol{r}) = \frac{4\pi c}{\kappa} \frac{\partial}{\kappa \partial y_0} G(\boldsymbol{r} - \boldsymbol{r}_0)$$

$$\varphi_{d_{xz}}(\boldsymbol{r}) = \frac{4\pi c}{\kappa} \frac{\partial^2}{\kappa^2 \partial x_0 \partial z_0} G(\boldsymbol{r} - \boldsymbol{r}_0)$$

$$\varphi_{d_{z^2}}(\boldsymbol{r}) = \frac{4\pi c}{\kappa} [\frac{\partial^2}{\kappa^2 \partial z_0^2} G(\boldsymbol{r} - \boldsymbol{r}_0) - \frac{1}{3} G(\boldsymbol{r} - \boldsymbol{r}_0)]$$

$$\varphi_{d_{x^2-y^2}}(\boldsymbol{r}) = \frac{4\pi c}{\kappa} [\frac{\partial^2}{\kappa^2 \partial x_0^2} - \frac{\partial^2}{\kappa^2 \partial y_0^2}] G(\boldsymbol{r} - \boldsymbol{r}_0)$$

The merit of the above expressions is that the probe wave function of different quantum numbers are now described using the same Green's function. This treatment will help the discussion on the tunneling matrix element.

According to the definition:

$$M = \frac{\hbar^2}{2m} \int_\Sigma (\varphi_\mu^* \nabla \varphi_\nu - \varphi_\nu \nabla \varphi_\mu^*) \cdot d\boldsymbol{S}$$

φ_ν is the sample state, and φ_μ the tip state. For the simplest case of s-wave tip:

$$M_s = \frac{2\pi c \hbar^2}{\kappa m} \int_\Sigma [G(\boldsymbol{r} - \boldsymbol{r}_0) \nabla \varphi_\nu - \varphi_\nu \nabla G(\boldsymbol{r} - \boldsymbol{r}_0)] \cdot d\boldsymbol{S}$$

$$= \frac{2\pi c \hbar^2}{\kappa m} \int_{\Omega_t} [G(\boldsymbol{r} - \boldsymbol{r}_0) \nabla^2 \varphi_\nu - \varphi_\nu \nabla^2 G(\boldsymbol{r} - \boldsymbol{r}_0)] \cdot d\tau$$

Using (2.34) and (2.36), one obtains

$$\varphi_\nu(\boldsymbol{r}_0) = \int_\Sigma [G(\boldsymbol{r} - \boldsymbol{r}_0) \nabla \varphi_\nu - \varphi_\nu \nabla G(\boldsymbol{r} - \boldsymbol{r}_0)] \cdot d\boldsymbol{S}$$

and

$$M_s = \frac{2\pi c \hbar^2}{\kappa m} \varphi_\nu(\boldsymbol{r}_0) \qquad (2.38)$$

This result is the same as the one obtained by Tersoff and Hamann [2.26]. One can further get

$$\frac{2\pi c\hbar^2}{\kappa m}\frac{\partial}{\partial z_0}\varphi_\nu(\boldsymbol{r}_0) = \frac{2\pi c\hbar^2}{\kappa m}\frac{\partial}{\partial z_0}\int_\Sigma [G(\boldsymbol{r}-\boldsymbol{r}_0)\nabla\varphi_\nu - \varphi_\nu\nabla G(\boldsymbol{r}-\boldsymbol{r}_0)]\cdot\mathrm{d}\boldsymbol{S}$$

$$= \frac{2\pi c\hbar^2}{\kappa m}\int_\Sigma\{\frac{\partial}{\partial z_0}[G(\boldsymbol{r}-\boldsymbol{r}_0)\nabla\varphi_\nu - \varphi_\nu\nabla G(\boldsymbol{r}-\boldsymbol{r}_0)]\}\cdot\mathrm{d}\boldsymbol{S}$$

$$= \frac{\hbar^2}{2m}\int_\Sigma [\varphi_{p_z}\nabla\varphi_\nu - \varphi_\nu\nabla\varphi_{p_z}]\cdot\mathrm{d}\boldsymbol{S}$$

$$= M_{p_z}$$

The following gives the transition element and the corresponding tip state (φ_ν represents the sample state).

State	Tip state	Transition matrix $M \propto$
s	$c(\kappa\rho)^{-1}\exp(-\kappa\rho)$	φ_ν
p_z	$c[(\kappa\rho)^{-1} + (\kappa\rho)^{-2}]\exp(\kappa\rho)\cos\theta$	$\dfrac{\partial\varphi_\nu}{\partial z}$
p_x	$c[(\kappa\rho)^{-1} + (\kappa\rho)^{-2}]\exp(\kappa\rho)\sin\theta\cos\phi$	$\dfrac{\partial\varphi_\nu}{\partial x}$
p_y	$c[(\kappa\rho)^{-1} + (\kappa\rho)^{-2}]\exp(\kappa\rho)\sin\theta\sin\phi$	$\dfrac{\partial\varphi_\nu}{\partial y}$

From the above results, one can estimate the magnitude of the transition matrix for various tip states. It can be concluded that higher quantum numbered states can enhance the detection sensitivity. It is worth noting that for widely used tip materials such as W and Pt, the high quantum numbered states indeed exist at the vicinity of the tip apex [2.27]. An example can be seen in Fig. 2.8 for the charge distribution of W(001) surface. At a distance of 2 Å from the surface, the charge consists mainly of (d_{z^2}) electrons.

2.2.3 Scanning Tunneling Spectroscopy (STS)

An advantage of the STM method is associated with the capability of tunneling spectra [2.24]. The mechanism is rooted in the expression of tunneling current, as in (2.3):

$$J = \frac{2e}{h}\int \mathrm{d}E\,[f(E) - f(E+eV)]\int \frac{\mathrm{d}^2 k_t}{(2\pi)^2} D(E, k_t)$$

$$= \int_0^\infty \mathrm{d}E_x D(E_x) N(E_x)$$

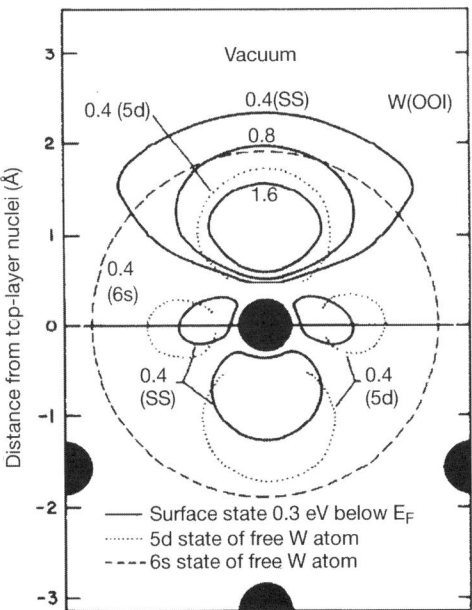

Fig. 2.8. The spatial distribution of density of states for the W atom at the W(001) surface. The states include the surface state, 6s, and 5d (extracted from [2.27])

Here, k_t is the momentum component parallel to the junction surface. The electron density of states $N(E_x)$ and transmission probability $D(E_x)$ are

$$N(E_x) = \frac{4\pi me}{h^3} \int_0^\infty [f(E) - f(E+eV)] \mathrm{d}E_t$$

$$= \frac{4\pi mek_\mathrm{B}T}{h^3} \ln\left\{\frac{1+\exp[(E_f - E_z)/k_\mathrm{B}T]}{1+\exp[(E_f - E_z - eV)/k_\mathrm{B}T]}\right\}$$

$$D(E_x) \simeq \exp\left\{-s\left[\frac{2m}{\hbar^2}\left(\overline{\phi} + E_F - \left|\frac{eV}{2}\right| - E_x\right)\right]^{1/2}\right\}$$

The normalized tunneling current is therefore [2.32]

$$\frac{\mathrm{d}I/\mathrm{d}V}{I/V} = \frac{\rho_s(eV)D(eV)}{\int_0^{eV} \rho_s(E)D(E)\mathrm{d}E/eV} + \cdots$$

$$\simeq \frac{\rho_s(eV)}{\int_0^{eV} \rho_s(E)\mathrm{d}E/eV} \qquad (2.39)$$

Note that it is assumed the electron density at the tip side is characteristic of free electrons. Therefore, the density of states of the sample can be explicitly reflected in the above expression. The consideration including the density of states of both tip and sample was given in the systematic analysis by

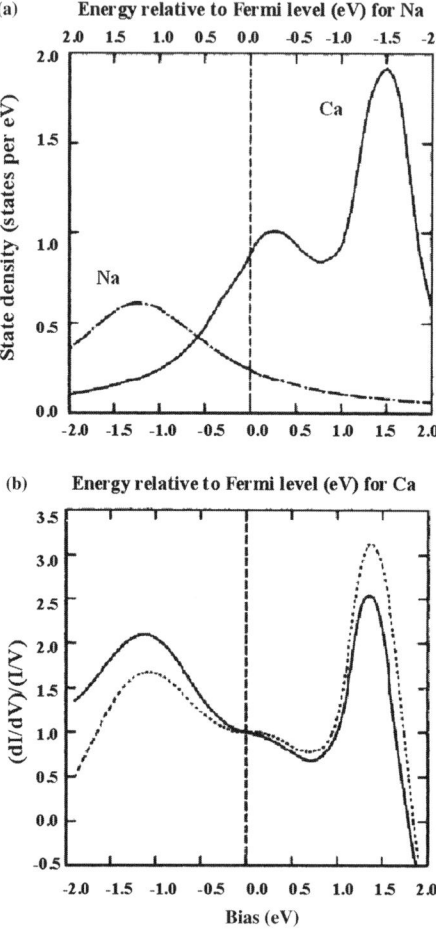

Fig. 2.9. a Distribution of density of states at a round Fermi level for Na and Ca atoms. **b** Calculated $dI/dV/(I/V)$ (or normalized conductance) for Ca/Na combination. (extracted from [2.31])

Lang [2.28, 2.31]. The results revealed that the tunneling current is jointly affected by the characteristics of density of state distribution at both sides of the electrodes (Fig. 2.9), suggesting that a rigorous discussion of STM/STS results may need to take into account the local density of states of the tip.

2.2.4 Measurement of Apparent Tunneling Barrier Height

The tunneling barrier is an important parameter in the study of tunneling phenomena. From traditional electron tunneling studies, it is known that the tunneling barrier is dependent on the electronic properties of the interface,

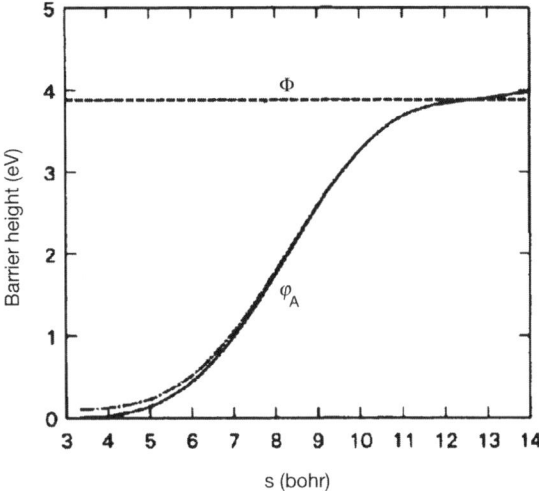

Fig. 2.10. Calculated behavior of apparent tunneling barrier at various tip-sample separations. A zero-barrier channel is depicted as a tip closing in to the sample surface (extracted from [2.33])

and the interaction between the tunneling electron and the insulating layer. The tunneling barrier can directly affect tunnel characteristics.

The thickness of the tunneling barrier in an STM junction is determined by the separation of the tip and sample. The range of the separation can be adjusted experimentally from point contact to the thickness of traditional tunneling junctions (tens of angstrom or larger). Such capability can provide a unique venue to study the effect of tunneling barriers. After taking into account the electrostatic potential and exchange correlation potential, Lang [2.33] concluded that the barrier height in STM junctions approaches zero at small separations (Fig. 2.10), suggesting the tunneling electron can experience little resistance at close tip-sample separations. Beginning from the expression of the tunneling current of STM (2.33):

$$I \propto V e^{-A\sqrt{\phi}s}$$
$$A = \frac{2\sqrt{2m}}{\hbar}$$
$$\ln I \propto \ln V - A\phi^{1/2} s$$

The barrier height can be deduced as

$$\phi^{1/2} = -\frac{1}{A}\frac{d \ln I}{ds}$$
$$\simeq -\frac{1}{A}\frac{d \ln I_{max} - \ln I_{min}}{s_{max} - s_{min}}$$
$$\simeq -\frac{1}{A}\frac{\ln(I_{max}/I_{min})}{\Delta s} \qquad (2.40)$$

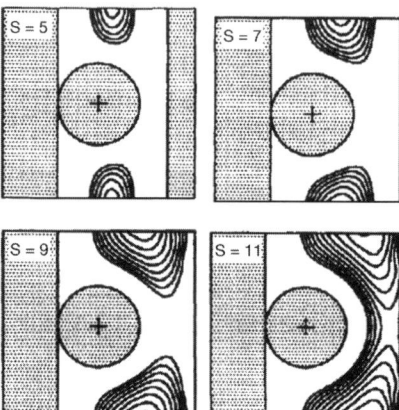

Fig. 2.11. Illustration of tunneling barrier distribution at various tip-sample separations. A zero-barrier channel is depicted as a tip closing in to the sample surface (extracted from [2.33])

The barrier height can have significant impacts on the topography and spectroscopy in STM measurements. As pointed out by Lang [2.33], tunneling electrons experience different barrier distributions as a tip approaches the sample surface. At certain separations, there is a finite-dimensioned tunneling channel that has no barrier within it. The general trend of a tunneling barrier as a function of tip-sample separation is illustrated in Fig. 2.11. This behavior has been qualitatively identified in measurements of layer compounds [2.34]. It was noted that in the case of low barrier height, which generally corresponds to small tip-sample separation, there is a risk of increased instabilities for both spectroscopy and topography measurements.

In addition, measurements of barrier height on metal surfaces revealed nearly constant behavior until the tip touches the sample surface. Olesen et al. [2.35] suggested that the tip-sample separation could be affected by the adhesion force. The tunneling junction resistance decreases as the tip approaches a sample surface. Considering that the resistance of an STM pre-amplifier is typically on the order of megaohms, the voltage across the junction is only part of the total applied bias. As a result, the assumption of constant bias is no longer valid. This effect can lead to an unchanged apparent barrier height.

In addition to the tip-sample separation-related barrier reduction, it has been shown that contaminants on the tip or sample surface could also result in anomalously low barrier heights, which is unfavorable for STM and STS studies. It is therefore desirable to have an appropriate barrier height for optimal STM/STS assessments. By adapting the barrier height measurement by means of a topography study, atomic resolution could be obtained on a sulfur adlayer on a Mo(001) surface (Fig. 2.12) [2.36]. This result could lead to

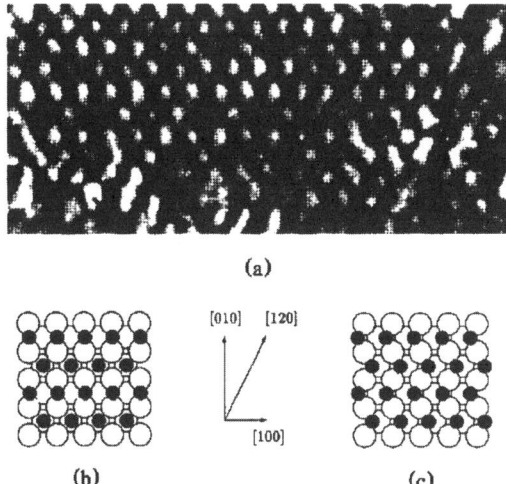

Fig. 2.12. a Atomic resolution of the barrier height distribution sulfur adlayer on a Mo(001) surface. **b, c** Two proposed models for the $p(1 \times 2)$ sulfur adsorbate structure on a Mo(001) surface (extracted from [2.36])

useful information about the microscopic distribution of local barrier height that is not available from other techniques.

The discussion in this chapter on various aspects of tunneling characteristics is largely derived from the knowledge of the ensemble average of molecules. As will be presented in the following chapters, studies at single molecule level have revealed a great deal of important insight that stimulated strong interest in the field of electron tunneling processes.

3 Single Molecule Structural Characterization

Single molecules, modeled as an isolated subject either in surface-bound state or in a three-dimensional environment (such as in solutions), experience non-periodic, heterogeneous interactions. It is conceivable that single molecules could adopt appreciably different properties, compared to the same molecule in the ensemble structures, because the surrounding environments are completely different. Such characteristics could develop into intrinsic merits of single molecule studies, such as high sensitivity to environmental variations, and uniqueness of dynamic properties, in many cases more complex in terms of theoretical analysis.

The motivation for single molecule structural studies is to understand the response of individual molecules to their immediate surrounding environment, such as supporting surfaces and solutions. This chapter is deals with the study of the characteristics of single molecules adsorbed to surfaces of metallic as well as organic monolayers. The knowledge of the principles behind molecular adsorption sites, adsorption configuration, etc., could lead to a fundamental understanding of intermolecular interactions, as well as molecule–surface interactions at microscopic scale. These issues also form the core topics of surface science studies. It is generally considered that acquired novel properties of single molecules will be discovered in the near future, and new technologies are expected in this domain.

On a very closely related topic, the exploration of the molecule assembling process has been a major scientific endeavor for several decades and represents one of the most productive scientific advances of our times. The assembled molecular structures are based on self-repeating units of individual molecule units, involving various intermolecular interactions. These interactions may be periodic and homogeneous. Single molecule studies can be seen as deeply rooted in the study of molecular assemblies, stimulated by fascinating advances in a range of experimental techniques in the past decade that facilitate single molecular-level experiments. Such increase in experimental capacity has spawned a vast and renewed interest to study molecules from the perspective of individual entities.

We could be witnessing the early stage of single molecule studies. The advance in both experimental techniques and theoretical insight must surely stimulate further studies in this field. The achievements could make important contributions to our understanding of molecular-based technologies.

3.1 Molecular Imaging Mechanisms of STM

Observing and identifying individual molecules forms the basis of practical investigations on single molecules. An important application of the STM method is to investigate the fine structures of various organic and biological molecules, in their surface-bound state. Many interesting observations have been reported in the past decade, such as high-resolution STM images of molecules (aromatic molecules, metal phthalocyanines or MPc, etc.) [3.1–3.4] adsorbed on metal surfaces, and the successful studies of a number of self-assembled molecular layers, liquid crystals [3.5–3.7] and long chain alkane molecules [3.8, 3.9] on inert surfaces of graphite and MoS_2. Another example is the high-resolution imaging of organic adsorbates on electrode surfaces under electrolyte-using electrochemical STM (ECSTM) [3.10–3.12]. These results have provided direct venues to examine the structural characteristics of molecules under a wide variety of experimental conditions, as well as the electronic properties of the molecules, such as the front orbital distributions and molecular polarizabilities. Furthermore, the adsorbate–substrate interaction is another important aspect. In general, the assembled structures of adsorbed organic molecules are stabilized mainly by the electrostatic multipole–multipole interaction, steric repulsion, intermolecular and surface forces.

The effort to resolve isolated single molecules by STM helps lay the ground for an extensive investigation of chemical and physical properties of single molecules in their surface-bound states. Related interests include adsorption geometry, understanding of submolecular contrast in STM images, spectroscopy characteristics of single molecules, and chemical reactions of single molecules. The results can be directly correlated to the interaction between molecules, and the impact of the environment, such as the substrate. As will be demonstrated in this chapter, STM is a powerful technique in imaging single molecules on metal, semiconductor and organic surfaces.

The accompanying theoretical efforts, as exemplified in this section, are crucial for our understanding of the physical nature of observed features for single atoms and molecules. The simulation of molecules observed by STM involves the modeling of both the tip states, as discussed in the preceding chapter, and the states of the adsorbed atoms. Several mechanisms have been assessed by various authors in this domain. One of the first mechanisms proposed is based on the perturbation theory by the formalism of Bardeen, Tersoff and Hamann, which was introduced in the preceding chapter [3.13, 3.14]. The application of scattering formalism to the interpretation of adsorbed molecules has also been extensively pursued [3.15].

The simulated profiles of molecules point to the possible chemical sensitivity of STM observations. The variation in the predicted contrast, such as from protrusion to depression, originates in the eigenstates of the tunneling current. Comprehensive reviews of STM imaging mechanisms of molecules can be found in the literature [3.13–3.15].

3.1.1 Molecular Orbital Model

Achieving submolecular resolution using STM is regarded as a major advantage of the technique, and also promotes identifying the electronic properties of single molecules. As generally interpreted, STM images represent the mapping of the local electronic density of states. The contribution from molecular orbitals, both lowest unoccupied molecular orbitals (LUMO) and highest occupied molecular orbitals (HOMO), is of key importance to establish the relationship between STM images and the predicted geometry. Such comparisons have been conducted for a number of molecules. The effects of, for example, substrate lattice and molecular orientation need to be taken into consideration to obtain acceptable results.

So far, much attention has been given to the understanding of contrast mechanisms associated with various functional groups. STM images of 4-n-alkyl-4'-cyanobiphenyls (nCBs), where n is the number of carbon units in the alkyl group, on graphite showed that the aromatic group or cyclohexane enhances the tunneling efficiency and appears as bright contrast [3.5–3.7]. Contrast differences are also prevalent in the derivative of linear alkanes [3.8, 3.9], and planar molecules [3.4].

Figure 3.1 is an STM image showing individual copper(II) phthalocyanine molecules (denoted as CuPc) with the molecular orbitals given in Fig. 3.2 [3.2]. The example shows that submolecular features can be directly correlated with the characteristics of the molecular orbitals. It should be noted that such correlations may reflect a grossly simplified signature. It can not be ruled out that a combination of front orbitals is the dominant contributor to the observed pattern.

On the other hand, the resonant tunneling mechanism, also proposed in terms of the positions of front orbitals (HOMO/LUMO) [3.5–3.7], provides a direct linkage between experimentally observed fine structures and intrinsic molecular orbitals. In many cases, the front orbitals are separated well away from the Fermi level of the substrate by more than one electron volt. Therefore, in order to account for the imaging at small tunneling bias voltages, one should consider the factors that could decrease the energy difference between the HOMO or LUMO of molecules and the substrate's Fermi level, such as the accompanying pressure within the STM gap, interactions between molecules and the substrate, and the applied electric field. It is worth noting that solid evidence is still needed to fully explain the contrast differences for organic and biological molecules. The results should be very beneficial for further studies aimed at recognizing individual functional groups within molecules.

The theoretical analysis of STM imaging of Xe (xenon) atoms also suggests a front orbital effect [3.16]. The electrical resistance of wires consisting of either a single Xe atom or two Xe atoms in series was measured and calculated on the basis of an atom–jellium model. Both the measurements and the calculations yielded a resistance of $10^5\,\Omega$ for the single-Xe atom system, and $10^7\,\Omega$ for the two-Xe atom system. These resistances greatly exceed

32   3 Single Molecule Structural Characterization

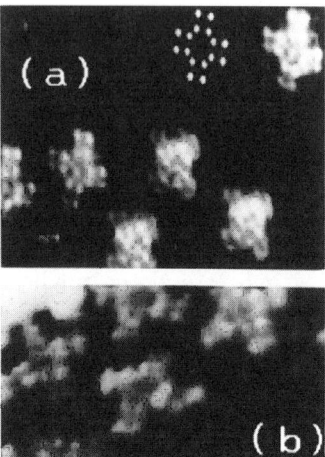

Fig. 3.1. STM image of CuPc molecules dispersed on a Cu(100) surface with tunneling conditions of −0.15 V and 2 nA at both **a** low and **b** high adsorption coverages. The internal structures represent the characteristics of the molecular front orbitals (extracted from [3.2])

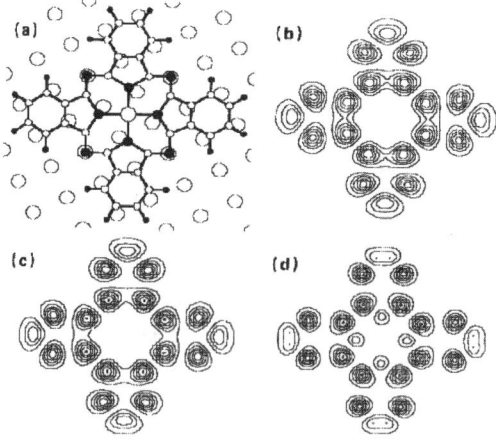

Fig. 3.2. Calculated molecular orbitals of a CuPc molecule corresponding to the experimental results in Fig. 3.1. **a** Structure model CuPc on a Cu(100) surface. *Small open circles* C, *small closed circles* H, *large open circles* Cu, *large closed circles* N. **(b)** and **(c)** are the distribution of charge density of HOMO and LUMO respectively at 2 Å above molecular plane. **(d)** is the charge density of HOMO at 1 Å above the molecular plane (extracted from [3.2])

the 12,900-Ω resistance characteristic of an ideal one-dimensional conduction channel, because conduction through the xenon atoms occurs through the tail of the xenon 6s resonance, which lies far above the Fermi level. Such conduction processes in a single atom-sized system can now be understood in terms of the electronic states of individual atoms.

3.1.2 HOMO-Ionization Potential Model

Further consideration involves the ionization potential (IP) to account for the observed contrast. The observed STM image contrast of alkane derivatives was shown to be affected by both topographic and electronic factors (Fig. 3.3) [3.17]. The contrast of the methylene regions is dominated by the H atom positions, as predicted by perturbation theory [3.18] and extended Hükel calculations [3.19]. As for the terminal functional groups, it has been found that the contrast depends on the magnitude of the applied bias voltage. The contrast strength was also shown to be consistently correlated to

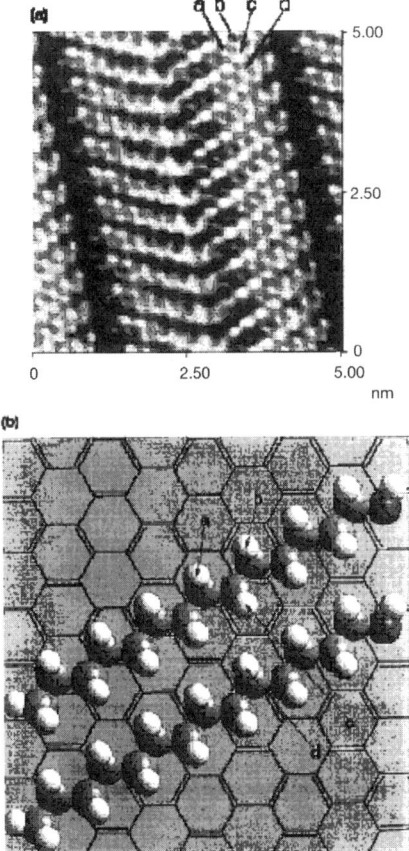

Fig. 3.3. a High-resolution STM image of tetradecanol molecules on a graphite surface (bias 1.127 V; tunneling current 650 pA). b Schematic of adsorbed tetradecanol molecules in trans-conformation (extracted from [3.17])

the ionization potential. The lower ionization potential leads to a higher contrast in STM images. This correlation could be caused by the contribution of HOMO to the tunneling current in STM. A lower ionization potential corresponds to closer energy levels between HOMO and the Fermi level of the tip, and more diffused molecular orbital structures. Both effects promote the enhancement of coupling between the adsorbate and the tip and sample. It has also been pointed out that the contrast of bromide could be accounted for by considering the contributions from both HOMO and LUMO orbitals [3.18].

3.1.3 Work Function Model

For saturated hydrocarbons and their derivatives, a different imaging mechanism has been proposed. This category of molecules have their orbitals far away from the Fermi level of the tip. The modification of the local work function due to the polarization of functional groups could be the factor contributing to the STM contrast. This has been qualitatively confirmed by systematically comparing the contrast of groups with various polarizability.

Several theoretical models have been suggested and have generated considerable interest. Spong et al. [3.20] suggested that local work function modification by the adsorbates is the origin for the observed contrast. By taking into account the polarizability of the adsorbed molecules, one can obtain the modified local barrier height:

$$\phi = \Phi - e\mu/\varepsilon_0 \qquad (3.1)$$

where ϕ is the barrier height for the adsorbate/substrate system, Φ the barrier height for a clean surface, μ the dipole moment of the adsorbates, and ε_0 the permitivity of free space. Thus, the dipole moment of organic molecules can reduce the barrier height of substrates such as graphite. As a result, the work function of the substrate surface will be periodically modified by the adsorbates and it becomes a function of position, such as in the case of linear alkane molecules. It should be noted that the terminal functional groups in the above studies have no net dipole moment in the direction perpendicular to the basal plane. The effect of polarizability due to the induced dipole moment alone was not sufficient to account for the observed contrast insensitivity to the bias polarity.

In a recent study of tridodecyl amine (TDA) molecules, which has a net dipole moment perpendicular to the adsorbate plane, a contrast reversal behavior was observed [3.21]. The asymmetry in contrast for positive and negative bias was associated with the total polarization of the amine group. The tunneling bias in this study was sequentially changed from negative to positive, and from positive to negative during the imaging process. It was observed that the amine group appears in higher contrast than does the alkane part when positive bias was applied, the reverse being the case when negative bias was used (Figs. 3.4a and b). The measured contrast variation is shown in Fig. 3.5. According to the molecular structure of TDA, the nitrogen

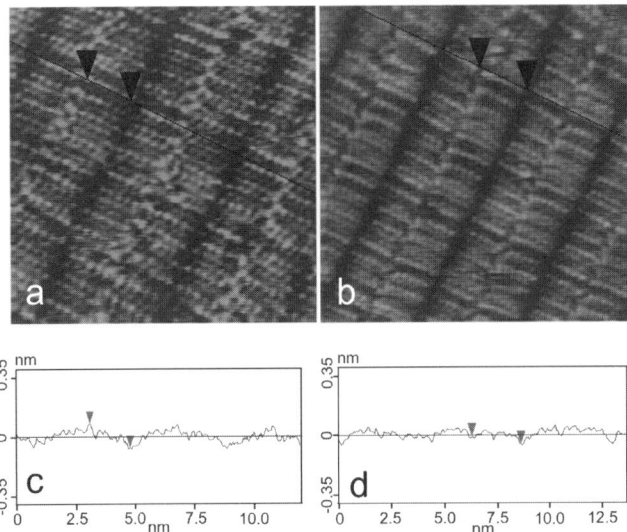

Fig. 3.4. STM images showing the effect of applied bias polarity on the STM contrast of amine groups. **a** 459 mV, 626 pA, **b** −405 mV, 626 pA. **c, d** are the corresponding cross-sectional profiles of the STM images of **a** and **b**, respectively. Note that the contrast of the central parts of the lamella is reversed for the two tunneling conditions (extracted from [3.21])

atom in the amine group is on one acme of the tetrahedron. Since the C–N bond is dipolar, the amine molecules are also dipolar, nitrogen being partially negatively charged. Thus, when amine molecules adsorb onto an inert surface of graphite, there exists an intramolecular electric field pointing either to the substrate or to the air. This intrinsic electric field can interact with the applied external field, and therefore the contrast of the nitrogen will be enhanced or diminished, depending on the polarity of the applied bias and the orientation of the C–N bond.

As proposed earlier for STM observations of organic molecules [3.20], the contrast of functional groups could be associated with the modification of local work function by the molecular dipole moment [3.22]. It was conjectured that the observed contrast reversal associated with bias polarity could be accounted for by the modification of local work function by the dipole moment of amine groups. The work function is expressed in (3.1), and the dipole moment of the molecule is:

$$\boldsymbol{\mu} = \boldsymbol{\mu}_0 - \alpha(\boldsymbol{E}_{\mathrm{dp}} + \boldsymbol{E}_{\mathrm{ext}}) \tag{3.2}$$

Here, μ_0 is the permanent dipole moment, α the polarizability of the molecule, e is electron charge. $\boldsymbol{E}_{\mathrm{dp}}$ the depolarization field associated with the adjacent dipole molecules [3.20], and $\boldsymbol{E}_{\mathrm{ext}}$ the external field between the tip and sample. Therefore, it can be seen that the induced dipole moment is dependant

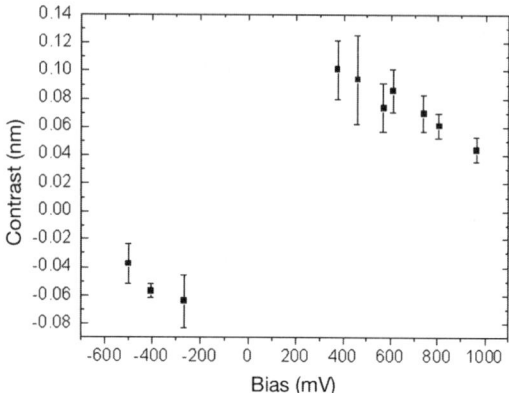

Fig. 3.5. The relationship of the STM image contrast of amine groups versus bias voltage (extracted from [3.21])

on the direction of the external field. The barrier heights for different bias polarities can be expressed in simplified form:

$$\phi_+ = \phi_0 - e\mu_0/\varepsilon_0 + \alpha(E_{dp} + E_{ext})/\varepsilon \quad (3.3)$$
$$\phi_- = \phi_0 - e\mu_0/\varepsilon_0 + \alpha(E_{dp} - E_{ext})/\varepsilon \quad (3.4)$$

By substituting the above expressions into that of the tunneling current I,

$$I \propto V \exp(-A\phi^{1/2}d)$$

where d is the separation between tip and sample.

One can rationalize the change in apparent contrast as a result of varying barrier height. As a comparison, when functional groups (such as amine) adsorb with a "flat" conformation where the molecular dipole is aligned perpendicular to the field direction, no contrast reversal is observed when bias polarity is changed [3.12, 3.23].

3.2 Single Diatomic Molecules on Metal Surfaces

The above section on imaging mechanisms has widened our view of the physical nature of the otherwise merely topography-based images of adsorbed molecules. In the subsequent discussions, we will present examples ranging from seemingly simple diatomic molecules to complex structures such as macrocyclic molecules. In this section, we show results on two representative diatomic molecules, i.e., CO as a heteronuclear molecule and O_2 as a homonuclear molecule. Their structures are inherently simple, compared to other molecules discussed in the subsequent sections. Additional examples of single diatomic molecules, dealing with dissociative adsorption, are given in the following chapter. The successful identification of CO and O_2 can also be viewed as representative in the pursuit of ultimate STM resolution of single molecules. These efforts are within the main themes of surface science to clarify adsorption configurations, and the STM results provide confirmative evidence for known information from other surface science techniques. The studies also provide a test ground for imaging mechanisms that are common to all STM studies of single molecules.

3.2.1 CO

CO molecule adsorption on metal surfaces has been one of the classical topics in surface science for decades. The interest in this molecule emerges from many aspects of surface chemistry. The pursuit of adsorption/desorption geometries, diffusion and reaction pathways has provided rich grounds for both theoretical and experimental investigations. The STM method provides the capability of imaging individually adsorbed CO molecules in real space, which is essential for studying the site specificity of molecular adsorption.

Adsorbate-induced restructuring of metal surfaces has been widely observed for a number of systems. Mass transportation and mobility of surface metal atoms are known to be caused by molecular adsorption through the formation of adsorbate–substrate species, weakening the interaction between substrate atoms. It was observed that CO adsorption at elevated pressures (>1 torr) could induce significant restructuring of Au(110) surfaces [3.24]. Micro-roughening of Pt(110) by CO adsorption is another example of the adsorbate-induced surface effect [3.25]. In another study, a dense CO adlayer on Ni(110) was observed at room temperature. The CO molecules are adsorbed at the short-bridge sites along the [1$\bar{1}$0] direction and form a 2×1 structure [3.29].

More detailed imaging results revealed that CO molecules appear with different characteristics. Pronounced shape variation of individual CO molecules adsorbed on Pt(111) was first reported by Stroscio and Eigler in 1991 [3.26]. It was proposed, and later confirmed by theoretical analysis [3.27], that the CO molecule at the top site appears as a clear protrusion, whereas the bridge site CO looks like a "sombrero", or an indentation within an enhanced circular boundary (Figs. 3.6 and 3.7). The simulation results suggest that the contribution from 3σ and 5σ orbitals of CO is site dependent. The molecule-mediated tunneling current at the top site is the cause for the protrusion-shaped CO image, whereas a much reduced contribution at the bridge site leads to the depressed contrast in images for bridge site CO.

Individually adsorbed CO molecules have been observed on Cu(211) surfaces as indentations, rather than protrusions as on other metal surfaces [3.28]. However, at high CO coverage, CO molecules are likely to appear as protrusions or enhanced contrast. This was postulated as the effect of a molecularly decorated tip. Such phenomena motivated more conclusive studies on imaging molecules with chemically specific tips [3.28, 3.30]. With the

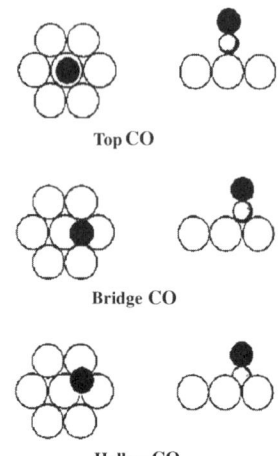

Fig. 3.6. Schematic of CO adsorption sites on a (111) surface (extracted from [3.25])

CO on platinum(111)

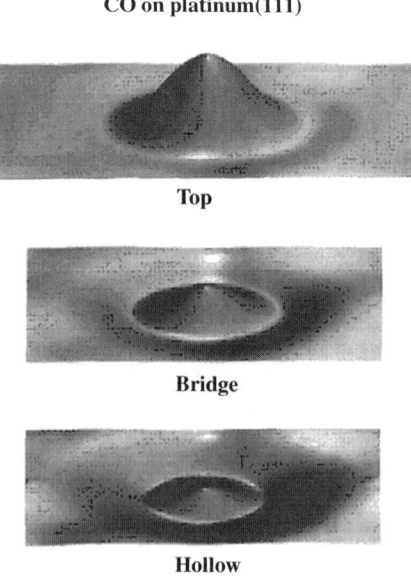

Fig. 3.7. Calculated STM image of CO molecules on a Pt(111) surface at the tunneling condition of 10 mV and 1 nA (extracted from [3.25])

capability of reliably transferring molecules between the tip and substrate, it was demonstrated that CO molecules, originally appearing as indentations on Cu(111) surfaces, can be turned into protrusions with a CO-decorated tip (Figs. 3.8 and 3.9) [3.30]. It has been observed that CO molecules appear as

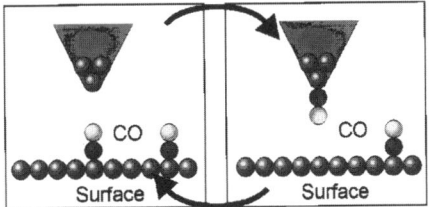

Fig. 3.8. Schematic of transferring CO molecules between the tip and substrate surface by ramping bias voltage (extracted from [3.30])

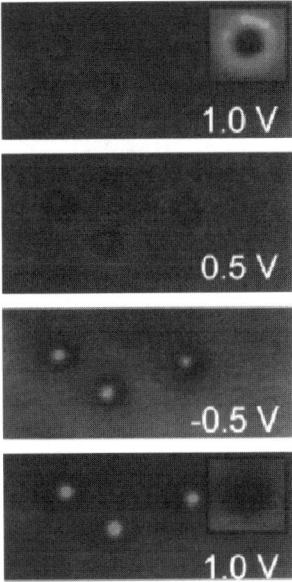

Fig. 3.9. Contrast variation of CO molecules using a CO-decorated tip for four different tunneling bias voltages at 15 K. The appearance of the CO molecules changes from indentation to protrusion as the bias voltage changes (extracted from [3.30])

indentations with a metallic tip, independently of the applied bias. On the contrary, coadsorbed O_2 molecules do not display contrast variation either with or without a CO tip [3.30].

It was shown that both topography resolution and the spectrum sensitivity can be improved by modifying the STM tip with single molecules. The adsorption sites of CO and O_2 are determined as atop and fourfold hollow on Ag(110) at 13 K (Fig. 3.10) [3.31]. CO molecules display a contrast reversal from indentation to protrusion when imaged with a bare metal tip and CO-decorated tip, respectively. The contrast variation associated with a C_2H_4

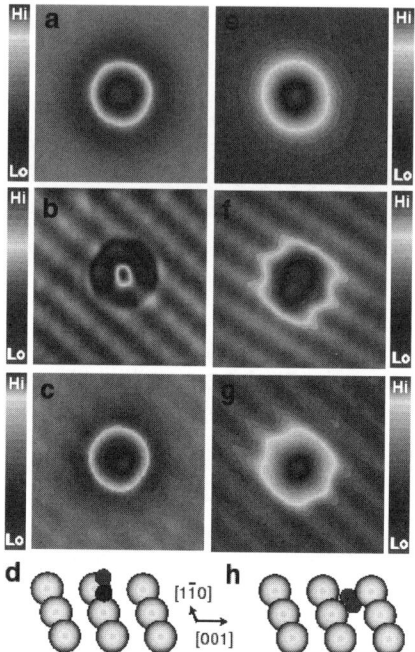

Fig. 3.10. STM images of CO (**a–c**) and O_2 (**e–g**) molecules on a Ag(110) surface using different tips, including bare metal (**a** and **e**), CO tip (**b** and **f**), C_2H_4 tip (**c** and **g**) (extracted from [3.31])

tip is less than that of a CO tip. The above results indicate the probable complexity of the STM tip in obtaining topography imaging.

In this same study, O_2 molecules appeared as indentations when observed with a bare tungsten tip, CO or ethylene (C_2H_4)-modified tips, with different apparent heights. By subtracting the spectrum background associated with the tip molecule, the hindered rotation mode of CO on Ag(110) (21 meV at positive bias and 19 meV at negative bias) was identified with a CO tip, and the hindered translation mode of CO on Ag(110) (7 meV) was recorded with a C_2H_4 tip (Fig. 3.11). A feature at around 26 meV was found with a CO tip on $O_2[1\bar{1}0]$, which was suggested to be the hindered rotation of $O_2[1\bar{1}0]$ parallel to the surface and/or the O_2–Ag stretch mode. It is interesting to note that the detection of the CO stretching mode (at about 240 meV) using the STM method remains a controversial issue. One possible reason might be the lateral disruption of the tip at the corresponding energy.

3.2.2 O_2 Molecules

O_2 molecules were found to adsorb at fourfold hollow sites on Ag(110) surfaces with the O_2 bond axis along the [001] direction, as shown in Fig. 3.12

3.2 Single Diatomic Molecules on Metal Surfaces   41

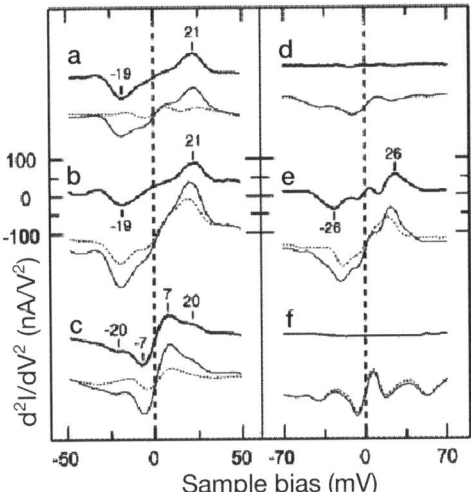

Fig. 3.11. Tunneling spectra of single CO molecules measured with *a* bare metal tip, *b* CO tip and *c* C–2H$_4$ tip. *Thick solid line* is the subtraction result of spectra at the molecular site (*thin solid line*) and clean Ag(110) surface (*dashed line*). *d–f* are parallel experiments with O$_2$ molecules (extracted from [3.31])

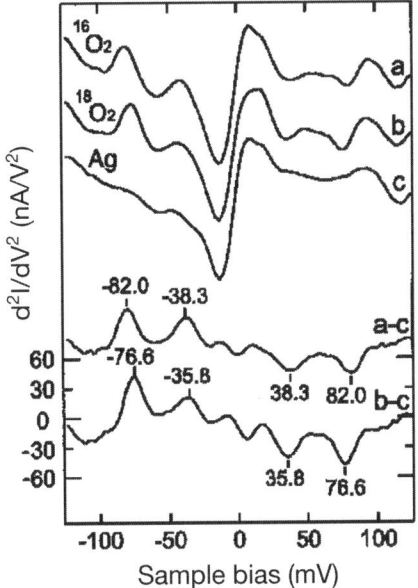

Fig. 3.12. Tunneling spectra of single O$_2$ molecules adsorbed on Ag(110) surface (extracted from [3.12])

[3.32]. IETS measurements were able to identify the O–O stretch mode at 82 meV (76.6 meV for $^{18}O_2$). It was noted in this study that the relevant vibrational features are representative of a decrease in tunneling conductance. The interpretation was provided by an earlier study that predicted the excitation of vibrational mode could be accompanied by suppression of elastic tunneling, as a result of interference between direct tunneling and the excitation of vibrational mode [3.33–3.35].

Olsson et al. [3.36] used density function theory and the Tersoff-Hamann approximation of the tunneling process to simulate the adsorption site and LDOS of O_2 on Ag(110). Consistent results were obtained using an s-wave tip and p_z-wave tip for imaging.

3.3 Aromatic Molecules and Macrocyclic Molecules

The STM technique has significantly advanced our capability of resolving internal structures of single molecules such as benzene and other planar and linear species. The adsorption configurations could be more complicated than that for small molecules such as CO, because of constitutive functional groups. Rigorous interpretation of observed high-resolution molecular structures has been a challenging topic that requires quantitative calculation of the molecule–substrate systems, and of the interaction between tunneling electrons and molecular orbitals. These investigations have become a vital part in the field of STM-related research.

3.3.1 Single Benzene Molecules Observed by STM

The first real-space STM image of benzene was obtained on Rh(111) surfaces when coadsorbed benzene and CO form compacted 3×3 [3.37] and c($\sqrt{3} \times 4$)rect [3.38] structures. In the 3×3 overlayer structure, both benzene and CO are on the hcp hollow sites based on LEED analysis. In the c($\sqrt{3} \times 4$)rect structure, benzene occupies the hcp site and CO is at the threefold hollow site, and both species can be resolved in the STM images. The existence of CO is an example of using coadsorbates to reduce later diffusion of the molecules of interest. A characteristic threefold pattern was observed, consistent with the symmetry of empty π orbitals of benzene located near the Fermi level of the substrate [3.15].

Distinctively different patterns were observed for single benzene molecules adsorbed on Pt(111) surfaces at 4 K. Thus, threefold patterns (with two configurations of 60° relative rotation), and cylindrical and simple protrusions were typically observed, as shown in Fig. 3.13 [3.39]. The subsequent analysis using the electron-scattering quantum chemistry (ESQC) method [3.40,3.41] concluded that these characteristic patterns correspond to different adsorption sites i.e., hcp hollow site, on-top site, and bridge site, respectively (Fig. 3.14a). These data suggest that the observed patterns of single molecules

3.3 Aromatic Molecules and Macrocyclic Molecules   43

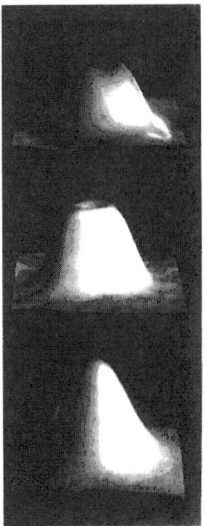

Fig. 3.13. Characteristic topography images of a single benzene molecule on Pt(111) (extracted from [3.39])

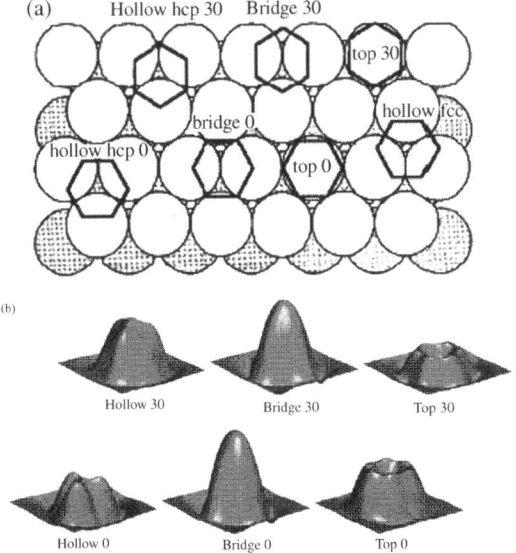

Fig. 3.14. a Schematic adsorption structure model of the adsorption site of benzene molecules on a Pt(111) surface. **b** Calculated STM images of benzene molecules at different adsorption sites (extracted from [3.40])

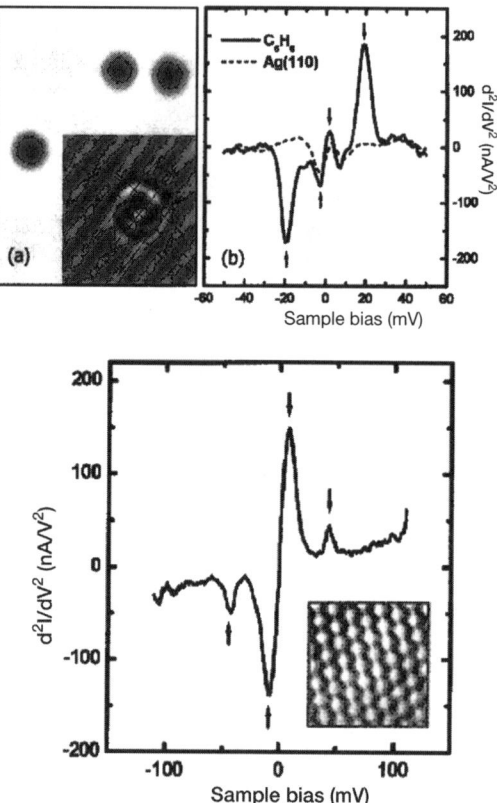

Fig. 3.15. a STM image of a single benzene molecule on a Ag(110) surface and the corresponding tunneling spectrum. **b** Tunneling spectrum for benzene molecules in a dense layer (extracted from [3.42])

could be a reflection of the interaction between single molecule and local substrate structures (Fig. 3.14b).

Site-dependent adsorption properties of single benzene molecules on Ag(110) were studied by Pascual et al. [3.42, 3.43]. Comparison of the adsorbate stabilities demonstrated that benzene is physisorbed on Ag(110) surfaces at 4 K and chemisorbed at temperatures below 66 K [3.43]. Benzene molecules were observed to adsorb at the atop sites in open terraces with molecular planes parallel to the surface at about 30 K (Fig. 3.15). In addition, an asymmetric shape for benzene molecules was observed at sites close to the corner of kink sites. The IETS study at 4 K revealed significantly different characteristics for molecules at the two types of sites (Fig. 3.15). The difference was attributed to the distortion of adsorption geometry at the kink sites that may modify the interaction between tunneling electrons and vibration excitation state [3.42].

In a similar study, benzene molecules were found to adsorb exclusively at the sites above step edges, possibly due to the well-known Smoluchowski effect [3.44]. The orientation of the compacted molecular domains developed at higher benzene coverage is subsequently determined by the step direction. In addition, the STM images were observed to be dependent on tip-sample separation. The images obtained with a given tunneling bias will vary from molecular modulation at large tip-sample separations to clearly Ag(110) lattices at small tip-sample separations [3.45]. This observation could be an example of STM imaging of varying decay behavior of sample wave functions. The wave functions of the benzene molecules extend beyond the tail of Ag states, and can be detected by an STM tip first when approaching the sample surface. With the Ag wave function being screened by the benzene overlayer, the Ag(110) lattice can be observed only at smaller junction distance.

Theoretical simulations by ab initio methods for the total energy and electronic structure of benzene molecules adsorbed on graphite and MoS_2 surfaces concluded that the molecular states are weakly mixed with the LDOS of the substrate [3.46]. The results support the interpretation of observed images of organic molecules as being the decayed tail of molecular orbitals situated far away from the substrate Fermi level.

Adsorption of benzene molecules on oxygen pre-covered Ni(110) surfaces causes structural transition from (3×1) to (2×1) as a result of compression of –Ni–O– rows [3.47]. The benzene molecules were observed to adsorb along the newly opened trough in a [001] direction. STM observations of adsorbed benzene molecules on Cu(111) surfaces at 77 K revealed two distinct states, i.e., two-dimensional solid state along step edges, and two-dimensional molecular gas state in open terraces [3.48, 3.49]. In the two-dimensional solid state, the benzene molecules are densely packed along step edges, and nucleate into small islands on terraces above step edges. The interface is formed at the boundary of the nucleated bands, which allows direct investigation of the dynamic behavior of individual molecules in two-dimensional gas and solid states.

Individual benzene molecules were observed to adsorb at the fourfold hollow sites of Pd(110) surfaces and evolve into a molecular superlattice of $c(4 \times 2)$ within a temperature range of 100–300 K [3.50]. A diffusion barrier of about 0.57 eV along the [001] direction was estimated based on the hopping rate of benzene molecules determined from STM images.

A variety of adsorption structures have been observed for iodobenzenes on Cu(111) at near-saturation coverage. As many as seven coexisting superstructures were identified with submolecular resolution [3.51]. The results point to the joint effect of molecule–molecule and molecule–substrate interactions for physisorbed species. The tip-induced dissociation effect of the adsorbed iodobenzene molecules will be discussed in the next chapter.

3.3.2 Phthalocyanines (Pc)

The first high-resolution image on "flat-lying" copper phthalocyanine (CuPc) molecules was obtained on a Cu(100) surface. A clearly resolved fourfold symmetry is consistent with the symmetry of molecular orbitals (HOMO and LUMO) [3.2]. The observed internal structures provide direct mapping of the density of states distribution of the molecule–substrate system. CuPc monolayers on graphite and MoS$_2$ surfaces prepared by the molecular beam epitaxial method were also studied by STM. Two distinct assembly phases were observed on MoS$_2$, and a quadratic, close packing pattern was observed on graphite surfaces [3.52].

The contrast of center ions of metal phthalocyanine (MPc) in STM images has been shown to be related to the metal ion valence configurations [3.53–3.55]. Specifically, the Co(II) d^7 and Fe(II) d^6 have larger d orbital contribution at the Fermi level than do Cu(II) d^9, and Ni(II) d^8 ions, corresponding to bright and dark contrasts in STM images, respectively (Fig. 3.16).

Individually adsorbed CuPc molecules were found immobilized on semiconducting surfaces of hydrogen-terminated Si(111) [3.56,3.57], Si(100) [3.57], GaAs(110) [3.58], InSb(100), and InAs(100) [3.59,3.60]. The CuPc molecules are positioned on top of the Si dangling bonds with symmetric configuration. CuPc molecules were observed to adsorb on terraces of Ag(110) surfaces and induce faceting [3.61]. In another study, monoatomic steps on Ag(111) surfaces were also observed to be modified by the adsorption of CuPc [3.62].

Apart from the STM studies under ultra-high vacuum conditions, it was demonstrated that peripherally coordinated alkane chains could substantially improve the adsorption and immobilization of phthalocyanines on inert surfaces such as graphite, in organic solutions or under ambient conditions.

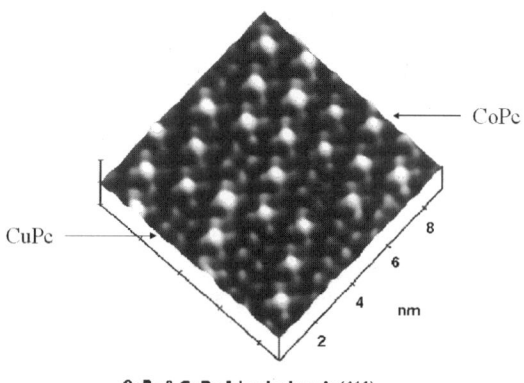

Fig. 3.16. Topography of a mixed assembly of CoPc and CuPc on a Au(111) surface. The *bright-centered* molecules are CoPc and the *dark-centered* molecules are CuPc (extracted from [3.53])

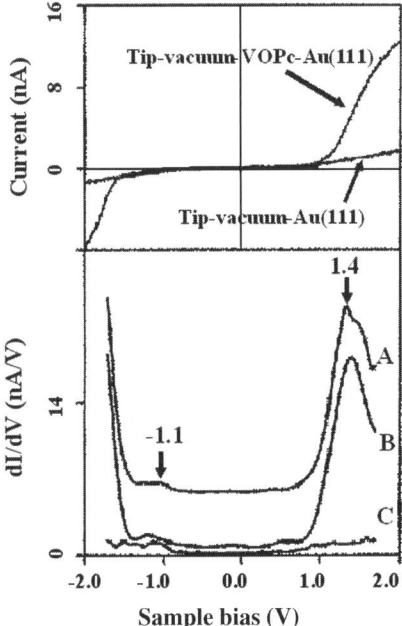

Fig. 3.17. Tunneling spectroscopy of VOPc adsorbed on Au(111). *Upper panel* is I versus V and *lower panel* is dI/dV versus V characteristics (extracted from [3.67])

Highly resolved submolecular features of the phthalocyanine cores and the interdigitated alkyl parts are clearly visible [3.63]. This approach can be adapted to studies of other molecules to observe submolecular features, and could also be helpful to obtain two-dimensional assemblies of mono-dispersed molecules, especially planar molecules.

For the non-planar type of MPc such as SnPc, STM images can help directly resolve the orientation of the molecular plane [3.64]. Tunneling spectroscopy measurements of SnPc adsorbed on graphite surfaces indicated an energy gap of around 3.4 eV [3.65]. For PbPc adsorbed on MoS_2, the Pb^{2+} ion is positioned out of molecular plane due to the large ion radius. Both bright and dark contrasts were observed as a result of molecular geometry [3.66]. Another non-planar phthalocyanine species is vanadyl phthalocyanine (VOPc), in which the oxygen atom is positioned outside the phthalocyanine molecular plane at a similar distance as that of Pb in PbPc. However, only dark-centered molecular patterns were observed [3.67, 3.68], in contrast to the results on PbPc and SnPc. This was attributed to the blocking effect of oxygen atoms that hampers the tunneling current. The accompanying spectroscopy measurements revealed an energy gap of around 2 eV (Fig. 3.17).

An example of electronic properties of phthalocyanines is found in their potential to act as a molecular rectification unit. Evidence of rectification

48   3 Single Molecule Structural Characterization

behavior of CuPc was observed in the asymmetric I–V characteristics measured by STS methods using acidified graphite as substrate [3.69]. In this case, the CuPc molecules were bonded to the substrate through N atoms in the outer ring, resulting in a tilted molecular plane relative to the substrate surface. There may be several factors that contribute to the reported rectification phenomena, including asymmetric tunneling junction configuration of substrate–molecule–tip, molecular front orbitals, etc. The capability of measuring tunneling characteristics at single molecule level should help advance our knowledge on the mechanisms of electron transportation in molecules.

3.3.3 Porphyrin

A number of porphyrin molecules have been subjected to STM studies. Cu-tetra[3,5 di-t-butylphenyl]porphyrin (Cu-TBPP) consists of a porphyrin core connected with four additional di-tertiary butylphyenyl(DBP) ligands as appendages [3.70]. High-resolution STM images on Cu-TBPP adsorbed on Cu(100), Au(110), and Ag(110) surfaces revealed characteristic molecular patterns for the molecule, which were attributed to the rotation of DBP groups as a result of different molecule–substrate interactions. The molecule interacts with the substrate surface through the hydrocarbon groups in DBP ligands. This interaction can vary according to substrate atomic corruga-

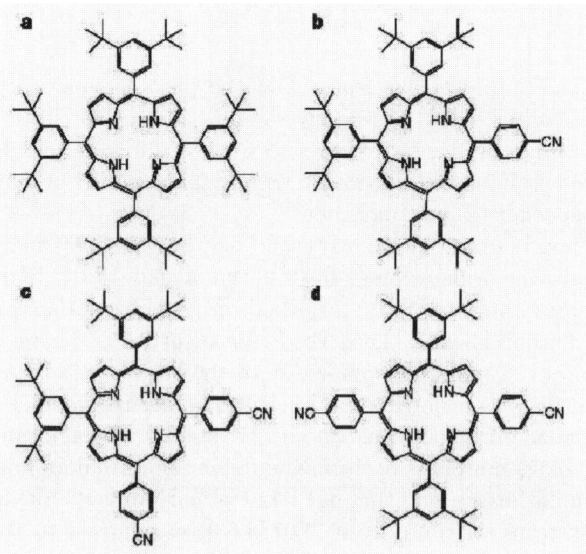

Fig. 3.18. Molecular structures of cyanophenyl porphyrin with differently numbered cyanophynyl substitution groups. **a** 5, 10, 15, 20-tetrakis-(3, 5-di-tertiarybutylphenyl) porphyrin (H_2-TBPP); **b** CTBPP; **c** *cis*-BCTBPP; **d** *trans*-BCTBPP (extracted from [3.72])

3.3 Aromatic Molecules and Macrocyclic Molecules   49

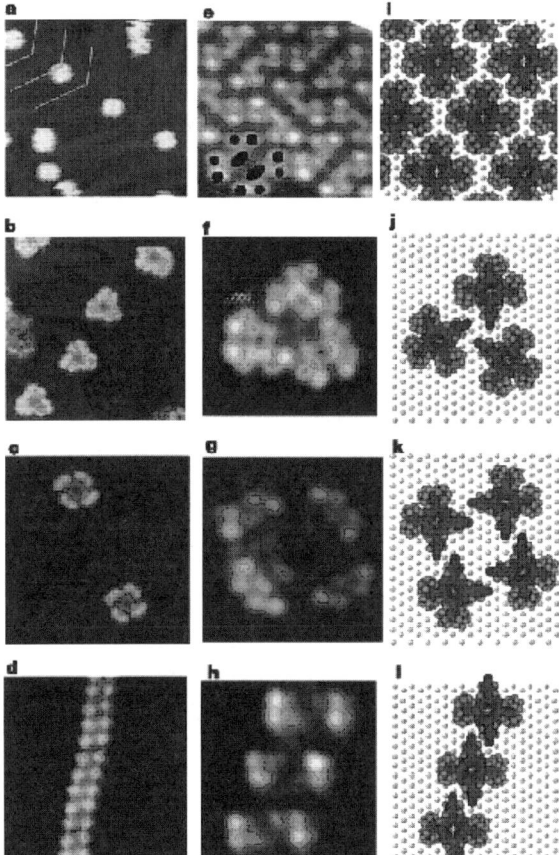

Fig. 3.19. a–d Characteristic molecular aggregate structures on a Au(111) surface for the molecular species **a–d** in Fig. 3.18, respectively. **e–h** are the high-resolution images of the molecular aggregates. **i–l** are the corresponding molecular models (extracted from [3.72])

tion and spacing. Such variations would be compensated by the rotation of phenyl-porphyrin bonds, and reflected in apparent changes in STM images. This work demonstrated that the observed molecular patterns could be a reflection of subtle changes in molecule–substrate interactions.

The Cu-TBPP molecules can also be laterally manipulated by an STM tip at room temperature [3.71]. The experiments on Cu(110) surfaces involved using an STM to mechanically contact the molecule to overcome the diffusion barrier due to interaction of the hydrocarbon groups with the substrate. The force motion of the molecule was observed to resemble a "slip-stick" type of pattern.

50   3 Single Molecule Structural Characterization

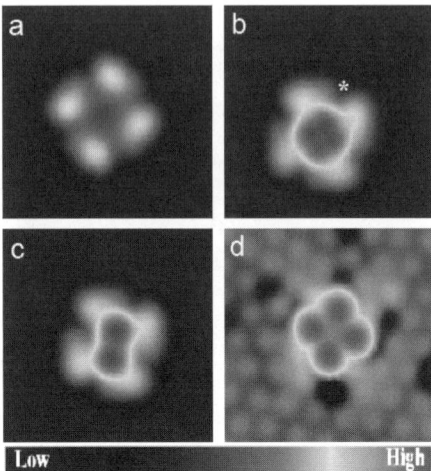

Fig. 3.20. STM images of **a** CuEtioI, **b** Xe@CuEtioI, **c** Xe$_2$@CuEtioI, **d** Xe$_4$@CuEtioI (extracted from [3.74])

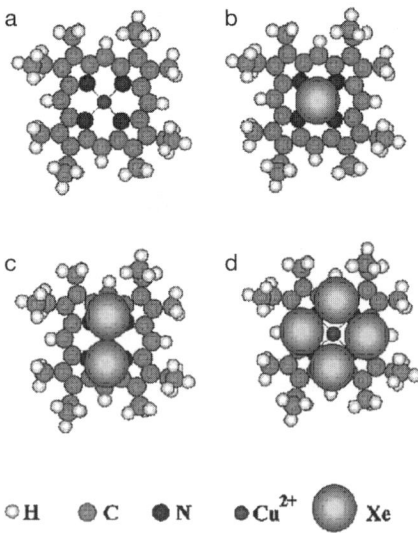

Fig. 3.21. Structural model of Xe–CuEtioI complexes (extracted from [3.74])

Intermolecular interactions can be adjusted by the dipole–dipole interaction via functional groups, and thus provide a venue to design and construct supramolecules [3.72, 3.73]. An example can be seen in cyanophenyl porphyrin aggregations (Fig. 3.18). The H$_2$-TBPP molecules were preferentially adsorbed at the kink sites on Au(111) surfaces (Fig. 3.19). The aggregated cluster geometry is shown to be adjustable by changing the number and location of cyano-substitution groups. As a result, monomer (H$_2$-TBPP, without cyano-substitution), trimer(CTBPP), and tetramer(*cis*-BCTBPP) as well as one-dimensional molecular wire structures (*trans*-BCTBPP) can be obtained.

Molecular complexes of xenon-Cu(II) etioporphyrin I (CuEtioI) have been demonstrated with precision manipulation of xenon atoms to various sites on the CuEtioI molecules on Cu(001) at 11 K (Figs. 3.20 and 3.21) [3.74].

3.3.4 Heterocyclic Molecules

STM and STS studies of five-membered heterocyclic molecules of pyrrole, thiolphene, pyrrolidine and tetrahydrothiophene adsorbed on Cu(001) surfaces at 9 K showed that the unsaturated heterocycles (pyrrole and thiophene) in-

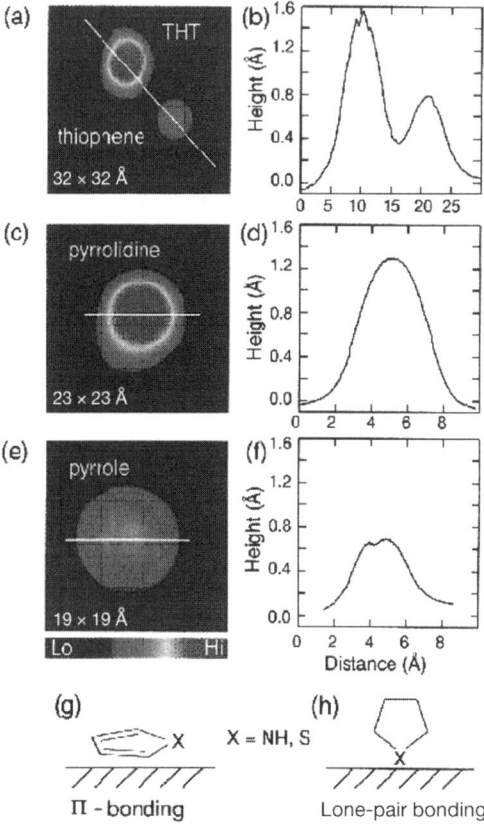

Fig. 3.22. STM images and cross-sectional profiles of **a, b** tetrahydrothiophene and thiophene, **c, d** pyrrolidine, **e, f** pyrrole molecules on a Cu(001) surface. **g** and **h** are two adsorption conformations (extracted from [3.75])

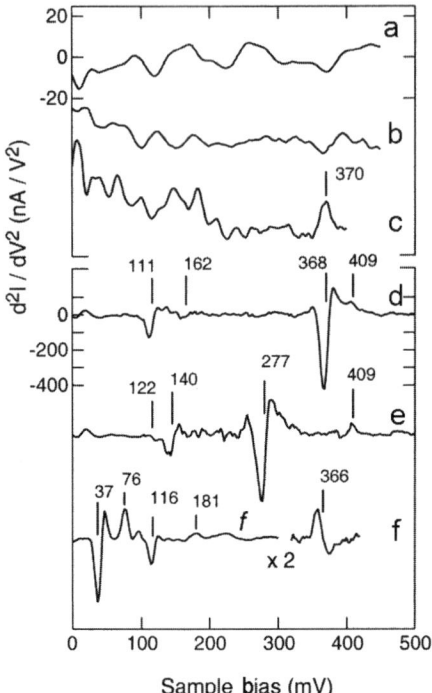

Fig. 3.23. Tunneling spectroscopy of molecules of *a* pyrrole on Cu(001), *b* thiophene on Cu(001), *c* tetrahydrothiophene on Cu(001), *d* pyrrolidine on Cu(001), *e* pyrrolidine-d8 on Cu(001), and *f* pyrrolidine on Ag(001) (extracted from [3.75])

teract with metal surfaces through π–electron interaction, whereas saturated heterocyles (pyrrolidine and tetrahydrothiophene) adsorb with the lone-pair electrons (Fig. 3.22) [3.75]. The conclusion is derived from tunneling spectroscopy on single molecules. The flat-lying molecules (pyrrole and thiophene) showed no vibrational modes whereas a number of modes can be identified with pyrrolidine and tetrahydrothiophene molecules (Fig. 3.23). In addition, pyrrolidine displayed features ascribed to the conformation switching between standing and bent orientations on Cu(001) and Ag(001) [3.76]. This effect was shown as negative differential resistance and could be explored as molecular switches.

In a separate study, six-membered heterocyclic molecules, pyrimidine molecules, were observed to preferentially adsorb on open terraces, rather than step sites on Pd(110) [3.78]. The pyrimidine (1,3-diazine, $C_4H_4N_2$) ring structure contains two nitrogen atoms. The lone pair of electrons associated with the nitrogen atoms can lead to preferential adsorption to the substrate and hydrogen bonding between molecules. The molecules were found to adsorb parallel to the surface and remain stable at room temperature. No significant variation in adsorption populations for the on-top sites of Pd[1$\bar{1}$0] row and midway sites was observed. The tendency of forming dimers was also identified as a result of attraction (most likely, the hydrogen bonding between the nitrogen atom and the hydrogen in the CH group) between pyrimidine molecules.

The above studies demonstrate that the interactions due to π-electrons and lone-pair electrons within heterocyclic molecular structures can significantly affect the adsorption configuration of single molecules, as well as the tunneling characteristics. Such studies fall into the wider category of designing functional molecular devices through molecule–substrate interactions.

3.3.5 Fullerene

Monolayers and multilayers of C_{60} molecules have been obtained on a number of surfaces, such as GaAs(110) [3.78], Si(111) [3.79], thiol SAM [3.80], and Pd(110) [3.81]. One-dimensional C_{60} chains have been constructed on Au/Ni(110) surfaces along Au chains [3.82], and thiol SAM templates on Au(111) surfaces [3.83]. Individual C_{60} molecules were observed on Si(100)-2×1 surfaces with submolecular resolution [3.84, 3.85], $TiO_2(100)$-(1×3) [3.86], and Au(110)-(1×2) surfaces [3.87, 3.88]. The high-resolution STM images revealed a variety of submolecular characteristics for single C_{60} molecules, depending on their orientations (Fig. 3.24) [3.89].

The internal fine structures of fullerene cages can also be revealed by using the STM technique. The spatial characteristics of a single Dy atom encapsulated within a C_{82} cage can be clearly resolved together with the molecular orientation of cage structure. In addition, the orbital hybridization and charge transfer effects between Dy and C_{82} were also identified [3.90].

Bimolecular arrays of C_{60} have been obtained on decanethiol lamella templates on Au(111) by annealing treatment. The C_{60} molecules were adsorbed atop thiol groups in the stripes of decanethiols [3.83]. In a separate study, by using a C_{60}-decorated tip, a threefold electron scattering pattern was directly imaged at the defect sites on graphite surfaces at room temperature [3.91]. This is ascribed to the sharpened local density of states from the molecular

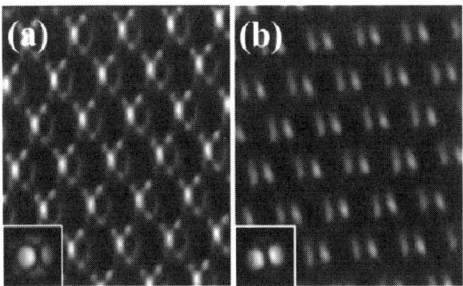

Fig. 3.24. a and b illustrate two examples of STM images for two different orientations of C_{60} molecules. Insets are the calculated STM images (extracted from [3.89])

orbitals of C_{60}, which is the effective tip. Similar patterns were also observed subsequently by using tungsten tips at 77 K [3.92].

3.3.6 Other Molecules

Submolecular resolutions have also been demonstrated in a wide range of molecular assemblies, such as naphthalene on Pt(111) [3.93], TTF-TCNQ [3.94], PTCDA [3.95–3.97], and NTCDT [3.98]. The evidence of interaction between electrons of molecular adsorbates with the substrate was subsequently found in the adsorption of single Lander molecules ($C_{90}H_{98}$) on Cu(110) surfaces by Rosei et al. [3.99]. The molecules were deposited on the surface at room temperature and were removed from their original adsorption sites at lower temperature (100–200 K). The authors observed that the adsorption sites previously occupied by the molecules were transformed into a characteristic nanostructure two Cu atoms wide and eight Cu atoms long. This observation pointed to the limited diffusion of Cu atoms at kink sites under room temperature that enables the restructuring of surface atoms and retained at lower temperature. The individual molecules were adsorbed at the lowest energy conformation at the step edges, and subsequently transformed the underlying atom arrangements at the adsorption sites.

In a similar study, molecular-induced imprints were identified on Cu(110) at the adsorption sites previously occupied by hexa-tert-butyl-decacyclene (HtBDC, $C_{60}H_{66}$) molecules (Figs. 3.25, 3.26) [3.100, 3.101]. The cause for the reconstructed local lattice was ascribed to a possible steric effect of the t-butyl groups that contacted the surface, or the enhanced reactivity due to the molecule–substrate interaction based on the relationship of bonding strength and the metal coordination number at the adsorption site [3.102].

The conformations of L-Lander molecules adsorbed on Cu(100) surfaces were characterized by the distance between the observed lobes using STM [3.103]. The interaction between the central molecular board and the substrate causes the rotation of the attached legs. The adsorbate-induced removal of substrate atoms was also observed by chiral molecules of cystein on Au(110) surfaces [3.104].

The preferential adsorption behavior of Lander molecules was observed on pre-oxidized Cu(110) surfaces [3.105]. The Cu–O/Cu surface consists of periodic one-dimensional strips of bare Cu and reconstructed (2 × 1)-O regions.

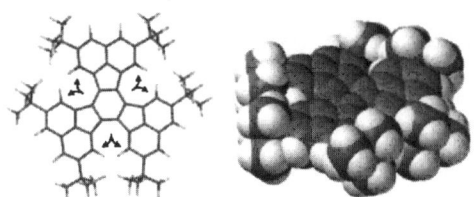

Fig. 3.25. Structure of HtBDC molecules (extracted from [3.100])

3.4 Single Hydrocarbon Molecules

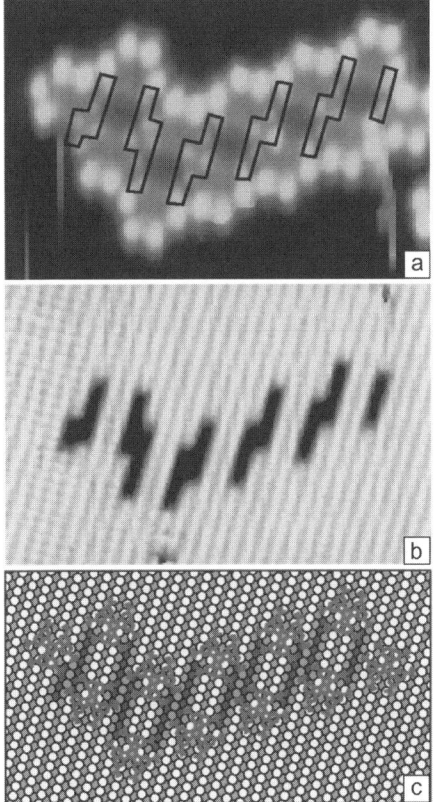

Fig. 3.26. a, b STM observations showing Cu atoms being removed from the surface after the adsorption of HtBDC molecules at 41 K. **c** Schematic of the adsorption of HtBDC molecules on a Cu(110) surface (extracted from [3.100])

The adsorption of Lander molecules is exclusively on the metallic part of the surface, suggesting the attractive interaction between the π board of the Lander molecules and the underlying charge density is the dominant source for immobilizing the adsorbates. In addition, the orientation of the Lander molecules could also be adjusted in order to fit in the metallic part of the surface.

3.4 Single Hydrocarbon Molecules

Remarkable resolution was demonstrated by Stipe et al. [3.106] for the tunneling spectroscopy of individual acetylene (C_2H_2) adsorbed on Cu(110) surfaces. This was the first time that STS was used to study vibrational characteristics of a single molecule. The evidence of C–H stretching mode was

56   3 Single Molecule Structural Characterization

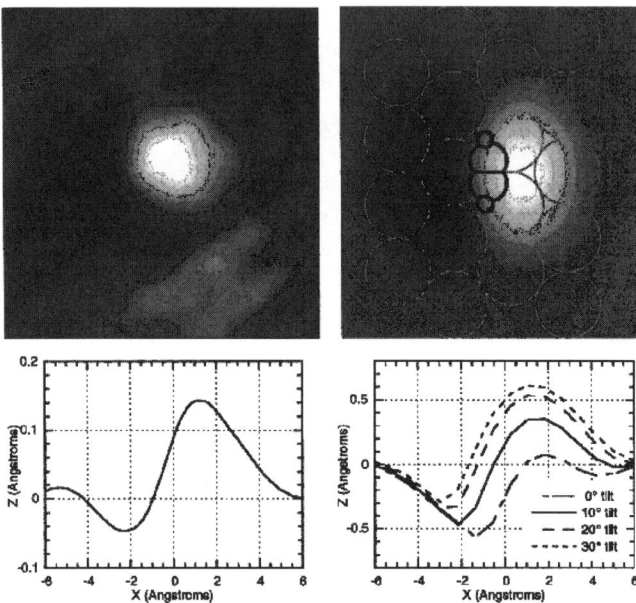

Fig. 3.27. STM image of single acetylene molecules on Pd(111) obtained at 40 mV and 400 pA (*left*), Calculated STM image(*right*) (extracted from [3.108])

obtained in the second-order derivative of the tunneling current, d^2I/dV^2 versus V characteristics. The acetylene molecule is stabilized on Cu surfaces through C–Cu bonds, with the C–H pointing outward from the Cu surface. A deuterium isotope effect was also confirmed in the tunneling spectrum. The results opened possible applications of the STS method for inelastic electron tunneling spectroscopy of single molecules. Theoretical analysis suggested that the coupling between tunneling electrons and the molecule is short-range and non-resonant, and further confirmed that the contribution from C–H or C–D stretch modes is dominant, compared with other vibrational modes [3.107].

The adsorbed single acetylene molecules on Pd(111) were recorded as individual protrusion–depression pairs by STM (Fig. 3.27) [3.108]. The molecules were found to rotate randomly in three equivalent orientations 60° apart at temperatures around 44 K. This is considered as evidence of threefold adsorption sites, i.e., hcp and fcc sites on the surface. The apparent shape of the molecule was confirmed by simulations to bind to both hcp and fcc hollow sites with the molecular axis (C–C bond) parallel to the surface. A slight tilt of the molecular plane was conjectured to achieve optimized overlapping between the π orbital of the molecule and the STM tip, thereby causing the asymmetric topographic appearance. It was also found that lateral diffusion of molecules begins at around 70 K [3.108]. In another study, ethylene molecules were identified to adsorb at on-top sites of Pd(110) along a [1$\bar{1}$0] direction and form a 3 × 1 overlayer structure [3.109].

3.5 Single Molecules Immobilized by Molecular Matrix

In order to achieve submolecular resolution with an STM, one needs to immobilize the single molecules sufficiently on the substrate. Such immobilization may not be obvious for assembled structures that could yield very high imaging qualities. However, success in imaging isolated, individual molecules requires more stringent conditions. The key is to reduce diffusion to sustain inevitable disturbances associated with the scanning probe.

One prevalent approach suitable for ultra-high vacuum STM studies is cooling to lower temperatures, which would reduce the diffusion mobility of adsorbed species on surfaces. Such a preparation approach by direct cooling is still widely adopted in current STM studies on single molecules.

Other approaches involve the assisted immobilization effect from different coadsorbate species. One of the first demonstrations of this approach was seen as benzene coadsorbed with CO on Rh(111) surfaces. CO molecules were introduced to inhibit the lateral diffusion of benzene molecules [3.37]. The existence of CO molecules can be seen in the c($2\sqrt{3} \times 4$) structure [3.38]. This immobilization scheme can be extended for studies carried out in ambient conditions and solvents, as shown in the following discussions.

Assembled networks with cavities of various sizes and geometries provide another viable approach to explore guest–host interactions by immobilizing guest species at single molecule level. Such efforts could be expanded into the generalized areas of host structure construction and physicochemical properties of the resulting complexes [3.110, 3.111].

3.5.1 Hydrogen-Bonded Networks and Single Molecule Inclusions

As one of the most useful interactions, hydrogen bonds require specific directions and interaction ranges, and exist in many molecular species. These interaction properties have been extensively studied and used in molecular self-assemblies. In a recent study, the hydrogen bond configuration of 1,3,5-benzenetricarboxylic acid (trimesic acid, TMA) adsorbed on Cu(100) surfaces was observed to depend on temperature [3.112]. At low temperatures (around 200 K), a honeycomb-like network with core diameter of about 20 Å is the stable structure, whereas the striped structure prevails at room temperature. The distortion of the hexagonal molecular lattice is due to the preferential adsorption at the hollow sites of the fourfold lattice of Cu(100) substrates [3.112]. In addition to the hexagonal honeycomb lattice, the "flower" structure is also stabilized by hydrogen bonding with two different cavity sizes [3.113]. Single TMA molecules can be observed as guest species in the

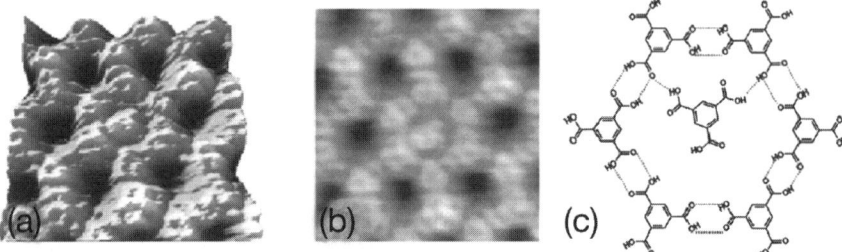

Fig. 3.28. a and b illustrate the inclusion of single TMA molecules in the TMA networks. c Structural model of a guest molecule trapped inside the host network. Reproduced with permission from [3.113]

cavities of existing networks. The entrapped guest TMA molecule was suggested to be stabilized by two hydrogen bonds to the host network, as shown in Fig. 3.28. Little effect on the host lattice geometry from the guest molecule can be observed.

Large areas of hydrogen-bonded molecular network-formed byperylene tetra-carboxylic-di-imide (PTCDI) and 1,3,5-triazine-2,4,6-triamine (melamine) were obtained on Ag/Si(111)-$\sqrt{3} \times \sqrt{3}$R 30° surfaces under UHV conditions [3.114]. This supramolecular network adopts a hexagonal pattern with a lattice constant of 34.6 Å. The open pores can be packed with up to seven C_{60} molecules. The adsorption registry of C_{60} heptomers is clearly resolved and is found different from that of C_{60} on Ag/Si(111)-$\sqrt{3} \times \sqrt{3}$R 30°. This is a reflection of the effect of the host lattice structure. In addition, the C_{60} can also adsorb directly on top of PTCDI and melamine molecules, leading to a replicated lattice structure.

As another example, a monolayer of 1,3,5-tris (carboxymethoxy) benzene (TCMB) shows two-dimensional hexagonal networks formed by hydrogen bonds, whereas the monolayer of 1,3,5-tris (10-carboxydecyloxy) benzene (TCDB) shows two-dimensional tetragonal networks on highly oriented pyrolitic graphite (HOPG) [3.115]. The inclusion effect of hydrogen-bonded two-dimensional networks of TCDB was demonstrated on the surface of highly oriented pyrolytic graphite (HOPG) in ambient conditions [3.116]. With the TCDB network as the host structure and copper(II) phthalocyanine (CuPc), coronene, decacyclene and pentacene as guest molecules, the host–guest architectures of CuPc/TCDB and coronene/TCDB were achieved when host and guest molecules coadsorbed on HOPG. An appreciable variation of the lattice dimension was observed as a result of the guest–host interaction. Control of adsorption site and geometry of organic molecules in self-assembled monolayers was accomplished by this method.

Self-assembled, second-generation Fréchet-type dendrons have been observed to assemble into nearly uniform, disk-shaped structures connected by hydrogen bonds at the focal points on a graphite surface [3.117]. Two kinds of cavity regions can be observed: one in the center of the disk, with a diameter of about 0.8 ± 0.1 nm, the other (which was surrounded by three disk units) in the inter-disk region, with a diameter of about 1.3 ± 0.1 nm. The measured values agree with those of a simplified structural model.

The above examples suggest there are ample molecular species that can be used for building molecular lattices via hydrogen bond interaction. The cavity symmetry and size can be designed by using different molecules. Such capability is certainly an expansion of using bare substrates as adsorption support, and will enable studies of novel composite molecular structures.

3.5.2 Molecular Networks Stabilized by van der Waals Interaction

In contrast to the highly directional hydrogen bond interaction, the van der Waals interaction is ubiquitous among molecules, wellknown as the important driving force for molecular crystallization and assembling processes. Characteristic hexagonal and quasi-quadratic structures derived from the first generation of n-alkoxy-substituted stilbenoid dendrimers have been observed. The assembly structures are stabilized by the interdigitated alkoxy chains, and

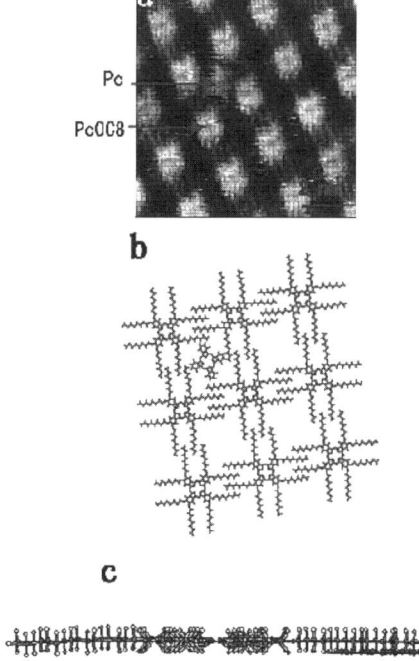

Fig. 3.29. a Immobilization of single molecules inside the quadratic lattices of alkylated phthalocyanines. No distortion of the host lattice geometry is observed. **b** Structural model of the guest molecule entrapment. **c** Side view of the molecular structure (extracted from [3.119])

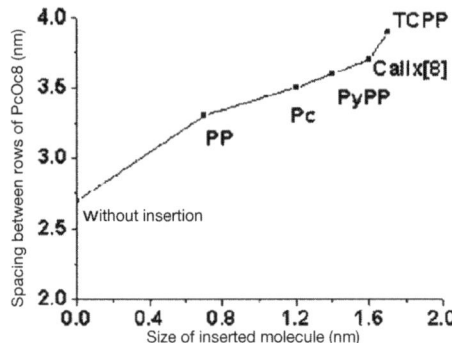

Fig. 3.30. Effect of guest molecule insertion on the lattice spacing (extracted from [3.119])

the lattice geometry changes appreciably as the alkoxy chain length increases to more than 16 carbon units [3.118]. The interaction between the constituent molecules is solely van der Waals force.

Quasi-quadratic molecular lattices can also be constructed by interdigitated alkylated copper phthalocyanines [3.63]. The trapping effect of two-dimensional assembly of octa-alkoxyl-substituted phthalocyanine (PcOC8) for individual molecules of phthalocyanine, prophyrins, and calix[8]arene has been observed (Fig. 3.29) [3.119]. It was shown that single molecules are trapped in quadratic, rather than hexagonal lattices, and domain boundaries are the preferential trapping sites, compared with sites within the domains [3.119]. The observed trapping behavior is, in some respects, analogous to the impurity segregation phenomena of point defects, dislocations, and grain boundaries in solid-state materials.

The molecular inclusion-induced spacing increase between host molecular rows is presented in Fig. 3.30. It is evident that, with an increase in intercalating molecular size, the distances of neighboring PcOC8 rows increase correspondingly. This could be an indication of the flexibility of the PcOC8 lattice. The insertion of molecules in the initially packed lattices leads to enhanced repulsive interaction between molecules. On the other hand, the overlapping alkyl parts could readjust to accommodate additional molecules and compensate the associated increment of repulsions. Molecular networks thus show a certain degree of flexibility in trapping differently sized and shaped single molecules.

3.6 Single Molecule Adsorption on Organic Substrates

The finite tunneling range (typically within 1 nm) in the STM method in principle can detect two layers of molecules, each a few angstrom high. It has been known from a number of reports that double molecular layers can be resolved with STM. This fact has opened the possibility of studing single molecule adsorption on top of monolayers of organic molecules. This is an additional advantage of the STM method that is readily applicable to studies on bare surfaces, as presented in the preceding sections. The theoretical analysis of single molecule adsorption, such as adsorption configurations and kinetics, is analogous to that of bare metal and semiconductor surfaces. The results can also be directly related to investigations of the interface of heterogeneous organic structures with various functionalities. The examples provided in this section illustrate some progress in this area.

Molecular adsorption-induced effects, such as molecular diffusion, and surface restructuring, are well documented in the case of molecular adsorption on metal and semiconductor substrates [3.120–3.123]. By contrast, only few experimental studies have been dedicated to the microscopic effects of molecular adsorption on an organic molecular supports [3.124–3.126]. The pancity of information on microscopic adsorption on organic substrates could be due mainly to the lack of experimental capabilities suitable for such systems. With the wealth of functionalities associated with molecular assemblies, it is conceivable that such studies could be highly rewarding. The information obtained should be of high importance to the construction of molecular nanostructures in both two and three dimensions.

Considering the different polarity and electronic properties of functional groups from the family of alkane derivatives, monolayers of alkane derivatives could provide ideal templates for the investigation of adsorption behavior of organic molecules. Compared with the nanofabricated surface, this alkane-derivative-modified surface is purely organic, rather than metallic or semiconducting. The relative polarity and the ratio of polar/non-polar areas could easily be modified by changing the functional groups and the chain length of the alkane derivatives.

The interlayer interaction is an important factor for determining the overlayer structures. The heterogeneous organic–organic interface is generally associated with weak interactions, typically van der Waals force. The intermolecular interactions are the dominant factors, such as in the case of PTCDA on thiol SAM [3.124]. The introduction of various functional groups to the buffer layer offers possibilities to further explore the effect of molecular interfacial interactions on the assembly structures.

3.6.1 Simple Alkane Lamella

When pre-covered on the support surface, molecular lamella structures inherently introduce the heterogeneous adsorption sites and anisotropic diffusion barriers associated with functional groups. The presence of heterogeneous adsorption sites can result in the selective adsorption of single molecules. The adsorbed species would also experience the anisotropic diffusion barrier, and would organize in a restricted manner.

62   3 Single Molecule Structural Characterization

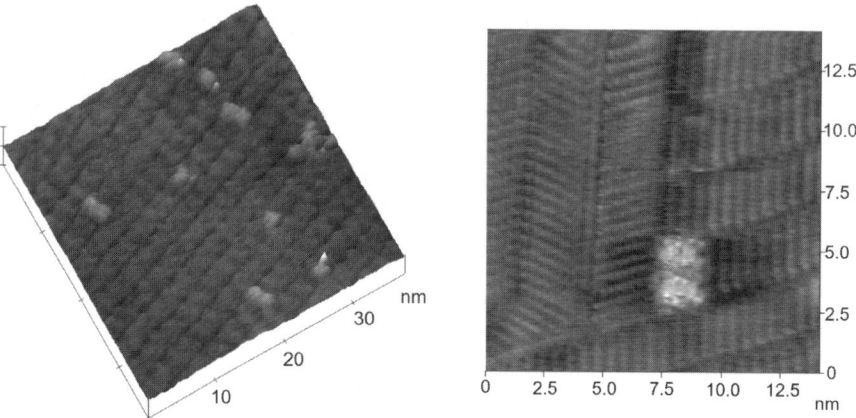

Fig. 3.31. Isolated CuPc selectively adsorbed on the hydrocarbon-chain portion of stearic acid (*left*), and CuPc dimer adsorb on the monolayer of 1-octadecanol (*right*) [3.127]. Copyright 2004 American Chemical Society

Site-selective adsorption of copper phthalocyanine (CuPc) has been observed atop organic surfaces of monolayers of various alkane derivatives (stearic acid, 1-octadecanol and 1-iodooctaecane) supported on the surface of HOPG. STM studies showed that the alkane derivatives form lamella templates, which direct the adsorption of the CuPc. This selective adsorption behavior is attributed to a preference of the CuPc for adsorption on the hydrocarbon-chain portions of the supporting layers [3.127].

When phthalocyanine (Pc) molecules coadsorb with stearic acid, isolated and paired Pc molecules can be detected on top of the stearic acid monolayer, as shown in Fig. 3.31 [3.127]. The Pc molecules are located on top of the alkane part of stearic acid lamellae, whereas no Pc molecules were observed at the location of carboxyl groups. This immobilization effect of alkanes may originate from the variation of adsorption barrier due to van der Waals interaction among molecules.

Two representative adsorption sites are considered in molecular mechanics simulations schematically shown in Fig. 3.32, i.e., site II in which CuPc adsorbs on the top of troughs linked by head-to-head functional groups, and site I in which CuPc adsorbs on the alkyl moiety [3.126]. The calculated results indicate that the system potentials at site I are higher than those at site II, by more than 21 kJ mol^{-1} for three alkane derivative systems. This may be caused by two factors. For one, the trough linked by head-to-head functional groups is about 3 Å in width and the concentration of atoms in the trough is less than that for other sites of the organic monolayers. Consequently, the van der Waals interaction between the adsorbed CuPc and underlying organic substrate decreases when CuPc adsorbs on the top of the trough. Moreover, seeing that the trough is linked by atoms and groups of

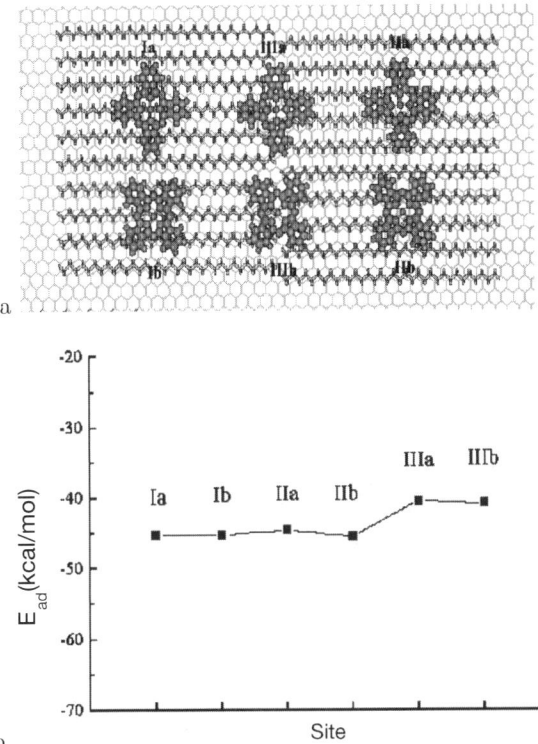

Fig. 3.32. Simulation of the adsorption energetics of CuPc molecules at different sites on top of alkane lamellae. **a** Schematics of the adsorption sites. **b** Adsorption energies of CuPc at different sites on alkane layers

large polarity, the electrostatic repulsion between the ring of CuPc and the functional groups will be stronger than that between CuPc and alkyl. These two factors may drive the selective adsorption of CuPc on the alkyl moiety to achieve the most stable adsorption state. The selectivity is dependent on the relevant functional groups and could vary among alkane derivatives. This analysis suggests that molecular surface decoration can be essential in achieving an optimized immobilization effect of guest species.

3.6.2 Alkylated Amino Acid Molecular Templates

Site-selective adsorption of urea molecules has been observed on the lamella templates of double-alkyl amino acids [3.128]. The unprotected amino acid groups were found to be the preferential adsorption sites, as illustrated in Fig. 3.33. Each amino acid group can adsorb either one or two urea molecules, as reflected by the local clustering of the adsorbates at some adsorption sites in Fig. 3.33b. By contrast, the lamella structure of single-alkyl amino acids did

64   3 Single Molecule Structural Characterization

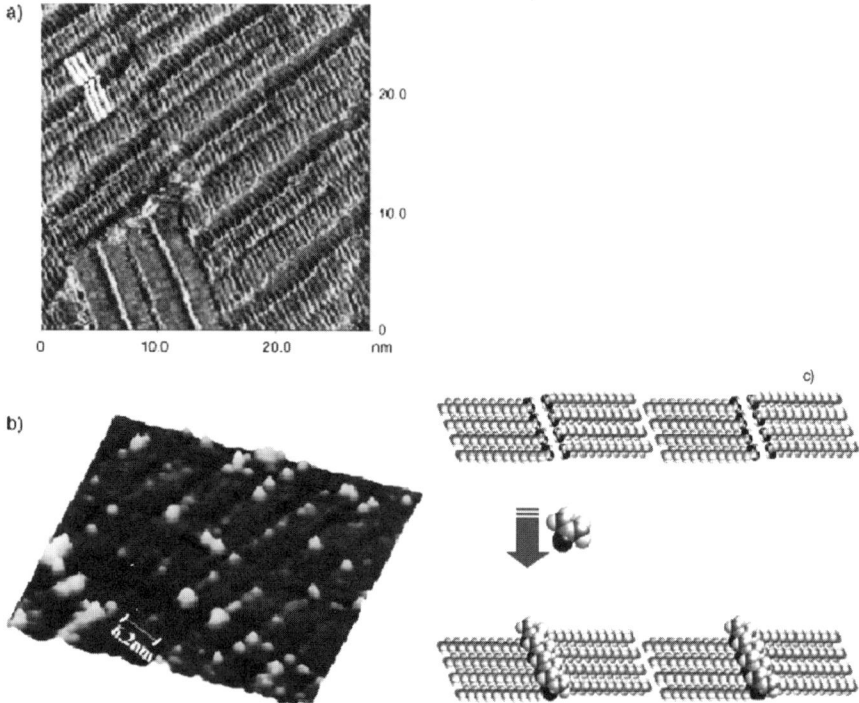

Fig. 3.33. a, b Selective adsorption of urea molecules at the sites of amino groups. **c** Schematic of site-selective adsorption of guest molecules atop molecular templates. Reproduced with permission from [3.128]

not show any templating effect, as a result of dimer formation that saturates the functional groups.

However, such saturation of the amino acid groups can be avoided by introducing fatty acids ($C_{23}H_{47}COOH$) as the matrix molecule [3.128]. The alkyl-substituted amino acids are randomly distributed within the matrix lamella of $C_{23}H_{47}COOH$, and the amino acid groups are available for interaction with adsorbates such as palladium(II) acetate and urea.

3.6.3 Tridodecyl Amine (TDA) Templates

An apparent template effect has also been illustrated with lamellae of tridodecyl amine (TDA) molecules on the adsorption, diffusion and assembly structures of copper phthalocyanine. The conformation of nitrogen atoms in the amine group of TDA is tetrahedrical, the nitrogen atom sitting on one acme of this tetrahedron (Fig. 3.34). Since the C–N bond is dipolar, amine molecules are also dipolar, the nitrogen being partially negatively charged. Thus, when amine molecules adsorb onto an inert surface of graphite, there exists a net dipole moment pointing nearly perpendicular to the surface.

3.6 Single Molecule Adsorption on Organic Substrates 65

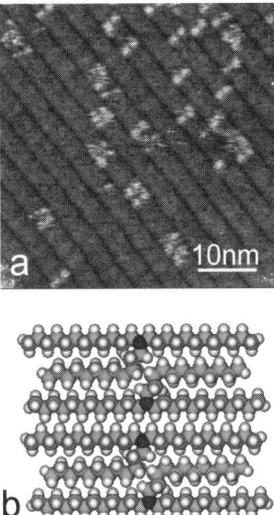

Fig. 3.34. a High-resolution STM image of the TDA lamella structure and b proposed packing model of the lamella structure [3.129]. Copyright 2003 American Chemical Society

Benzoic acid was found exclusively on the TDA assembly adsorbed through hydrogen bonding at the sites of amine groups [3.129]. This shows the possibility of using alkane derivative lamellae as templates to direct adsorption and assemblage of other organic molecules.

By co-deposition of TDA with copper phthalocyanine (CuPc), isolated CuPc molecules and clusters can be stabilized in the alkane part of the lamellae (Fig. 3.35) [3.130]. When coadsorbed with TDA at low CuPc:TDA ratio, CuPc dimers are most commonly observed, with a smaller population of quadrimers and hexamers. From the large-scale view in Fig. 3.34, one can observe two CuPc molecules located on both sides of the amine group of the TDA lamellae. The CuPc dimers appear to adsorb on top of the alkane part of the TDA lamellae.

The lateral diffusion of the single CuPc molecules as well as clusters of adsorbed CuPc molecules was found exclusively along the direction of the TDA lamellae. Such highly directional diffusion behavior can be considered direct evidence of the one-dimensional template effect of TDA lamellae. Such effects have never been observed on lamellae of simple alkanes, possibly due to the lack of functional groups that could establish sufficient diffusion barriers for the overlayer adsorbates. This concept may be generalized to the construction of molecular templates for novel molecular nanostructures.

There are many aspects of molecular templates yet to be explored in a wide range of single molecules and molecular ensembles, such as sensor or catalytic behavior of molecular assemblies, molecular devices based on

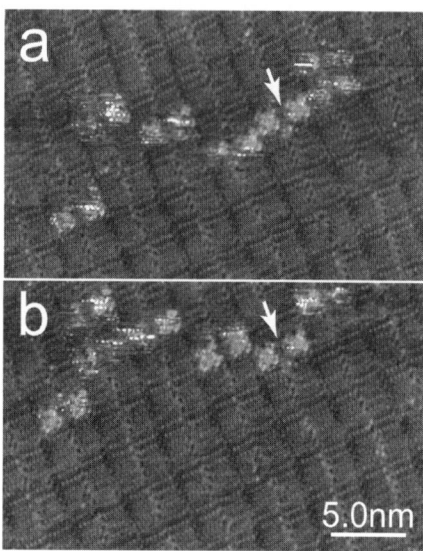

Fig. 3.35. Isolated CuPc molecules observed on lamellae of TDA. *Arrows* indicate the migration of the molecules in consecutive scans in **a** and **b** [3.130]. Copyright 2004 American Chemical Society

heterogeneous assemblies (donor, acceptor, *p*-, and *n*-types), and assemblies of molecular magnets. It is conceivable that with the rich variety of functional groups that can be incorporated into molecular structures, the pursuit of templates with novel physicochemical properties could be fruitful. Efforts in theoretical analysis of the assembling processes are important in gaining deeper insight of the driving mechanisms.

3.7 Electron-Spin Resonance Study of Single Molecules

Along with the effort to immobilize individual molecules on various surfaces, it is recognized that chemical identification of individual molecules is a great challenge to the STM method. Incorporating chemical specificity into the structural resolution capability of STM is the leading demand in the development of STM techniques. The tunneling spectroscopy of the electronic structures and vibrational modes of molecules sets examples for such endeavors. The detection of the electron-spin resonance effect using STM discussed in this section represents an effort from a different perspective.

Electron-spin resonance from single molecules is an important signature for identifying the chemical nature of the molecule. Efforts at single molecule recognition using the electron-spin resonance effect can be seen from electron tunneling, force and optical detection approaches. The latter two experimen-

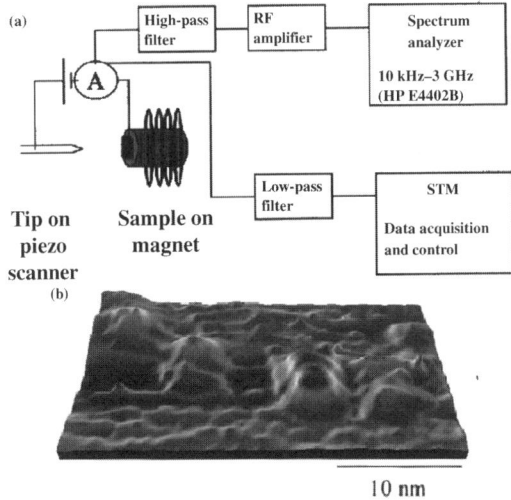

Fig. 3.36. a Schematic of the setup for measuring STM-ESR spectra. **b** Individual BDPA molecules on a graphite surface (extracted from [3.134])

tal schemes will be presented in the following chapters. In the electron tunneling approach, the spin centers can interact with the tunneling electrons through the magnetic dipole field localized at the spin centers, or affect the local density of states due to exchange interactions [3.131–3.134]. It should be noted that there are ongoing discussions on the nature of the interaction between the spin centers and the tunneling electrons. In either of the above two interacting mechanisms, the magnitude of the interaction strength is dependent on the orientation of the spin that can be modulated by an external magnetic field (either DC or AC). Such modulation can be described by the Larmor angular processing frequency ν:

$$\nu = g\mu_B B/h$$

where g is the g-value of the spin center, μ_B the Bohr magneton, and B the applied field.

As a result, the tunneling current may contain an ac component in the radio-frequency (rf) regime that corresponds to the spin-resonance effect. The Larmor precession frequency modulation can be measured from the tunneling current using a spectrum analyzer (Fig. 3.36).

There are three reported types of spin centers studied by the STM method, namely, Si radicals on partially oxidized silicon surfaces [3.131], iron-doped Si systems [3.133], and molecular radicals [3.134]. As an example, in the study on single molecular radicals, the magnetic field was introduced by a small Sm/Co permanent magnet of field strength 190–300 G at the sample surface. The measured STM-ESR spectra of single molecules of α, γ-bisdiphenylene β-phenylallyl (BDPA) confirmed the linear dependence of the

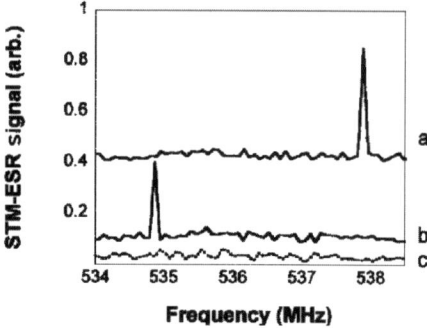

Fig. 3.37. a, b STM-ESR spectrum of BDPA molecules and c the control spectrum on a bare graphite surface (extracted from [3.134])

peak frequency position (within the range of 533–539 MHz) on the applied external field strength (as expected from the above equation) and the g-factor was determined as 2 ± 0.1 (Fig. 3.37) [3.134]. The molecules displaying the spin-resonance effect are active only for a short time and recovery can occur after certain delays. In addition, the spectra peak position and linewidth showed a certain degree of variation that could be related to the molecular orientation within the tunneling junction, and the aggregation of molecular clusters. The pursuit of the single molecule sin-resonance effect has advanced the study of single paramagnetic atoms, molecular radicals and defects using STM. It can be expected that, with more rigorous evidence on the mechanisms of the spin-resonance effect on tunneling electrons, and the improvement of experimental capabilities, the STM-ESR technique will deliver very useful data on single spin centers, especially the interactions between the spin centers and the surrounding environments.

The observations of single molecules presented in this chapter are based on electron tunneling principles. These results have enriched our general knowledge of molecular adsorption, as well as opened up new fronts in investigating the physical and chemical properties of individually adsorbed species with very high structural resolutions, as will be demonstrated in the following chapters. The topography resolution of STM will undoubtedly continue to reveal interesting insights of single molecules as well as molecular assemblies. Probably a more interesting and challenging improvement of the STM technique is its chemical specificity, possibly through improved spectroscopy functions. The accomplishment of this task may lead to further breakthroughs in single molecule studies.

4 Single Molecule Diffusion and Chemical Reactions

This chapter continues the study of surface-bound molecules using the STM method. The focus is shifted from the adsorption configuration to the dynamic properties of adsorbed species, i.e., diffusion, tip-controlled movements, and more importantly the chemical reactions of adsorbed single molecules. The interest in these processes is related to many fundamental mechanisms in heterogeneous catalytic reactions, material surface properties, etc. The success of observing individual molecules on various surfaces presented in the preceding chapter provides us with a solid basis for exploring the two-dimensional kinetics of single molecules on surfaces.

4.1 Molecular Diffusion on Surfaces

Diffusion processes are important to a wide range of topics in surface physics and chemistry, such as adsorbate-induced mass transportation, and heterogeneous catalytic reactions. In many surface-mediated reactions or heterogeneous catalytic reactions, the promotion and inhibition of the reactions are closely associated with the mobility of the reactants [4.1].

The STM method offers a real-space approach to investigate the diffusion behavior of single adsorbates on atomically resolved substrate structures. This represents a useful complement to existing in-depth knowledge of diffusion processes from a range of experimental techniques such as field ion microscopy (FIM) [4.2], and laser-induced thermal desorption [4.3–4.5], etc.

4.1.1 Thermal-Activated Single Molecule Diffusion

The diffusion of a particle on surfaces is characterized by stochastic motion or random walk on a two-dimensional lattice [4.6]. The diffusion barrier is caused by variations of adsorption energy minimum and maximum, as schematically shown in Fig. 4.1. The mean square distance of a particle travelling from its original position at time t can be expressed as:

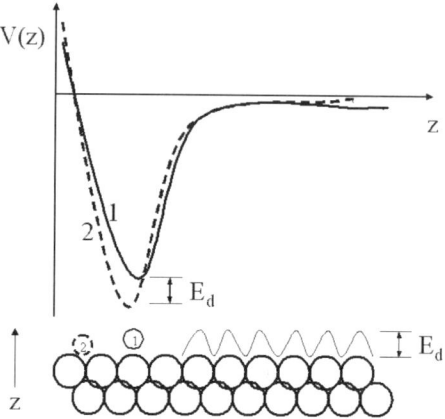

Fig. 4.1. Schematic of diffusion energy barrier resulting from the difference of adsorption energy at different adsorption sites. The adsorption potentials for sites denoted *1* and *2* are schematically illustrated as *solid* and *dotted curves*, respectively. The difference of the equilibrium energy is the origin of the diffusion barrier

$$<r^2> = 4Dt$$
$$D = D_0 e^{-E_d/k_B T} \quad (4.1)$$
$$D_0 = \frac{1}{4}a^2 \nu$$

where a is the jump length, ν the effective jump frequency, and E_d the activation energy. The activation energy of diffusion can be provided by a thermal source, light irradiation, or an external field. Thermal-activated diffusion is the most common driving mechanism.

Experimental measurements of the diffusion coefficient D can be achieved by observing the variation of macroscopic coverage of adsorbates at different temperatures. Rigorous derivations of the formalism can be seen in a number of reviews and books [4.7]. Much success has also been shown with STM in studying the diffusion behavior in real space at single molecule level, as presented in this chapter.

The diffusion properties of single CO molecules and multimers on Cu(110) were directly obtained observing time-lapsed STM images [4.8]. CO molecules are known to adsorb to the atop sites on the Cu surface. The molecular motion is exclusively along the $\langle 1\bar{1}0 \rangle$ direction, i.e., along Cu atom chains. The results in the temperature range of 42 K–53 K suggest that multimeric CO molecule chains display higher mobility than that of single CO molecules. The observed difference in mobility has been attributed to the effect of entropy of the diffusion pre-factor, and the entropy of a molecule in an adsorbed dimer is lower than that of an isolated one. This observation provides an example of adsorbate–substrate interaction in the mobility of adsorbate molecules.

Increased mobility was also observed for multimeric CO on Cu(111) surfaces, and was ascribed to the attractive interaction between CO molecules [4.9].

As demonstrated in the preceding chapter, a CO-decorated tip can lead to much enhanced resolution of the adsorbate structures. Another example of vibration-stimulated hopping was shown for CO on Pd(110) at 4.8 K [4.10]. A threshold bias voltage of about 350 mV was observed to induce lateral movement of CO on Pd(110), and the value is ascribed to the C–O stretching mode. It was also suggested that discontinuous motion or hopping of CO molecules could be tuned by the internal vibrational mode [4.11]. The excitation can be initiated by dosing electrons from an STM tip.

The enhancement of molecular mobility of H_2O clusters, compared with single H_2O molecules, was observed on Pd(111) [4.12]. In this system, the mobility is increased by several orders of magnitude for molecular clusters. The enhancement is due to the mismatch of molecular clusters to the substrate lattice, leading to a reduced diffusion barrier.

The above studies illustrate the difference between single molecules and molecular aggregates from the perspective of diffusion properties. The single molecule behavior is clearly sensitive to its surrounding environment, in this case, the substrate and adjacent molecules.

4.1.2 Laser-Activated Single Molecule Diffusion

Enhanced anisotropic diffusion of CO on Cu(110) was achieved with femtosecond laser pulse irradiations (200 fs with repetition rate of about 1 kHz). The STM observation was performed at 22 K, which is lower than the temperature for typical thermal-originated diffusion [4.13]. In addition to the dominant diffusion along Cu atom chains of $(1\,\bar{1}0)$, a minor diffusion channel (with branching ratio of 0.34) across the Cu atom chains along (001) was also identified. Considering that the temperature is well below the threshold for thermally driven diffusion, and that such anisotropic diffusion behavior has never been observed under thermally initiated diffusions, it was suggested that the diffusion is driven by an electronic effect. As a matter of fact, the estimated electronic temperature increase is about 3,000 K for a time period of about 1 ps under these irradiation conditions, in contrast to the lattice heating temperature rise of about 140 K under the same conditions.

4.1.3 Field-Induced Diffusion of Single Atoms

In addition to thermal- and laser-induced diffusions, the local electric field at the tip apex of an STM (typically on the order of a tenth of V/Å) can also significantly alter the local barrier and stimulate diffusion. Under the experimental conditions of STM, the adsorbed atom or molecule can be polarized under the influence of a strong electric field at the tip apex. The induced dipole will show a parabolic-shaped potential, the minimum coinciding with the tip position [4.14].

The potential profile due to the local electric field at the tip apex can be expressed as:

$$U(r) = -\mu E(r) - 0.5\alpha E(r)^2 + \cdots \quad (4.2)$$

where μ is the permanent dipole, and α the effective polarizability of the adsorbate. The gradient of the potential is the force experienced by the adsorbate:

$$\frac{dU(r)}{dr} \simeq -(\mu + \alpha E)\frac{dE}{dr}$$

Therefore, the inhomogeneous distribution of field strength may result in a directional diffusion behavior. Field-induced diffusion can be observed as directional diffusion of adsorbates either toward or away from the tip position. Therefore, the concentration of the adsorbates should be dependent on scanning conditions such as speed, bias, and polarity.

The field-induced diffusion of Cs adatoms on p-GaAs(110) and n-InSb(110) has been reported in STM observations. Under positive tip bias, directional diffusion of adatoms toward the tip position has been observed and interpreted in terms of the above-mentioned arguments [4.15]. Tip-induced movements of Sb dimers were observed on Si(001) with high bias (+3.5 V) (Fig. 4.2) [4.16]. Another observation of the tip-induced effect is the reversible rotation of Sb dimers on Si(001) [4.17]. By counting the population of Sb dimers on Si(001) surfaces annealed at various temperatures, the diffusion barrier of Sb across the substrate dimer row was determined as 1.2 ± 0.1 eV, and the corresponding pre-factor as $10^{-4 \pm 1}$ cm^2 s^{-1} (Fig. 4.3) [4.16]. The dissociative adsorption of Sb$_4$ on Si(001) surfaces was observed as five distinctive types of Sb dimer configurations. By annealing the sample at several temperatures and observing the population variation for each type of Sb dimer, the energy barrier can be estimated in each case [4.18].

The diffusion of Si adatoms on Si(001) surfaces was identified as being predominantly along surface dimer rows with an activation energy of 0.67 ± 0.08 eV [4.19]. Details of the diffusion of Si dimers on Si(001) surfaces have been studied by STM with atom-tracking capability [4.20]. The STM tip is locked onto the target atom and the migration of the target is followed by dithering the tip around the target. The STM is used mainly for measuring the kinetics of the target rather than scanning the large area. Such improvement enhances the capability of STM to study dynamic events by 3 orders of magnitude. Using this approach, the diffusion barrier for Si dimers on Si(001) surfaces was determined as 0.94 ± 0.09 eV.

In a separate study, a tunneling-current-stimulated migration mechanism was proposed for inducing diffusions of adsorbed Br atoms on Cu(100) surfaces [4.21]. It was observed that, under the same tunneling bias, the current density is the determining factor for the migration of Br atoms. This mechanism could be due to the localized thermal activation of the adsorbates resulting from inelastic scattering of the injected electrons. Since the thermal activation to the adsorbates can be dissipated through the adsorbate–substrate

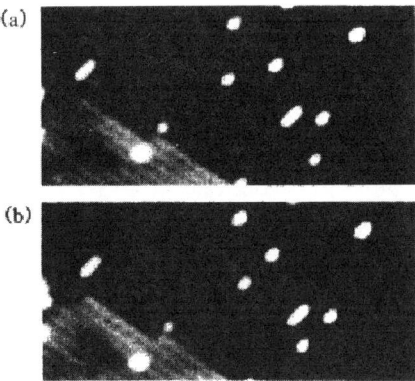

Fig. 4.2. STM image of Sb dimers **a** before and **b** after annealing at 479 K for 2 min (extracted from [4.16])

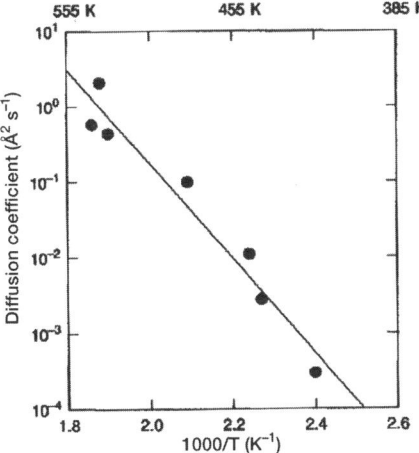

Fig. 4.3. The Arrahnius plot of temperature dependence of diffusion coefficients for Sb dimers (extracted from [4.16])

interaction, a threshold tunneling current is needed to trigger observable adsorbate migrations.

In practice, there are certain limitations for studying the diffusion process of individually distributed adsorbates using STM. The limited scanning speed of STM is obvious here, and the unavoidable interaction between the tip and adsorbate would introduce a force on the order of 10^{-9} N [4.22]. The latter effect could be reduced somewhat by increasing the tip-sample separation, yet it can not be completely eliminated. If the adsorbates diffuse too fast, they will not be observable by STM, this being the case also if the adsorbates interact too weakly with surfaces (cf. many physisorbed molecules).

4.2 Single Atom and Molecule Manipulations

The manipulation of single atoms and molecules represents an important application of tip-substrate interaction. By optimizing the tip-sample interaction strength, STM can be used to move adsorbates such as metal particles, atom clusters, or single atoms from one place to another. So far, precision manipulations by STM have been demonstrated for a variety of atoms and molecules, and this section provides only a short summary of some representative cases. This manoeuvering capability of single molecules could be useful for building devices made out of small particles, perhaps from particles of different materials. It could also be useful for studying interactions between particles, and between particles and substrates. Eventually, it may become possible to build molecules in an atom-by-atom manner. Even though such a construction approach is inherently slow, it nevertheless can provide a great deal of insight into processes of intermolecular interaction, and possibly also novel reaction pathways. The in-depth exploitation of the tip-sample interaction can lead to permanent restructuring of the surface, such as in the study of direct writing, lithography, electron-beam-induced deposition, and etching.

4.2.1 Controlled Manipulation of Single Xe Atoms

The simplest adsorbate species on surfaces is single atoms. It is possible to move these atoms according to predetermined paths using STM [4.23, 4.24]. Due to the strong field strength, the tip of an STM always exerts finite force on an adsorbate atom [4.22]. This force has both van der Waals and electrostatic contributions. By adjusting the tip-sample separation and the voltage of the tip, the magnitude and direction of this force can be tuned. Seeing that it generally requires less force to move an atom along a surface than to pull it away from the surface, this makes it possible to optimize these parameters such that the STM tip can pull an atom across a surface although the atom nevertheless remains bound to the surface.

Eigler and Schweizer [4.24] were the first to transfer Xe atoms back and forth repeatedly between the tip and the substrate surface. The electrical conductance between the tip and the substrate depends on the position of the Xe atom, which results in a switching device with a low-conductance state when the Xe atom is on the substrate, and with a high-conductance state when the Xe atom is on the tip. This atomic switch is a bistable element, and this type of components is important to microcircuit constructions. A threshold tip height was identified in the above study. Only when the tip-sample distance was lower than the threshold, i.e., the tunnel current was high enough, were the adsorbed xenon atoms moved. Simple investigations showed that neither the magnitude nor the sign of the applied voltage had significant effects on the threshold tip-sample distance. This suggests that the dominant force between the tip and the xenon atom is due to van der Waals interaction.

The notion that atoms of a noble element such as Xe should be observable by STM is not a foregone conclusion. According to the theory of Tersoff and Hamann [4.25], for small bias and constant current an STM image corresponds to a map of constant local density of states at the Fermi level. The extent to which an adsorbate will be "visible" by STM depends on how it locally changes this density of states. Lang [4.26, 4.27] has shown that for single atom adsorbates on metal surfaces, the crucial parameters in determining the apparent height of an atom in a low-bias STM image are predominantly the s-state and p-state densities due to the adsorbate at the Fermi level. In the case of noble gas species of Xe, this is not expected to contribute significantly to the density of states at the Fermi level when adsorbed on a metal surface. The highest filled state of Xe is $5p$, which is about 6–8 eV below the Fermi level, and the lowest unfilled state is $6s$, which is about 4.5 eV above the Fermi level. However, the calculated $6s$ state density displays a broad distribution that spans the Fermi level. This residual $6s$ density of state at the Fermi level is considered as the origin for the observed contrast of Xe atoms in STM images [4.28].

STM studies indicated that the Xe atom appears as a nearly cylindrically symmetric protrusion of 1.53 ± 0.02 Å from the Ni(110) surface. Similar images for Xe adsorbed on Pt(111) surfaces at 4 K have also been obtained [4.29]. Xe atoms were also shown to be transferable between a Ni(110) surface and tungsten tip by pulsing bias voltage [4.30]. Calculations of Xe atoms adsorbed on metals indicate that there should be substantial vibrational heating for tunneling current >100 pA. The observed sideway motion at smaller tip-sample separation may be due to the increased van der Waals attraction to the tip as the tip is brought closer to the surface.

4.2.2 Si Atoms

In addition to adsorbed species, chemically bonded atoms and atomic clusters can also be manipulated with STM. An example of such experiments is for Si(111) surfaces [4.31]. As the tungsten tip is brought to the sample surface, the measured apparent barrier collapses at about 3 Å from the surface. At even closer distance to the sample surface, the chemical interaction between the tip and sample becomes more pronounced. The Si atoms and clusters can be removed from the basal plane of the Si(111) surface by applying a positive voltage pulse of 3 V at the tip-sample separation of approximately 1–3 Å. By this approach, as many as tens of atoms can be successfully removed from the Si(111) surface and transferred to the tip. The Si clusters that adsorbed to the tip apex were shown to be transferable back to the surface with negative voltage pulses.

Several mechanisms have been proposed to account for the reported manipulation of single atoms, including possible field evaporation of negative ions, and electromigration of atomic adsorbates [4.32, 4.33]. Ionization followed by field evaporation has been suggested [4.31] for the reversible transfer of Si atoms between the tip and Si surface by using STM. The required threshold field strength for tip-induced Si atom manipulation is estimated at around 1 V/Å, which is much lower than the typical value for evaporating atoms in field-ion microscopy (about 3 V/Å). Such reduction of threshold field strength can be attributed to the strong chemical and mechanical interaction between the tip and sample surface, as well as to high current density during manipulation bias voltage pulsing.

4.2.3 Gold Atoms

Gold atoms adsorbed on NiAl(110) surfaces have been manipulated to construct one-dimensional gold chains [4.34]. Tunneling characteristics revealed a strong change of the electronic states as the gold atoms were added to the chain one by one. The measured conductivity of the Au atom chains can be considered as resulting from the delocalized electron transport within the one-dimensional quantum well structure and the wave function of the Au atoms. The sequential evolvement of the electronic structure, from single atom resonance, coupling between adjacent Au atoms, to the appearance of a one-dimensional electronic band, has been identified. The development of the one-dimensional band structure was confirmed by the dependence of energy position of lowest conductivity peaks versus chain lengths of Au_3 to Au_{20}. An effective mass of $0.4 \pm 0.1\ m_e$ has been estimated from the energy dispersion relationship of an 11-atom chain [4.35]. More rigorous theoretical treatment of the electronic properties of the one-dimensional metal atom chains should include possible coupling with the substrate. This experimental demonstration can be further developed to study the electronic characteristics of complex structures, especially artificially constructed ones.

4.2.4 CO Molecules

A number of artificially constructed atomic and molecular structures have been successfully demonstrated, such as for xenon atoms and CO molecules, and nanometer atom rings of iron atoms. Recent achievements in moving Cu atoms from different sites on Cu(211) surfaces have led to the possibility of constructing metallic clusters from individual atoms. The shift of Cu atom positions at the step edge with adsorbed CO molecules was shown to reveal the adsorption registry of CO [4.36]. The lateral manipulation of CO molecules on Cu(211) was found to be achieved via repulsive interaction between the tip and CO due to the confinement of CO molecules to the step edges [4.37]. It should be noted that attractive interaction between the manoeuvering tip and CO prevails for controlled movements of CO on low-index surfaces.

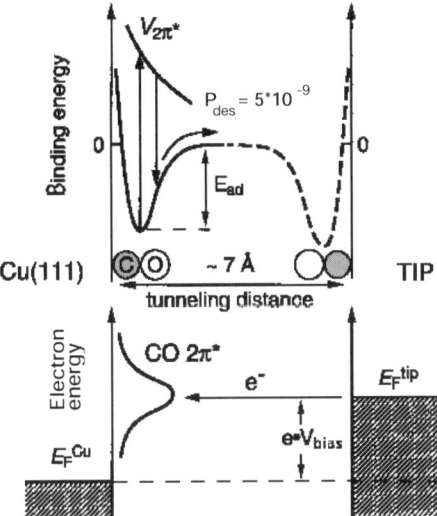

Fig. 4.4. Energy diagram of CO transfer between Cu(111) and the tip (extracted from [4.38])

Bartels et al. [4.38] studied the vertical transfer mechanism for single CO molecules on Cu(111) using STM at 15 K (Fig. 4.4). It was suggested that the adsorption potential for the tip and substrate is well separated, and that molecular hopping is initiated by the electronic transition of CO from ground state to the excited state $V_{2\pi^*}$, followed by a desorption step. A threshold voltage of about 2.4 V was found for the transfer of CO from the Cu(111) substrate to the tip.

Individual CO molecules were imaged and manipulated by using STM on Ag(110) surfaces at a temperature of 13 K at the presence of coadsorbed Fe atoms [4.39]. It was demonstrated that, by moving CO molecules to the sites of Fe atoms, new complexes could be formed such as Fe(CO), and Fe(CO)$_2$, and identified by single molecule inelastic electron tunneling spectroscopy (IETS) of the C–O stretching mode. The tilt angle of the CO ligand was also determined from the topography image. This result illustrated that by picking up CO molecules at the tip apex, one could appreciably enhance the resolution of STM. The substrate lattice could be well resolved, together with the molecular adsorbates. This approach is very promising for the identification of adsorption sites of single molecules on metal substrates.

4.2.5 C$_{60}$ Molecules

C$_{60}$ molecules adsorbed on Si(001) 2 × 1 surfaces have been subjected to manipulation while the lateral force was simultaneously measured (Fig. 4.5) [4.40]. The lateral force experienced by the STM tip is monitored by the "H"-shaped reflector shown in Fig. 4.5. The reflector is supported by two thin Si

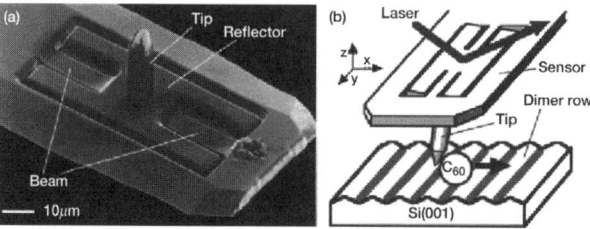

Fig. 4.5. Lateral force sensor for studying molecular manipulation in lateral directions (extracted from [4.40])

beams, resulting in a lateral force constant of 68 N m^{-1} and a normal force constant of 26 N m^{-1}. The migration of the adsorbate is preferentially in the dimer row direction. The migration of C_{60} molecules across dimer rows can be initiated by adjusting the tunneling current, and recorded as a jump in measured lateral force. A force in the range of 2–3 nN was required for the lateral movement of C_{60}. In addition, from the reported stiffness values of C_{60} molecules (40–100 N m^{-1}), an elastic energy of 0.1–0.7 eV was estimated in association with adsorbed single C_{60} molecules [4.40]. Such experiments may help gain direct evidence on the nature of the C_{60} adsorption on surfaces.

The single molecule manipulation by STM is still a state-of-art technique that can be achieved only under optimized conditions. Similarly to the uncertainties encountered in the imaging of single molecules by STM, the tip conditions (geometry, chemical specificity, etc.) remain largely poorly predictable, which will greatly impact the efficiency of the tip-controlled molecular movements. Development of techniques for fabricating precision tips will definitely help extend the scope of this very interesting field.

4.3 Single Molecule Chemical Reactions on Metal Surfaces

The understanding of the basic physics and chemistry of surface reactivity is an important topic for catalyst designs and chemical engineering applications. The investigations of surface-initiated reactions at single molecule level will assist the effort in establishing the principles of catalysis phenomena. The following sections in this chapter summarize findings of single molecule chemical reactivities on electronically different surfaces, namely, metal, semiconductor, and oxide surfaces. These surfaces are representative of a wide variety of catalysis studies. The STM tip is shown to be an important tool to induce the dissociation and association of reactants at single molecule level. Such capability could lead to useful insights into novel reaction pathways at unprecedented precision levels. It should be noted that although the studies so far are restricted to low temperatures, vacuum conditions and crystalline

4.3.1 Single Molecule Oxidative Reaction on Metal Surfaces

Oxygen adsorption can cause dramatic changes in surface structures and properties for a wide variety of samples. The impact of studying oxidative reactions could be significant for many surface science topics. In this section, we provide a list of examples that cover a range of metal surfaces subjected to oxidations. Many of the observations have been made at single molecule levels that enable detailed comparison of the reactivity associated with various surface structures.

O_2 adsorption on Ru(1000) surfaces at 300 K results in both compacted domains of oxygen atoms with a (2×2) structure, and highly mobile individual oxygen atoms [4.41, 4.42]. The residence time of the oxygen atoms as a function of separation distance from neighboring oxygen atoms can be correlated with interatomic interactions. An STM capable of rapidly scanning up to 20 frames per second was developed to enable studies on the dynamics of molecular adsorbates. In a study of oxygen atoms on Ru(1000), a hopping rate of $14 \pm 3\,\mathrm{s}^{-1}$ was determined [4.43].

STM observations of O_2 adsorption on Al(111) surfaces revealed that the separation between O atoms is exceedingly large, in comparison with the O–O bond length. This result led to the proposition of a so-called cannonball mechanism to account for the abstractive adsorption of O_2. Molecular oxygen adsorption on Pt(111) was found to be highly dependent on temperature. Thermal dissociation of molecular species into atomic oxygen proceeds at above 95 K. The adsorption studies of O_2 on Pt(111) at low temperatures revealed that molecular O_2 is mobile at 60 K on Pt(111). At 160 K, dissociative adsorption is evident with paired O atoms distributed on the surface (Fig. 4.6). The separation between the oxygen atom pairs is between one and three lattice constants, with a maximum population at two lattice constants (Fig. 4.7) [4.44]. This can be considered as evidence of a non-thermal process of dissociation of molecular oxygen in which hot atoms are generated [4.44, 4.45].

Further dissociation of oxygen atom pairs into individually separated oxygen atoms occurs at 205 K. The diffusion barrier of oxygen atoms has been determined as 0.43 eV, with a pre-exponential factor of $10^{-6.3}\,\mathrm{cm}^2\,\mathrm{s}^{-1}$. In addition, the dissociation of single oxygen molecules at fcc sites of Pt(111) surfaces can be induced by injection of low-energy electrons (0.2–0.4 V) from a tunneling tip (Fig. 4.8) [4.45]. This was achieved through activation of the intramolecular vibrational mode due to the resonant inelastic electron tunneling process. Direct comparison of the dissociation process for molecules at various adsorption sites showed that O_2 at bridge sites could be either dissociated or desorbed by the injected tunneling electrons, whereas the molecules at step edges would always disappear (possibly be desorbed) under voltage

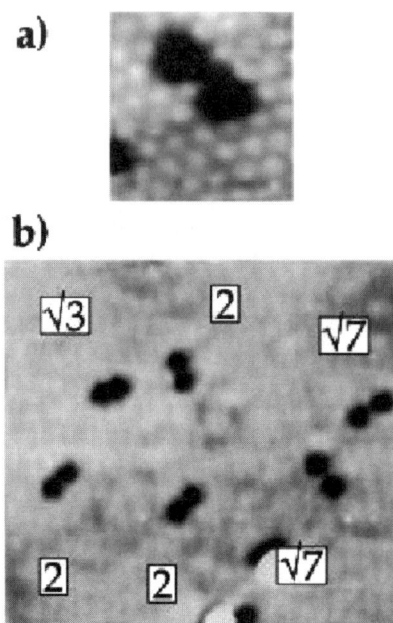

Fig. 4.6. STM image of an O_2-dosed Pt(111) surface at 160 K (extracted from [4.44])

pulses. The dissociation efficiencies at different bias voltages may indicate the quantization of vibrational energy levels, resulting in variable dissociation efficiency behavior. Supportive evidence was provided from molecular dynamics simulations [4.46].

The dissociative adsorption of O_2 molecules on Pt(111) surfaces was found to develop temperature-dependent aggregation patterns observed by STM. At about 160 K, only isolated pairs of dissociated adsorbates were observable [4.47]. By contrast, between 100 and 140 K, small chains of oxygen clusters prevail and grow at decreasing temperature, indicative of a higher dissociation probability at the chain ends. At even lower temperatures (less than 80 K), compact molecular oxygen islands, typically in the shape of triangles, become the dominant feature on the surface, suggesting a high mobility of precursors on the surface.

Consecutive stages of the oxidation process of Pd(111) and Pd(100) resulting from the adsorption of O_2 and NO_2 have been revealed by STM [4.48, 4.49]. The oxidation process is preceded by the fast chemisorption of oxygen to the surface. Time-elapsed observations revealed that NO_2 adsorption can cause continuous growth of step edges on Pd(100) (or the growth of oxidized islands) and form new surface structures [4.49]. The adsorbate-induced surface reconstruction can be subtly affected by adsorbate–adsorbate interaction. The observed evolution of characteristic adsorbate patterns in

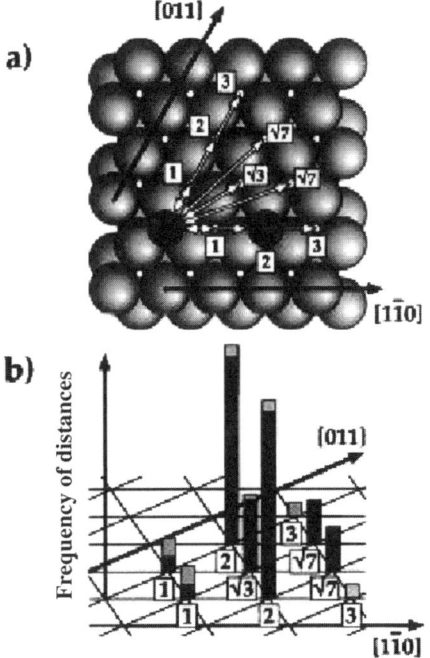

Fig. 4.7. Distribution of the separations between O adatoms (extracted from [4.44])

such studies demonstrates the joint effects of mobility, spontaneous dissociation, and thermal desorption of individual precursor molecules. Such effects have been widely observed with STM as well as other surface techniques.

O_2-induced reconstruction of Rh(110) is characterized by (1×n) structures with every nth row missing in the $[1\bar{1}0]$ direction [4.50]. This differs from the observed oxygen-induced (2×1) structures on Cu(110), Ni(110) and Ag(110) surfaces.

Oxygen-induced (2×1) structures on Ni(110) are manifested by a missing Ni atom row in the $[1\bar{1}0]$ direction [4.51]. The reconstructed Ni(110) surface undergoes a series transformation once H_2S is introduced [4.52]. The (2 × 1)O structure breaks down and is replaced by randomly distributed troughs and islands. This behavior can be explained by considering the removal of adsorbed O atoms by H_2S:

$$H_2S(g) + O(ad) \longrightarrow H_2O(g) + S(ad)$$

The released Ni atoms from the –Ni–O– chains aggregate into islands, and are subsequently covered by sulfur atoms. The sulfur overlayer develops a c(2 × 2) structure.

The oxidation of Cu(110) surfaces has been studied both at room and higher temperatures [4.53–4.56]. Similarly to Ni(110) surfaces, Cu(110) also develops a (2 × 1) structure in the presence of oxygen adsorbates. The added

82 4 Single Molecule Diffusion and Chemical Reactions

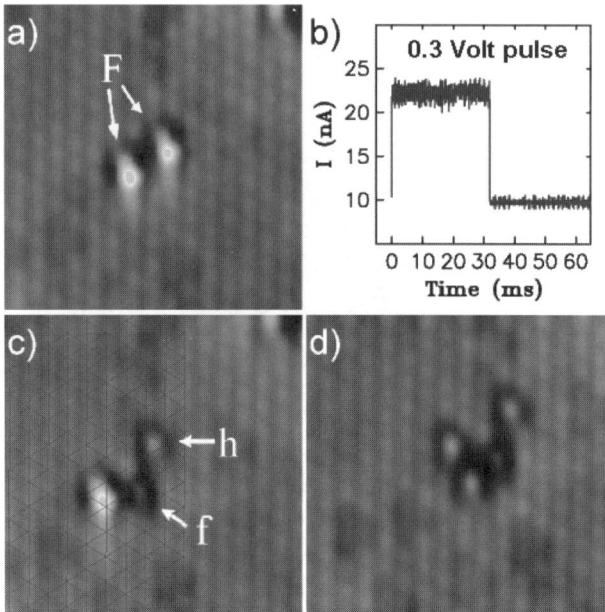

Fig. 4.8. Electron injection-induced dissociation of two O_2 molecules adsorbed on a fcc site Pt(111) surface. The surface was dosed with O_2 at 85 K and cooled to 50 K for STM studies (extracted from [4.45])

–Cu–O– chain is oriented along the (001) direction. The formation of the –Cu–O– chain was proposed to result from the condensation of mobile substrate adatoms and chemisorbed oxygen atoms [4.57, 4.58].

4.3.2 Dissociative Adsorption of H_2

The impact of vacancy dynamics on hydrogen dissociation on Pd(111) surfaces at low temperatures was examined in real time. According to the statistics of annihilation events of vacancy clusters due to H_2 adsorption, it was concluded that molecular hydrogen dissociates at the cluster sites of three or more hydrogen vacancies (Fig. 4.9) [4.59]. This effect was attributed to observations that each stable pair of vacancies takes up three adsorption sites, and one H adatom is rapidly exchanged between these three sites. As a result, the vacancy pairs always appear as triangular-shaped patterns in STM images. In addition, a three-vacancy cluster includes six fcc sites with three H atoms moving inside. A four-vacancy cluster contains 10 fcc sites with six moving H atoms inside. This observation is supported by the free energy-based consideration that bridge sites are associated with lower diffusion barriers than is the case for top sites.

The dissociative adsorption of H_2 causes abrupt changes in the topological appearance of vacancy clusters. The individual vacancy sites reappear after

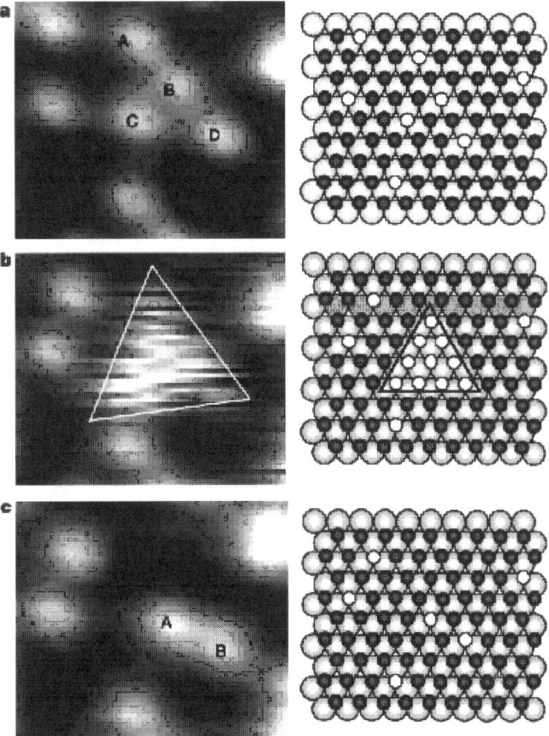

Fig. 4.9. STM images of the vacancy sites of H on a Pd(111) surface. The vacancy pairs appear in triangular shape as a result of fast exchange of a H atom across the bridge sites (extracted from [4.59])

occupation by additional H atoms. The dissociation of H_2 can be accommodated only by the vacancy clusters containing more than three vacancy sites.

In a study of H_2 adsorption on Pt(111) surfaces, a strong enhancement (factor of 500) of diffusion rate was observed for the Pt–H complex [4.60]. The density functional simulation suggests the atomic hydrogen in the Pt–H complex is sitting on top of the Pt atom, causing a reduction of the diffusion barrier by about 0.09 eV, which is in good agreement with the experimentally estimated value of 0.16 eV.

4.3.3 Dissociative Adsorption of NO

Reduction of NO on catalyst surfaces is an important step in automotive exhaust treatment. The dissociation of NO on Ru(0001) surfaces has been identified preferentially along monatomic step edges [4.61]. The adsorbed NO molecules migrate to the step edges from both sides, and dissociate into N and O atoms. The oxygen atoms diffuse away from the reaction sites to form small clusters, whereas N atoms diffuse slowly and are randomly distributed in the vicinity of the reaction sites (step edges). The population of N(ad) displays a linear relationship with the width between step edges, suggesting that the diffusion length of NO is larger than the terrace width in this case. Such reactivity can be attributed to the local electronic structures of the step edges. By quantitatively analyzing the distribution of the dissociated products of NO, it was proposed that the dissociation efficiency is higher at metal atom sites that have two neighboring atoms at the bottom of the step. The activation energy for NO dissociation was estimated as 1.28 eV at sites on open terraces, and 0.15 eV at step edges, based on density function theory calculations [4.62]. The reactivity of the step edges is also affected by the presence of oxygen adadoms.

Three adsorption structures, c(4 × 2), c(8 × 2) and p(2 × 2), of NO on Pd(111) have been identified by STM in the range 170–300 K [4.63]. An additional disordered structure has also been observed. At room temperature, the c(4 × 2) structure prevails at high NO pressure ($> 3 \times 10^{-6}$ torr), and the Pd(111) surface becomes disordered at low NO pressures. There are two adsorption sites in the the c(4 × 2) structure for NO molecules, based on measured topography height. The adsorption registry of NO has been assigned, based on density function theory simulation, to the atop sites with tilted orientation. c(8 × 2) and p(2 × 2) structures can be observed only at low temperatures, coexisting with the c(4 × 2) structure.

4.3.4 Dissociation of NH_3

The injection of electrons was shown to effectively induce desorption and migration of individual NH_3 molecules on a Cu(100) surface at 5 K (Fig. 4.10) [4.64]. The excitation of the vibrational states is revealed as either translational motion or desorption of the NH_3 molecules. At low tunneling currents (smaller that ∼0.5 nA), the observed threshold voltage value (400 mV for NH_3 and 300 mV for ND_3) coincides with the stretching frequency of N–H(N–D) for NH_3 (ND_3) molecules (Fig. 4.11). At higher tunneling currents (greater than ∼0.5 nA), an additional threshold voltage value (270 mV for NH_3) was identified, which can be attributed to the umbrella mode of NH_3.

The vibrational excitation due to inelastic scattering was proposed as driving mechanism. Based on existing evidence from ultraviolet and infrared experiments, the desorption of NH_3 from Cu(100) surfaces is mainly through the excitation of the umbrella mode. From the measured dependence of desorption events on the tunneling current, it was suggested that the desorption is a three-electron process, in which three inelastic electrons are needed to overcome the desorption barrier of ∼600 meV.

The translation motion of NH_3 is initiated by the anharmonic coupling between the N–H (N–D) stretching mode and the translational mode. In addition, a two-electron excitation mechanism was identified at higher tun-

4.3 Single Molecule Chemical Reactions on Metal Surfaces 85

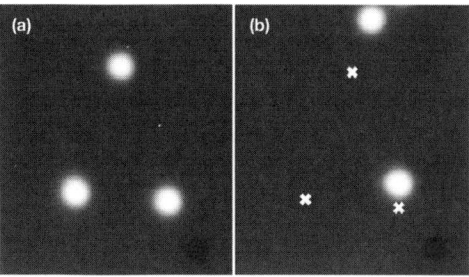

Fig. 4.10. Single NH$_3$ molecules can be desorbed or laterally moved by applying a voltage of about 420 mV (extracted from [4.64])

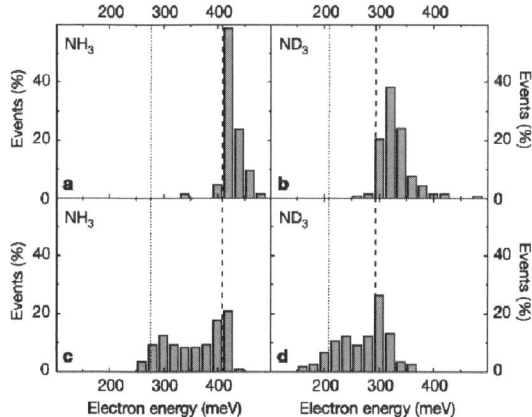

Fig. 4.11. Statistics of threshold voltage required to induce reactions for NH$_3$ and ND$_3$. **a** ND$_3$ at tunneling current of 0.5 nA. **b** ND$_3$ at tunneling current of 0.5 nA. **c** NH$_3$ at tunneling current >1 nA. **d** ND$_3$ at tunneling current >1 nA (extracted from [4.64])

neling current regimes. Such studies opened up possibilities of mode-selective excitation of single molecules, and exploration of reaction pathways at single molecule level.

4.3.5 CO Oxidation

Direct observation of CO oxidation on oxygen pre-covered Pt(111) by STM in the temperature range 237 K–274 K revealed that the reaction initiates preferentially from the (2×2)-O domain boundaries, manifested as the compressing of initially random-sized (2×2)-O domains into large ones [4.65]. The characteristic c(4×2) CO structures are observable under STM when CO domains are sufficiently large. The surface become fully occupied by CO at the end of the oxidation reaction. The measured reaction rate is proportional to the length of the (2 × 2)-O domains, and the corresponding Arrhenius plot leads to an activation energy of 0.49 eV. This value is very close to the value of 0.51 eV reported earlier for CO_2 formation on Pt(111) surfaces.

The observation is supported by the theoretical analysis suggesting a reduced activation energy barrier for CO adsorbed at bridge sites [4.66], in analogy to the likely situation at domain boundaries. Several reconstruction patterns have been observed on oxygen pre-covered Rh(111) surfaces, including c(2 × 2) for low oxygen coverage and c(2 × 6), c(2 × 8), and c(2 × 10) for high oxygen coverage [4.67]. This variety of reconstructed structures provides an opportunity to compare their relative reactivity. The oxygen atoms are not visible in STM images. Upon introducing CO, the oxygen atoms are reacted in the (011) direction. The step edges of c(2 × 8) show highest reactivity.

The CO oxidation reactions on Cu(110)-p(2 × 1)-O surfaces are preferentially along –Cu–O– chains, starting from the defect sites at the chain ends. The released Cu atoms have been observed to incorporate into the existing terraces, rather than form new islands [4.68].

4.3.6 Dehydrogenation of Single Molecules

Dehydrogenation reactions are an important category of catalytic processes. Since the reaction conditions (pressure and temperatures) are not adaptable to STM experiments, the study of C–H bond dissociation at single molecule level can be explored by applying a voltage pulse through the STM tip. The threshold voltage for breaking the C–H bond in acetylene was identified as 2.8 V, and C–D bond in DCCD as about 3.9 V (Fig. 4.12) [4.69]. Similar exercises can be performed by removing H (or D) atoms from pyridine and benzene (Fig. 4.13).

Isolated molecules of *trans*-2-butene(C_4H_8) and 1,3-butadiene(C_4H_6) were observed on Pd(110) surfaces [4.70]. With the aid of a molecular tip, the substrate lattice could be resolved and thereafter the adsorption site for C=C of the molecules was determined (Fig. 4.14). The adsorbed molecules appear in distinctively different shapes. The *trans*-2-butene(C_4H_8) appears as pairs of bright protrusions or dumb-bells, and 1,3-butadiene(C_4H_6) takes the form of single oval-shaped spots. By dosing the tunneling electrons for *trans*-2-butene(C_4H_8), a conversion of molecular topography to that of 1,3-butadiene(C_4H_6) was observed at a bias higher than about 365 mV (Fig. 4.15). This was attributed to dehydrogenation, whereby the C–H bond in *trans*-2-butene(C_4H_8) is activated. Figure 4.15 illustrates that single *trans*-2-butene(C_4H_8) molecules (T_A and T_B in Fig. 4.15a) are transformed into oval-shaped 1,3-butadiene(C_4H_6) molecules (P_A and P_B in Fig. 4.15c). However, the threshold voltage is much lower than that expected for the excitation state of C–H antibonds, and could be associated with the excitation of C–H stretching modes that should be in the range of 350–370 meV. In another report, dehydrogenation experiments using single ethylene molecules on Ni(110) and STM yielded a threshold voltage of 1.1–1.5 V, with a contrast

4.3 Single Molecule Chemical Reactions on Metal Surfaces 87

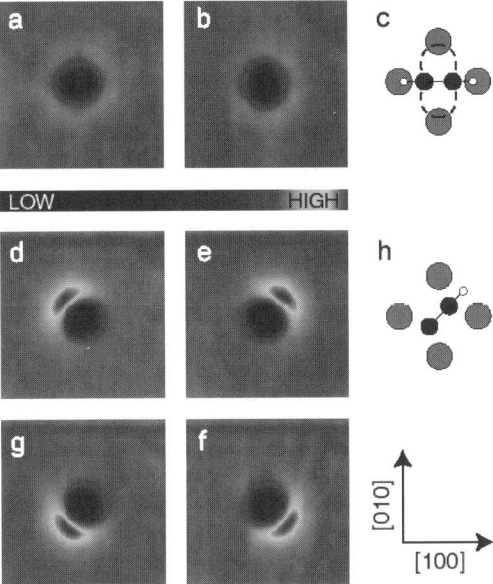

Fig. 4.12. STM images of adsorption and electron stimulated dissociation of single acetylene molecules. **a** and **b** are two adsorption orientations of dueterated acetylene (DCCD) molecule as schematically shown in **c**. **d**, **e**, **g** and **h** are four adsorption configurations of a CCD molecule illustrated in **f**. The scan conditions are 50 mV and 10 pA under temperature 9 K (extracted from [4.69])

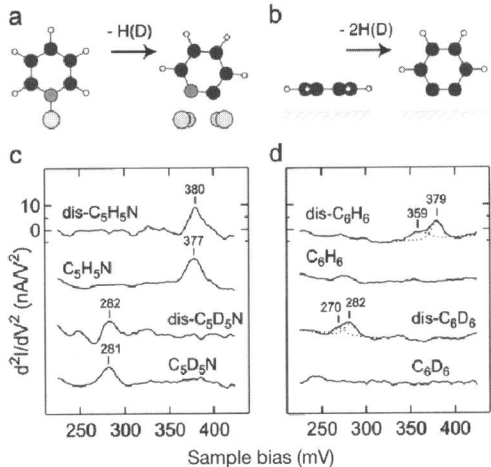

Fig. 4.13. Electron-induced dissociation of **a** pyridine and **b** benzene on a Cu(001) surface at 9 K. **c** and **d** are the corresponding tunneling spectra (extracted from [4.69])

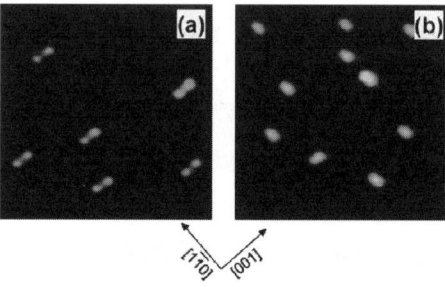

Fig. 4.14. Characteristic topography of **a** dumb-bell-shaped *trans*-2-butene (C_4H_8) and **b** oval-shaped 1,3-butadiene (C_4H_6) on Pd(110) (extracted from [4.70])

inversion from protrusion (ethylene) to depression (acetylene) [4.71]. Further dehydrogenation of acetylene to carbon atoms can be achieved by applying a voltage pulse of 1.0–4.8 V in STM. The identification of the products can be obtained by monitoring the C–H stretch feature in the tunneling spectrum of d^2I/d^2V versus V.

An artificial molecular structure through controlled dedydrogenation (as much as eight hydrogen atoms) of cobalt phthalocyanine molecules adsorbed on Au(111) surface was recently demonstrated by using voltage pulse at the magnitude of 3.3 V to 3.5 V [4.72]. Appreciable Kondo resonance effect was clearly identified in the tunneling spectroscopy on the individual dehydrogenated cobalt phthalocyanine molecule. It was proposed that the observed high Kondo temperature of the dehydrogenated cobalt phthalocyanine molecule is associated with the Coulomb repulsion and large half-width of hybridized d-level.

4.3.7 Tip-Induced Reactions of Single Iodobenzene on Cu(111)

A multiple-step approach for studying chemical reactions at single molecule level was demonstrated by Hla et al. [4.73]. The process is comprised solely of steps of controlled manipulations of single iodobenzene molecules using an STM tip at 20 K. At first, the iodobenzene immobilized at the step edges of the Cu(111) surface was stripped of iodine by the injection of tunneling electrons of 1.5-eV energy (Fig. 4.16). The dissociation rate was found to depend linearly on the tunneling current, suggesting that C–I bond breaking is caused by single electron scattering events. It was noted that electrons with 1.5 eV are not sufficient to break either C–C or C–H bonds. The dissociated components are automatically separated on the Cu(111) surface. The resulting phenyl reactants were subsequently brought together under the conditions 0.53 nA and +70 mV sample bias. The association of two adjacent phenyls into a biphenyl was accomplished by applying 0.5-V bias for 10 s (Fig. 4.17). The reaction process may involve the activation of σ_{C-Cu} bonds by in-plane rotation of phenyls, which leads to the formation of σ_{C-C} bonds. A similar

4.3 Single Molecule Chemical Reactions on Metal Surfaces   89

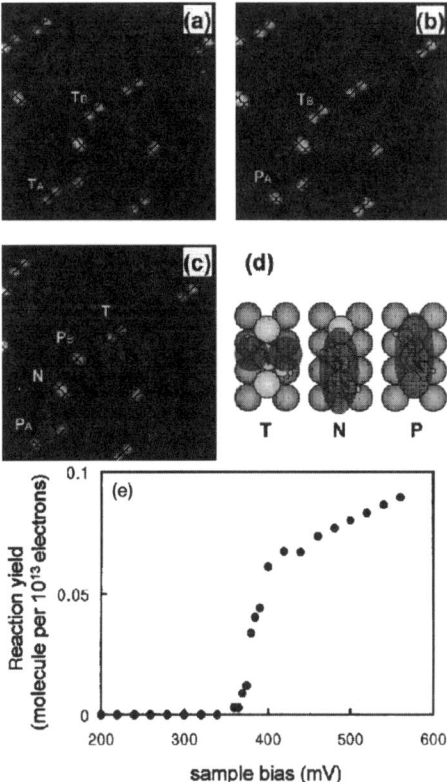

Fig. 4.15. Sequential STM image of the reaction process of *trans*-2-butene(C_4H_8) to 1,3-butadiene(C_4H_6). Two isolated *trans*-2-butene(C_4H_8) molecules marked as T_A and T_B in **a** are transformed into oval-shaped 1,3-butadiene (C_4H_6) molecules marked as P_A and P_B in **b** and **c**. **d** is the adsorption configuration of *trans*-2-butene(C_4H_8, T), 1,3-butadiene(C_4H_6, N), and the dehydrogenation product (P). Note that there is a half lattice shift of adsorption sites between the 1,3-butadiene(C_4H_6, N) and the reaction product (P). **e** Observed reaction yield from STM images (extracted from [4.70])

manipulation-assisted reaction was demonstrated with an iodine-decorated tip at improved resolution [4.74].

4.3.8 Formation of Metal Ligand Complexes

In addition to the metal ligand complex formed by CO and Fe introduced in Section 4.2.4, here we present another type of molecular clusters constructed from molecular adsorbates and diffusive metal atoms through metal ligand bonding. The Cu(100) surface is known to have highly mobile Cu adatoms released from the step edges at room temperature. These Cu adatoms can stimulate the deprotonation of adsorbed trimesic acid (tma) molecules. As a

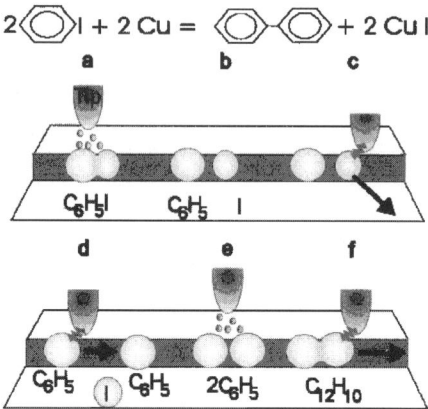

Fig. 4.16. Schematic of STM-induced dissociation of iodobenzene and synthesis of biphenyl molecules (extracted from [4.73])

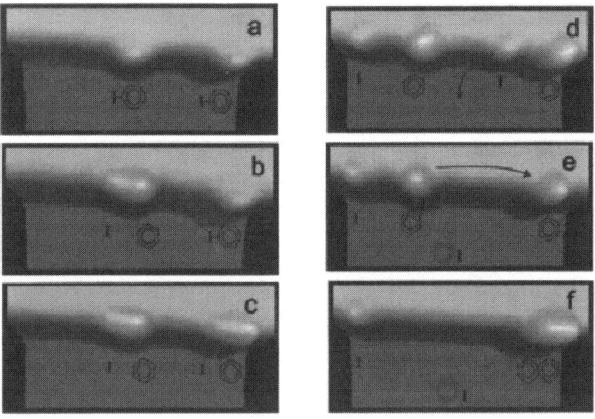

Fig. 4.17. STM images of the dissociation and synthesizing process shown in Fig. 4.16 (extracted from [4.73])

result of deprotonation, the COO$^-$ ligands of the tma molecules are available for further reactions. The product of the complexation between tma and Cu atoms forms an assembled overlayer that can be directly observed by STM. The trimesic acid molecules were observed to form a stable complex of $[Cu(tma)_4]^{n-}$ and $[Cu_2(tma)_6]^{m-}$ on Cu(100) surfaces (Fig. 4.18) [4.75]. Since the reaction is restricted to the basal plane of the Cu(100) surface, the effects of substrate lattice geometry and electric properties may be reflected in measured coordination structures in which the coordination bond Cu$\cdots$O distance is 3 Å, rather than 2 Å for three-dimensional copper carboxylates. In addition, the dominant ligand complex products closely match the substrate registry excluding other types of ligand structures.

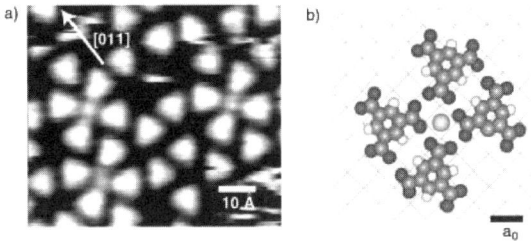

Fig. 4.18. a STM image of individual $[Cu(tma)_4]^{n-}$ on a Cu(100) surface. **b** is the proposed structural model for the complex consisting of four tma molecules and a Cu atom. The mobile Cu adatoms are not visible in the images (extracted from [4.75])

The stability of the complex species is dependent on the adsorption site and size of the assemblies [4.75]. The isolated single complex molecules on open terraces clearly have the shortest lifetime, the molecules with one neighbor or more display much longer lifetimes. The steric hindrance effect is attributed to the enhanced stability for molecules at step edges and in molecular aggregates. The temperature dependence of the individual $[Cu(tma)_4]^{n-}$ between 260 and 300 K yielded a dissociation energy of $E_D = 0.31 \pm 0.08$ eV.

4.3.9 Other Reaction Model Systems

The scope of studies on single reaction events on metal surfaces is much broader than the examples discussed in the above sections. This growing effort has rekindled interest in searching for quantitative and microscopic theories that could provide better description of heterogeneous chemical reactions, such as oxidation and hydrogenation/dehydrogenation. This is a field that will continue to be very dynamic and rewarding.

Significant restructuring of Cu(110) surfaces was observed by acetic acid adsorption [4.76], causing steps dominantly bunched in the $(1\bar{1}2)$ direction. Similar bunching effects were observed for formic acid and benzoic acid adsorption on oxygen pre-covered Cu(110) surfaces [4.77]. The study of the decomposition reaction behavior of formic acid on Cu(110) in the presence of oxygen revealed that the active sites are the step edges and domain boundaries [4.78].

Molecular adsorption and assembling behavior on model catalyst surfaces have been actively pursued using STM and other techniques. It was found that adsorption of methylacetoacetate on R,R-tartaric acid-covered Ni(111) surface could lead to the formation of two-dimensional co-crystals, which could help the study of the enantioselective hydrogenation mechanism of the system [4.79].

Ethylene (C_2H_4) adsorption and reaction on Pt(111) was studied by STM at full surface coverage [4.80]. Ordered molecular arrangements for both ethylene and ethylidyne (after annealing at high temperature) have been observed. No evidence of preferential reactions at Pt step sites was found in this case.

The reaction process of methanol to formaldehyde (CH_2O) on oxygen pre-covered Cu(110) surfaces at 270 K shows gradual replacement of (2×1)O added-row structure by a methoxy-induced (5×2) structure consisting of zigzag chains separated by $c(2 \times 2)$ subunits [4.81]. The disappearance of the –Cu–O– row is initiated from the edge of the islands along the (001) direction. The methoxy-induced islands disappear after an extended time, indicative of decomposition of methoxy into formaldehyde and hydrogen according to the following:

$$2CH_3OH + O(ad) \longrightarrow 2CH_3O(ad) + H_2O$$
$$2CH_3O(ad) \longrightarrow 2CH_2O(ad) + H_2$$

Anisotropic reaction behavior was observed for CO oxidation on Cu(110)-$p(2 \times 1)$-O surfaces [4.82]. CO molecules were found to react with oxygen atoms in the –Cu–O– chain direction, rather than in the perpendicular ($1\bar{1}0$) direction, as reflected from the fluctuation in –Cu–O– chains in STM images. The reaction probability at the end of an island was estimated as 500–1,000 times faster than that at the side.

NH_3 molecules interact with Cu(110)-$p(2 \times 1)$-O surfaces and dissociate into NH(ad) at 300 K. The NH(ad) species form linear-shaped structures in the $[1\bar{1}0]$ direction, perpendicular to the –Cu–O– chains. The reaction was found to initiate from step edges, and the NH(ad) arrays terminate at the –Cu–O– rows. This observation indicates that the reaction is inhibited along –Cu–O– chains, suggesting the "site-blocking" effect of the pre-adsorbed oxygen atoms [4.82].

The dissociative adsorption of NH_3 on Ag(110)-$p(2 \times 1)$-O develops into randomly distributed islands of NH(a) and Ag. The Ag atoms are released from –Ag–O– rows due to the reaction of NH(a) with O(a), and form islands of monoatomic height. The NH(a) induces a (2×3) restructuring of the surface [4.82].

The exploration of using nanometer-sized STM tips in localized catalytic reactions has been performed by means of $Pt_{0.8}Rh_{0.2}$ tips in rehydrogenation of carbonaceous deposits on Pt(111) surfaces [4.83]. The tip can be activated by pulsing the bias voltage in a hydrogen environment. It was suggested that the tip could be the site for hydrogen dissociation, which is the key to the rehydrogenation process.

Oxygen adsorption on Ni(110) surfaces leads to several restructured surfaces, (3×1), (2×1) and (3×1), consisting of added rows of –Ni–O–. NH_3 molecules were found to interact with the terminating atoms of –Ni–O– rows on oxygen pre-covered Ni(110) surfaces [4.84]. The dissociation proceeds in the (001) direction in an anisotropic way. As a result, molecular rows of OH evolved in the (001) direction, and the dissociation product NH_2(ad) forms a $c(2 \times 2)$ structure. The Ni atoms released from the –Ni–O– rows are aggregated to form islands.

Coexisting adsorption structures on Cu(110) surfaces have been identified as (2×1)O and (2×3)N, (2×3)N and $c(2 \times 2)$S and (2×3)N, (2×3)N and $c(2 \times 2)$ formate [4.85].

NO_2 adsorption on Ag(110)-p(2×1)-O surfaces results in p(4×1) and $c(6 \times 2)$ structures at low and high coverages respectively, reflected in STM images as straight and jagged rows in the (001) direction [4.86].

4.4 Single Molecules on Semiconductor Surfaces

The study of single molecule adsorption and reaction on semiconductor surfaces is of practical importance in semiconductor device processing and evaluations. There may be additional implications in the current trend of device miniaturization to nanometer scale. This also provides us with a test ground for understanding the possible effect of electric structures on molecular reactivities. The few examples in this section are given as a reflection of the effort in this direction.

4.4.1 Single H_2 Molecules on Si(100)

The effect of Si–Si dimer structures on the adsorption/desorption of H_2 on Si(100) 2×1 surfaces has been studied by STM [4.87, 4.88]. The Si–Si dimers can be generated by desorption of hydrogen (or deuterium) from Si(100) 2×1 surfaces. Both paired and unpaired Si–Si dimers coexist at low temperatures (below 620 K). Above 660 K, the unpaired Si–Si dimers are completely dissociated. The random-walk-like migration behavior and recombination of unpaired Si–Si dimers has been directly observed by STM [4.87]. The diffusion of unpaired Si–Si dimers is strictly one-dimensional in the dimer row direction

Two types of Si–Si dimer structures, tilted and untilted with respect to the surface plane, were seen to have very different adsorption efficiencies [4.88]. The untilted Si–Si dimers were shown to be generated by introducing submonolayers of hydrogen atoms and briefly annealing at 600 K. The untilted Si–Si dimers are those reacted with two hydrogen atoms, with Si–Si bonds parallel to the surface. The local strain field associated with untilted Si–Si dimers was seen to extend to 15–20 lattices. This conclusion was derived from the fact that each untilted Si–Si dimer is accompanied by 10–20 tilted Si–Si dimers at either side.

It was identified that such hydrogen-filled dimers are associated with much higher reactivity, with small adsorption barriers. According to the population variation of untilted dimers upon H_2 adsorption, the enhancement of the sticking coefficient has been estimated on the order of 10^9 at room temperature. This serves as a vivid example of self-promoted autocatalytic processes. It was also postulated that such catalytically active dimer sites could be generated by adsorption of other species such as halogens.

4.4.2 Single NH_3 Molecules on Si Surfaces

The adsorption of NH_3 on Si(001) surfaces at room temperature results in the formation of monohydrides with the Si atom positions unaffected [4.89]. The Si–H bonding state is detected from spectroscopy measurements, and is associated with the π_a dangling-bond state of clean Si(001) surfaces.

The dangling bonds within the (7×7) unit cell of Si(111) surfaces are correlated with different reactivities in dissociative adsorption of NH_3 [4.90]. The reaction was manifested in a strong change in topography in STM images and in the corresponding spectroscopy results (Figs. 4.19 and 4.20). By counting the affected surface sites in STM images, it was concluded that the rest atom sites are most reactive, followed by the center adatoms, and the least reactive corner adatom sites. The spectroscopy measurements proved helpful in elucidating the observed variation of reactivity for surfaces sites. For the clean Si(111) surface, the rest atom has a higher density of states than does the adatom site. This reduced density of states of adatom sites indicates a strong bonding between the adatom and the underlying Si atoms, leading to reduced reactivity. The reaction at the rest atom site is accompanied by a reverse charge transfer between the rest atom and the adjacent adatom, as evidenced by the change in tunneling spectra (curves A in Figs. 4.19 and 4.20). The difference in the reactivity of adatom sites (center and corner) can be accommodated by geometric arguments. Since the dissociation products (H,

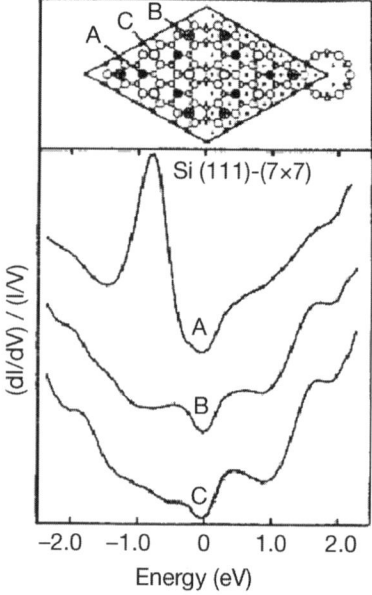

Fig. 4.19. Normalized conductance at different sites on a Si(111)-7×7 surface. *A* rest atom site, *B* corner adatom site, *C* center adatom site (extracted from [4.90])

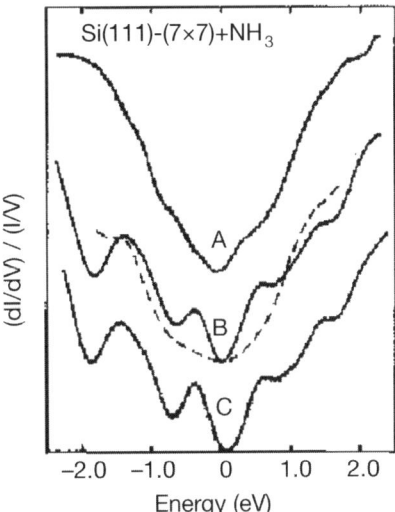

Fig. 4.20. Effect of adsorbed NH_3 on the conductance spectra. Curves A and B (*solid line*) correspond to the reacted sites. Curves B (*dashed line*) and C correspond to the unreacted sites (extracted from [4.90])

and NH_2) will react with the dangling bonds surrounding the reacted rest atom, the center atoms outnumber the corner atoms by a factor of 2, and therefore a high reaction probability can be observed for the center adatom sites.

4.4.3 Single O_2 on Ge(111), Si(100) and Si(111)

The oxidation process of semiconductor surfaces is an important issue in device fabrication. The understanding of this process is key to building devices of atomic precision. A study at single molecule level on the oxidation of Ge(111)-c(2 × 8) surfaces indicated that oxidation started from defects of dislocation in the c(2 × 8) structure [4.91]. Single atomic reactive sites can be generated by removing Ge adatoms from the basal plane. The oxidation reactivity of the single atom defect can be evaluated by direct visualization of the appearance of the defect sites. The hopping of adjacent Ge adatoms of the atomic defect sites was observed to cause diffusion of the defects, and lead to secondary atomic defects that could be oxidized again. This process may evolve further into a chain reaction behavior at single atom level. It was shown that the observed diffusion rate can be correlated with the reaction kinetics of defects with different numbers of surrounding adatoms [4.91].

The continued exposure of O_2 revealed the development of the oxidation process from single atomic defect sites to nanometer-sized domains (Fig. 4.20) [4.82]. In addition, the defect sites associated with adsorbed hydrogens were observed to develop oxidation behavior similar to that for adatom defects [4.92].

The oxidation process appears to evolve inhomogeneously around the initial reaction sites. The dangling bonds in (7×7) cells on Si(111) surfaces display different reactivity with O_2. The adatoms are most reactive at these sites. In addition, two types of long-lifetime molecular oxygen have been observed, at the adatom sites and the second-layer rest atom sites. Molecular oxygen dissociated at high sample bias voltages of positive polarity [4.93, 4.94]. Si(100) surfaces display much less oxidation reactivity, compared with Si(111) 7×7 surfaces. The reaction appears to be preferentially associated with C-type defects of dimer vacancy. The difference of reactivity between Si(100) and Si(111) surfaces correlates with the electronic properties of the surfaces [4.95].

4.4.4 Other Molecules on Si Surfaces

A variety of organic species have been studied with STM at single molecule resolution levels on semiconductor surfaces. The individual decaborane molecules are adsorbed exclusively at the defect sites on Si(111). Dissociation and fragmentation of single decaborane ($B_{10}H_{14}$) molecules adsorbed on Si(111) surfaces were demonstrated with bias voltages higher than 4 V, whereas no effects were observed with clean Si lattices under similar conditions [4.96]. The dissociated fragments of decaborane react with Si dangling bonds to form new dark sites under STM. In addition to dissociation effects, other effect such as lateral displacement, and cluster formation were also observable as a result of electron bombardment.

The dissociation of disilane molecules on Si(001) surfaces has been monitored by STM at atomic resolution level [4.97]. The study is related to the epitaxial growth of silicon on Si(001) surfaces, following the reaction pathway:

$$Si_2H_6(ad) \longrightarrow Si_2H_6(precursor) \longrightarrow SiH_3(ad) \longrightarrow$$
$$SiH_2(ad) \longrightarrow SiH(ad) \longrightarrow Si(epitaxial) \quad (4.3)$$

STM studies on the adsorption of iodochloride (ICl) molecules on Si(111) 7×7 surfaces revealed a high selectivity to form Si–I bonds, rather than Si–Cl bonds [4.98]. The quantitative ratio of Cl to I was determined by thermal desorption experiments. The molecular orbital argument was invoked to account for the observed high selectivity. The HOMO of ICl ($\pi_{x,y}*$ antibond) is dominated by the I $5p$ orbitals. The interaction between the HOMO of ICl and the partially filled Si dangling bond is the driving force for the preferential adsorption of the I end of ICl, and subsequently the abstraction reaction that results in I-rich coverage. The adsorption of halogen molecules on Si(111) 7×7 surfaces can be described using abstraction or dissociative models [4.99–4.101]. It was demonstrated by using molecular beam techniques

4.5 Single Molecule Reactions on Metal Oxide Surfaces

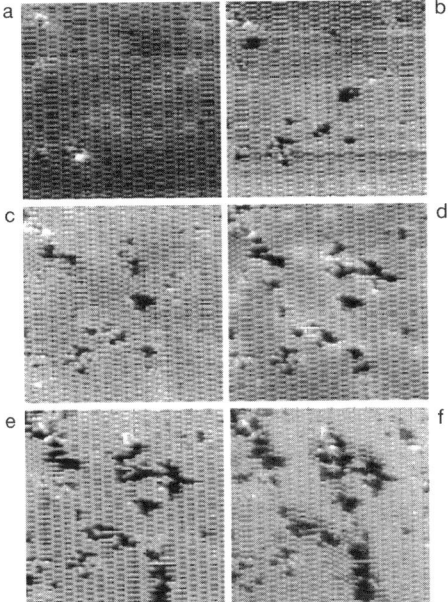

Fig. 4.21. STM images of the oxidation process of a Ge(111)-c(2 × 8) surface **a** Clean surface, **b–f** sequential exposure to 10, 20, 30, 40, and 50 L of O_2. The variation of the oxidized areas is evident (extracted from [4.92])

that the adsorption of halogen molecules of low translational energy is dominated by abstraction, in which only one of the halogen atoms chemisorbes at the surface and the other atom is ejected from the surface. In the case of molecules of high incident energy, adsorption of both atoms of the dissociated halogen molecules occurs. A detailed STM study identified that the center atom in the unit cell of Si(111) 7 × 7 is more reactive than other sites in the unit cell [4.101].

4.5 Single Molecule Reactions on Metal Oxide Surfaces

Application of STM in assessments of metal oxide surfaces has made much progress in the last several years. Metal oxide surfaces are different from pure metal surfaces in many aspects, considering the localized chemistry registry, and electronic structures. As a result, complex adsorption and catalytic behaviors should be expected. A number of oxide surfaces have been successfully analyzed at atomic resolution level [4.102, 4.103]. Though the intrinsically low conductivity of oxide materials still sets certain restraints to the generalized applications of STM techniques, the possibility of studying several important oxide surfaces demonstrates the promising prospects of STM studies.

The progress is directly related to many interesting reaction effects on oxide surfaces [4.104].

The empirically acquired estimate for minimum conductivity for STM observation is around 1 $\Omega^{-1}\text{cm}^{-1}$. Several types of natural metal oxides of high conductivity, such as ReO_3, WO_2, Fe_3O_4, and high-T_c oxide superconductors, could be readily studied by STM. The difficulty of low conductivity of metal oxides generally encountered with STM could be dealt with, at least partially, by the following [4.101]: (1) UHV annealing to generate defect states that alter the Fermi level; (2) appropriate doping to adjust the Fermi level; (3) high-temperature STM in which oxide conductivity can be enhanced; (4) epitaxially grown oxide layers on metal substrates; and (5) oxidation of alloy surfaces (such as NiAl(110) to obtain Al_2O_3).

4.5.1 TiO_2

Reconstructed TiO_2(110) surfaces are related to bulk stoichiometry. STM observations of bulk terminated TiO_2(110)-(1 × 1) surfaces reveal Ti cations as bright features at positive bias. The TiO_2-(1 × 1) surface is characterized as bright rows consisting of Ti atoms in the (001) direction, separated by O bridge-bonded oxygen atom rows [4.105, 4.106]. Two types of TiO_2(110)-(1 × 2) reconstruction structures were suggested as additional Ti_2O_3 rows

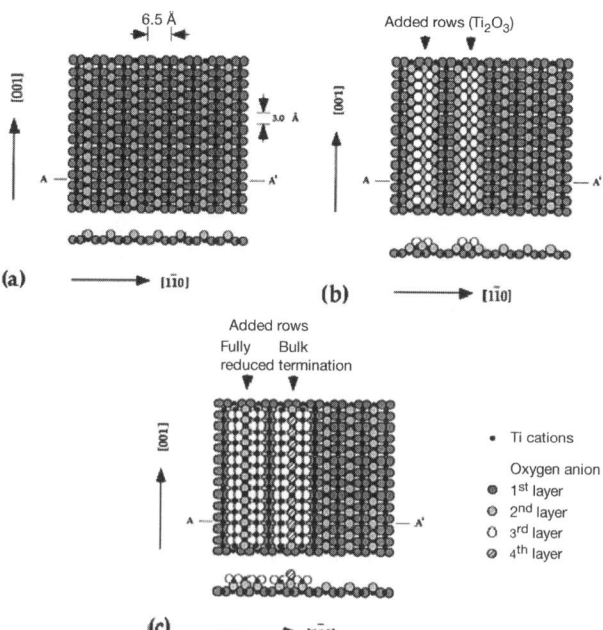

Fig. 4.22. Structural model of TiO_2(110) surface. **a** TiO_2(110)-(1×1) [4.105,4.106]; **b** TiO_2(110)-(1×2) [4.107]; **c** TiO_2(110)-(1×2) [4.108,4.109] (extracted from [4.111])

4.5 Single Molecule Reactions on Metal Oxide Surfaces

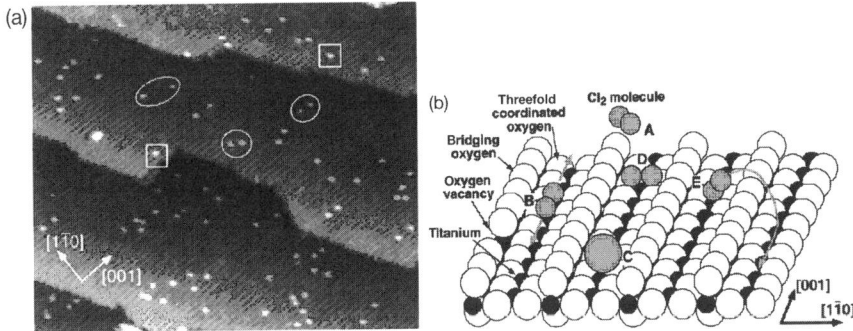

Fig. 4.23. a STM image of a TiO$_2$(110)-(1×1) surface decorated with Cl$_2$ molecules at room temperature. **b** Proposed model for adsorption of Cl$_2$ molecules (extracted from [4.114])

on top of (1 × 1) structures (Fig. 4.22b) or the reduced-row model/added-row model in (1 × 1) surface structures (Fig. 4.22c), corresponding to near-stoichiometric surface and heavily reduced surface respectively [4.107–4.109]. The spectroscopic measurements confirm the n-type semiconducting nature of the material with an energy gap around 3 eV. A summary of the titanium dioxide surface structures and related chemical properties can be found in the literature [4.110, 4.111].

The reactivities of TiO$_2$(110) surfaces to a number of molecular species have been studied at single molecule level. TiO$_2$(110) is known to have oxygen vacancies (or missing O atoms in the bridging O row) as a result of annealing treatment. These vacancy sites are very reactive to adsorbates. It was found that the oxygen vacancy population can be increased with the induction of atomic oxygen to the surface, due to the recombinatory reaction of O adatoms with the bridging O atoms [4.112]. By contrast, the adsorption of molecular O$_2$ can reduce the oxygen vacancy population. The migration of the oxygen vacancies was shown to occur only in the ($1\bar{1}0$) direction (perpendicular to the bridging O row) when molecular O$_2$ is present on the surface. This process is made possible by the abstractive adsorption of O$_2$, in which an O atom of molecular O$_2$ is lost to a vacancy site, and the other O atom is recombined with a neighboring bridging O atom.

In a related study, O$_2$ diffusion on TiO$_2$(110) was found surprisingly to depend on the oxygen vacancy density [4.113]. A proposed mechanism attributes the formation of transient species of charged O$_2^*$ as the key step in the hopping process of molecular O$_2$. The population of oxygen vacancy is directly correlated to the donor level, and thus chemical potential. As a result, electron density in the conduction band (n_c) is affected through the relationship:

$$n_c = N_c e^{-(\varepsilon_c - \mu(N_d))/k_B T}$$

N_c depends on temperature and the effective mass of electrons in the conduction band, and ε_c is the conduction band minimum, and N_d the va-

100   4 Single Molecule Diffusion and Chemical Reactions

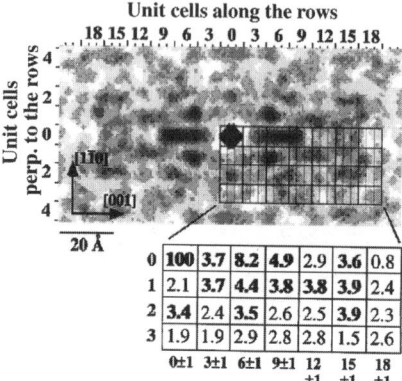

Fig. 4.24. Autocorrelation of Cl–Cl distances. (extracted from [4.114])

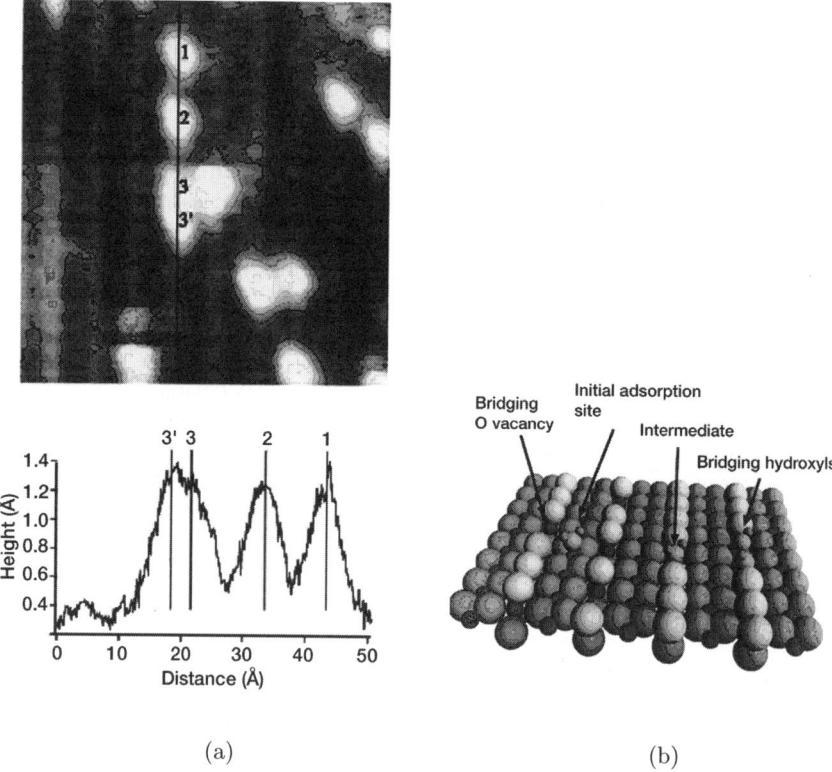

(a) (b)

Fig. 4.25. a High-resolution STM image of H_2O adsorbates on a $TiO_2(110)$-(1×1) surface. The distances between the markers in the cross-section profile are integer multiples of the lattice constant along [001]. **b** Schematic of adsorption sites for H_2O on $TiO_2(110)$ (extracted from [4.116])

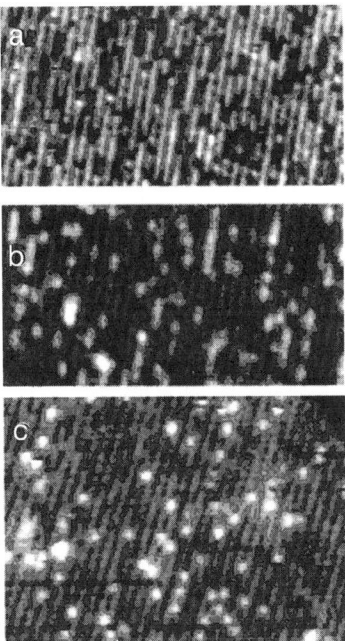

Fig. 4.26. Temperature effect of the H_2O adsorption. **a** 150 K, 24 nm × 12 nm, **b** 195 K, 24 nm × 12 nm, **c** 290 K, 20 nm × 14.5 nm (extracted from [4.116])

cancy concentration. Therefore, the rate of charge transfer between the O_2 adsorbate and the electrons in the conduction band is also affected. This observation suggests that vacancies can behave as a promotor to adsorbates by affecting the donor level in the electronic structure.

Dissociative adsorption of Cl_2 molecules on $TiO_2(110)$ surfaces results in separated Cl adatom pairs with an average distance of 26 Å (Figs. 4.23 and 4.24) [4.114]. In addition, single Cl adatoms were observed with smaller populations compared with paired Cl adatoms, due to abstractive adsorption. The observed large separation between Cl–Cl adatom pairs is attributed to the non-thermal transient process during dissociation, described as "cannonball-like" trajectories. Cl atoms were found to adsorb on the rows of fivefold coordinated surface Ti atoms, and no preferential adsorption was observed for step sites and oxygen vacancy sites. The reaction sites for molecular dissociations could be the undercoordinated Ti sites. A smaller population of adsorbates was found at bridging oxygen sites [4.115]. By contrast, adsorption at 200 and 300 °C results in Cl atoms at bridging oxygen sites only, possibly due to the replacement process, as indicated by the UPS method [4.115].

The adsorption of molecular H_2O was found predominantly on the row of fivefold Ti atoms on $TiO_2(110)$ surfaces at 150 K (Figs. 4.25 and 4.26) [4.116]. Upon heating the surface to 290 K, molecular H_2O is dissociated into bridge-bonded hydroxyls, suggesting the dissociation is activated at the vacancy sites along bridge oxygen rows (Fig. 4.26).

Adsorption of pyridinecarboxylic acid and benzoic acid molecules on $TiO_2(110)$ surfaces was also studied by STM. Dimeric (or larger multimeric) adsorption of isonicotinic acid molecules was found to involve the dominating species at low molecular coverages [4.117].

Organometallic complexes of $Me_2Au(acac)$ ($Me = CH_3$, $acac = CH_3COCH_2COCH_3$) and $Ir(acac)_3$ were found to adsorb on the added rows of Ti_2O_3 on $TiO_2(110)$ terraces [4.118]. These molecules are used in gas-phase grafting (GG) methods for preparing metallic nanoparticles.

The reconstructed $TiO_2(110)$-(2×1) structure was found to be the preferential adsorption site for SO_2 at low coverage [4.111]. Adsorption at $TiO_2(110)$-(1×1) sites becomes significant at high SO_2 coverage, along the fivefold Ti atom rows.

4.5.2 CO on $RuO_2(110)$

$RuO_2(110)$ surfaces can be obtained by epitaxial oxidation of $Ru(0001)$ surfaces (Fig. 4.27) or by growth from RuO_2 powders at elevated temperatures [4.119–4.121]. Single crystal surfaces of RuO_2 have been studied by STM in both UHV and ambient conditions. Figure 4.28 is an example of the ordered atomic structure of RuO_2 crystal surfaces obtained by STM in ambient conditions [4.121].

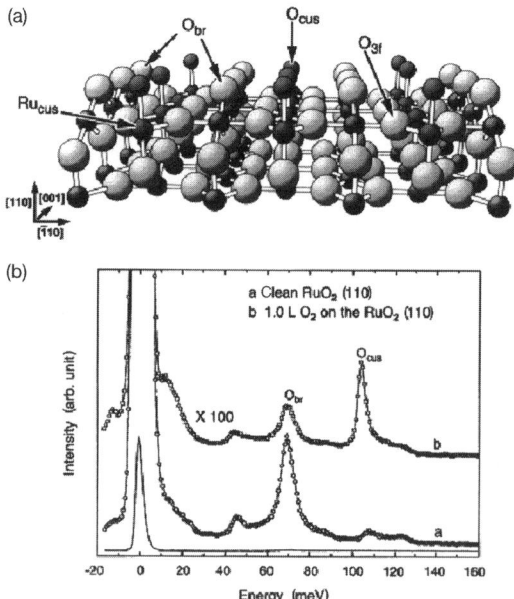

Fig. 4.27. Structural model of $RuO_2(110)$ surface. O atoms can be found at bridge sites (O_{br}) and coordinatively unsaturated sites (O_{cus}) (extracted from [4.120])

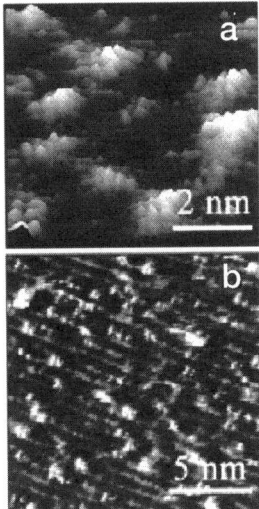

Fig. 4.28. a, b STM image of $RuO_2(110)$ surface in ambient conditions showing atomic resolution (extracted from [4.121])

Under UHV conditions, the individual CO molecules were found to adsorb to ruthenium atoms at the coordinatively unsaturated sites (cus) of RuO_2 surfaces [4.119, 4.120]. The CO molecules interact with the $RuO_2(110)$ substrate through coupling of 5σ electrons with the substrate, and the $2\pi^*$ of CO and the d electron of the substrate. The bonding strength therefore provides an indication of the local density of d electrons at the adsorption sites. The adsorbed CO on RuO_2 is immobile [4.119], suggesting a high diffusion barrier. The accompanying density function theory simulation suggested a diffusion barrier of about 1 eV, which is much higher than the typical diffusion barrier on clean metal surfaces (a few tenths of an eV). This enhanced diffusion barrier on oxide surfaces could be due to the fact that highly localized bonds form during chemisorption.

The oxidation of CO to CO_2 can be achieved by reacting with lattice oxygen atoms [4.119, 4.120]. The driving mechanism for the enhanced reactivity was explained by the reduced bonding strength of bridge site oxygen atoms (about 1.6 eV), which matches the adsorption barrier of CO anchored to the coordinatively unsaturated (cus) Ru atoms (1.2 eV). This result provides direct evidence for the catalytic activity of the cus mechanism for oxide surfaces [4.122].

4.5.3 Fe Oxide Surfaces

STM has been applied to obtain high-resolution images of single crystalline iron oxide films FeO(111), $Fe_3O_4(111)$, and α-$Fe_2O_3(0001)$, prepared by epitaxially grown on Pt(111) surfaces [4.123]. Site-specific dissociations of CCl_4

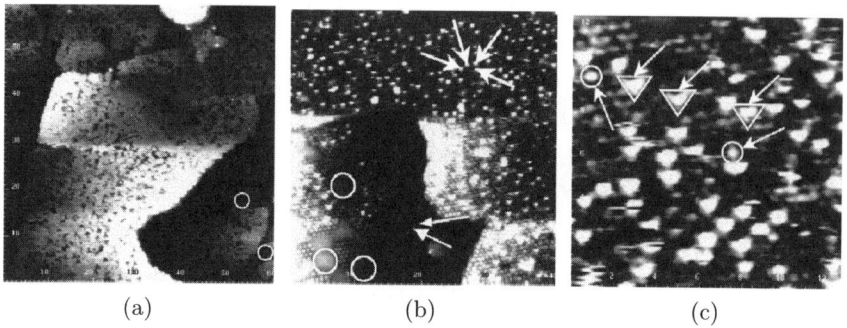

Fig. 4.29. $Fe_3O_4(111)$ surface **a** before and **b** after 0.1 L CCl_4 adsorption at room temperature. Tunneling conditions are 1.2 V and 0.6 nA. **a** 60 × 60 nm, **b** 40 × 40 nm. **c** High-resolution STM image of CCl_4 on Fe_3O_4 (extracted from [4.124])

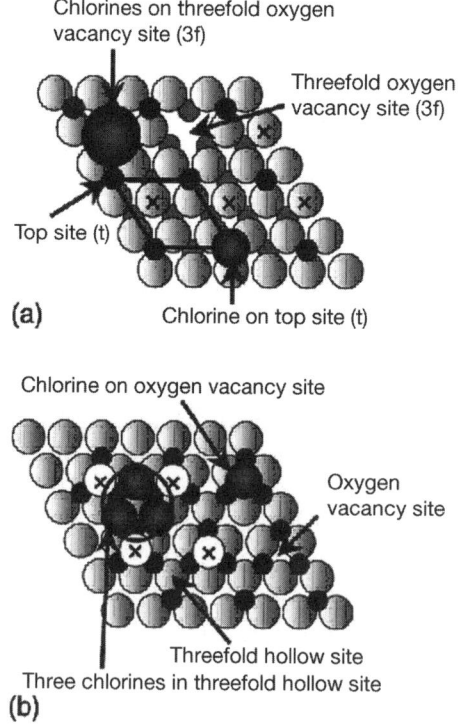

Fig. 4.30. Schematics of CCl_4 adsorption on Fe_3O_4 (extracted from [4.124])

and Cl_2 molecules on α-Fe_2O_3(0001) surfaces have been identified by STM (Figs. 4.29 and 4.30) [4.124]. The (2 × 2) structure for Fe_2O_3(111) was observed as the dominant feature on clean oxide surfaces, and is the active structure for dissociation reactions. The low-energy-electron-diffraction (LEED) pattern and the accompanying I-V spectroscopy result collected on the surface suggested the (2 × 2) surface is terminated with a 1/4 monolayer of Fe atoms. It was found that Cl atoms are preferentially attached to the atop sites of Fe atoms, and bring substantial changes in tunneling characteristics. In addition, the minority structure of (1 × 1) has tentatively been assigned to oxygen terminated regions from tunneling characteristics, and is inactive to the adsorbates. The threefold hollow sites are the additional adsorption sites for dissociation of CCl_4 molecules, attributed to the binding of $COCl_2$ to the second-layer Fe atoms (Fig. 4.30).

4.5.4 Other Oxide Surfaces

A growing number of STM studies dedicated to metal oxide surfaces have been seen in the literature, such as those dealing with WO_3(001) [4.125, 4.126], polyoxometalate films [4.127], and vanadium oxides [4.128]. The γ-WO_3 surface has been examined in the oxidative dehydration of alcohols at molecular resolution by STM [4.126]. The WO_3(001) surface develops a c(2 × 2) reconstruction pattern, terminated with a 1/2 monolayer of oxygen atoms. After annealing product of 1-propanol at 400 K, dispersed features correlated to alkoxide adsorbates were observed, and the binding sites were attributed to the exposed W^{6+} sites. Such efforts could contribute to the understanding of many important, heterogeneous catalytical reactions at single molecule level. Several oxide structures have been identified on thin-film vanadium oxide layers grown on Pd(111) using HREELS, including V_5O_{14}, V_6O_{14}, and s-V_2O_3 [4.129].

The silver oxide surface Ag(111) was prepared by exposing Ag(111) surfaces to large doses of NO_2. STM observations in UHV and low temperatures of this oxide surface revealed hexagonal, honeycomb structure of p(4 × 4)-O [4.130, 4.131]. Theoretical simulations by Green's function scattering formalism of the STM image suggest that the surface is actually a trilayer structure of $Ag_{1.83}O$ on top of Ag(111). Ethene molecules are found to adsorb exclusively on the Ag_3 sites of the silver oxide surface at 77 K (Figs. 4.31, 4.32) [4.131]. In addition, calculations based on density function theory point to the significant displacement of Ag_3 sites to optimize the adsorption of ethene molecules.

STM results on reaction systems are rather preliminary at this stage, and much more work should be performed to provide complementary and useful data to the body of information obtained from other traditional surface chemistry research. With the continued development of STM techniques, it is expected that this technique could facilitate studying practical heterogenous reactions under high pressures and temperatures, as well as disordered surfaces. There have been several generations of designs of STM instruments

106   4 Single Molecule Diffusion and Chemical Reactions

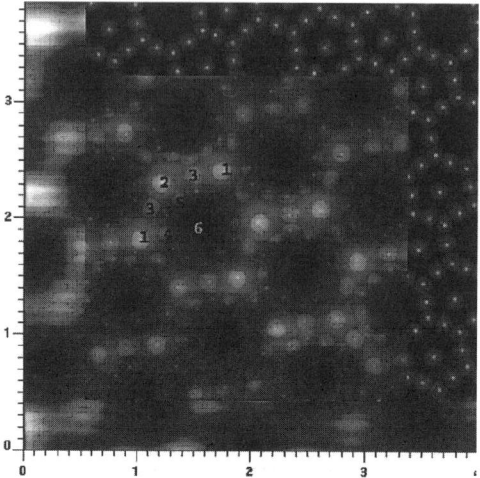

Fig. 4.31. Structural model of silver oxide (extracted from [4.131])

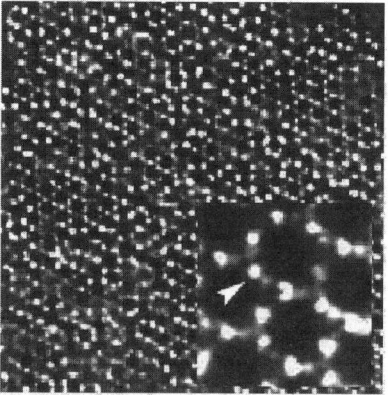

Fig. 4.32. STM image of ethene-covered silver oxide surface. The *arrow* marks the single ethene molecular adsorbate (extracted from [4.131])

toward this goal, with some progress. Until such time, more rigorous investigations can be performed to reveal details of heterogeneous catalytic reactions in real time and space, and the results will be surely highly important.

5 Molecular Scale Analysis
Using Scanning Force Microscopy

5.1 Basic Principles of Atomic Force Microscopy (AFM)

It is important to note that the "tunneling" phenomenon utilized by STM introduced in the preceding chapters requires that this instrument be used only for electrical conductors or semiconductors. In order to study an electrically non-conductive material with an STM, its surface must be covered with a thin conductive layer. However, the existence of conductive films often reduces the resolution and thus limits the usefulness of the STM. It was mainly this limitation that prompted the invention and development of atomic force microscopy (AFM) in 1986 [5.1, 5.2], which can be applied to image both conductors and non-conductors in air, liquids or vacuum. The force microscopes have demonstrated excellent resolution capabilities at nanometer scale, and the experimental conditions are greatly expanded. Since the invention of AFM [1], the concept has led to widespread interest in many branches in surface science studies [5.2]. As a result, several highly effective microscopy techniques have appeared, with the aim of developing new chemical mapping methods for surfaces [5.3, 5.4].

The principles behind various force detection schemes require detailed knowledge of the interaction force, both for long and short ranges, especially those interactions between the probe and various functional end groups of the surface. Extensive studies have provided ample examples of retrieving nanometer-scale surface compositional information based on mechanical, physical, chemical and magnetic characteristics that form the content of this chapter. The pursuit of nanometer-scale mapping of surface compositions using AFM-related techniques has produced many illuminating results. In view of the high spatial resolution, such efforts should provide insightful information complementary to that from other surface characterization techniques.

5.1.1 Introduction of Instrumentation

The operational principle of AFM is schematically illustrated in Fig. 5.1. The cantilever, which is extremely sensitive to even weak forces, is fixed at one end. The other end of the cantilever has a sharp tip that gently approaches the sample surface. The designs of AFM scanners can be identical to that used in STM. When the sample is being scanned in the x-y direction, because of the

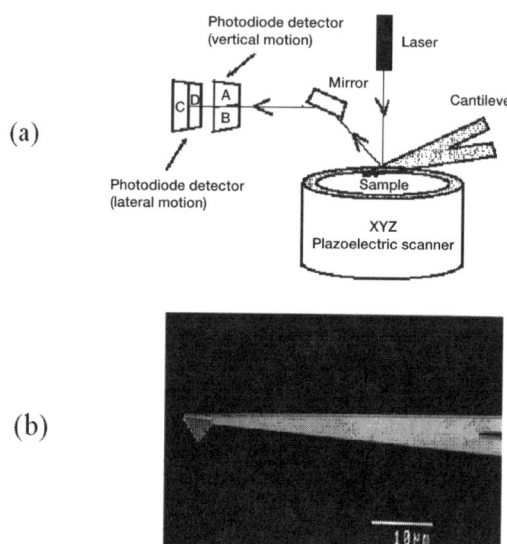

Fig. 5.1. Schematics of **a** AFM cantilever, and **b** detection configurations (extracted from [5.4])

ultra-small repulsive force between the tip atoms and the surface atoms of the sample, the cantilever will move up and down in the direction perpendicular to the sample surface, corresponding to the contours of the interaction force between the tip and surface atoms of the sample. The topographic images can be obtained either by recording the deflection of the cantilever at each point (variable-deflection mode) or by keeping the force constant using an electric feedback circuit and recording the z-movement of the sample (constant-force mode). The STM yields images related to the surface electronic densities near the Fermi level. The AFM tip may be kept in direct contact with the surface (contact mode) or it may be vibrated above the surface (non-contact mode). For high-resolution imaging and most routine topographic profiling, the repulsive force (10^{-9}–10^{-8} N) or contact mode is usually used. In the non-contact mode, van der Waals, magnetic, or electrostatic forces can be detected.

5.1.2 Cantilever

High lateral stiffness in the cantilever is desirable to reduce the effects of lateral forces in the image-acquiring process by AFM. Frictional forces can cause appreciable lateral bending of the cantilever, leading to associated image artifacts. Investigations have indicated that choosing a "V" shape for the lever can yield substantial lateral stiffness. In a recent study, it was suggested that a rectangular-shaped cantilever can actually lead to better lateral stability than is the case for the "V"-shaped lever [5.5]. The analysis was performed

5.1 Basic Principles of Atomic Force Microscopy (AFM)

by considering the ratio between the lateral spring constant and the normal spring constant, using typical parameters for V-shaped cantilevers. The theoretical analysis was confirmed by experiments using centimeter-scaled cantilevers [5.6].

For high-resolution topographical imaging, the cantilever stylus used in the AFM should satisfy the following criteria:

(1) Low force constant
(2) High resonant frequency
(3) High mechanical Q
(4) High lateral stiffness
(5) Short lever length
(6) Incorporation of a mirror or electrode for deflection sensing
(7) Sharp protruding tip.

For imaging atomically flat samples, one can simply use the end of a cantilever as an effective local tip. For imaging rougher samples, however, a sharp, protruding tip of known shape is preferable, so that the interaction between the sample and the cantilever can be characterized more precisely.

Fabrication processes have been developed to produce SiO_2 cantilevers with integrated conical tips or tetrahedral tips [5.7]. The low mass of these cantilevers allows them to have high resonant frequencies (10–100 kHz), with spring constants (typically 0.0006 N m^{-1}) small enough to detect forces less than 10^{-8} N. As an AFM can image both conductors and non-conductors, in air, liquids or vacuum, it has found a wide range of applications with atomic and molecular resolutions of surfaces. Images obtained using the cantilevers described above convincingly resolve the atomic periodicity of several types of crystalline samples, including graphite, MoS_2, WTe_2, $TaSe_2$, and boron nitride (BN) [5.8].

5.1.3 Cantilever Deflection Detection

Several designs of deflection detection have been demonstrated during the development of AFM techniques, and some can be found in commercially available instruments. The effectiveness has been widely tested under UHV, low-temperatures, ambient-temperature, and liquid conditions.

Optical detection of the cantilever deflection can be exemplified by the designs based on beam deflection or beam interferences. To measure the cantilever displacement, the direction of a laser beam reflected off the back side of the cantilever is monitored with a position-sensitive detector (PSD). In the case of surface topography, the bending of the cantilever is recorded with a segmented PSD, typically a bi-cell, which consists of two separated photoactive (e.g., Si) segments (anodes) with a common cathode. Additionally, lateral forces induce a torsion of the cantilever, which in turn causes the reflected laser beam to undergo a change in direction perpendicular to the surface. Thus, with a simple combination of two orthogonal bi-cells, i.e., a quadrant PSD, one is able to measure the deflection of the cantilever independently and simultaneously in two orthogonal directions [5.9–5.11]. This capability is unique to the optical beam deflection method that, due to its very nature, measures the orientation of the cantilever together with its displacement. The beam interference methods using optical interferometry have also been discussed in the literature [5.12, 5.13].

The electrical detection methods are derived from tunneling effects [5.14–5.16], the capacitance method [5.17, 5.18], and the piezoelectric method [5.19–5.21]. The inclusion of a piezo-resistive strain sensor to the cantilever enables direct measurements of the deflection at high resolution both in ambient conditions [5.22], UHV [5.23], and at low temperatures [5.24, 5.25]. In addition, a setup was proposed for studying single molecule forces based on magnetic stimulation. Detection of the AFM cantilever was achieved through an attached magnetic particle [5.26].

5.1.4 Cantilever Calibration

The primary difference between the scanning method of an AFM and an STM is that it is the sample that is scanned, rather than the delicate, and sometimes bulky, force sensor (composed of tip and cantilever). Much of the technology for optimized AFM performance is now well developed. Techniques for vibration isolation, scanning, sample approach, feedback control and image processing are adopted with little modification from those of STM. Construction of the force sensing cantilever stylus, and measurements of the deflection of cantilevers still need careful consideration.

The force sensor is a crucial component for AFM, determining its sensitivity and resolution. Whereas the force sensor senses the force, the role of the cantilever is to communicate this information to the controller. When the AFM is operated in the contact mode, in order to register a measurable deflection with small forces, the cantilever must flex with a relatively low force constant. The data acquisition rate in the AFM is limited by the mechanical resonant frequency of the cantilever. To achieve an imaging bandwidth comparable to that of an STM, AFM cantilevers should have resonant frequencies greater than 10 kHz. Fast imaging rates are not simply a matter of convenience, because the effects of thermal drifts are more pronounced with slow scanning speeds. If the scanning rate is too high or the cantilever resonance is too low, due to the inertia of the cantilever, the stylus tip may not track the steep downward slopes on the sample surface. The combined requirements of a low spring constant and a high resonant frequency can be met by reducing the mass of the cantilever stylus assembly.

Quantitative determination of spring constants of AFM cantilevers has been carried out following several methods. The thermal noise method utilizes the frequency spectrum generated from fast Fourier transformation (FFT) of the cantilever displacement spectrum to obtain the modulus value k [5.27]. An example of the power spectrum of cantilever fluctuation is given in Fig. 5.2.

5.1 Basic Principles of Atomic Force Microscopy (AFM)

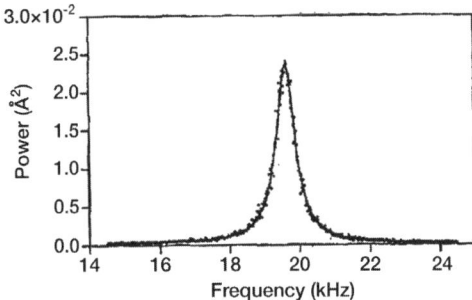

Fig. 5.2. Power spectrum of cantilever deflection fluctuations recorded by the detector (extracted from [5.27])

This approach can be understood by considering that the oscillation of the cantilever is a harmonic motion, and mean kinetic energy is equivalent to thermal energy expressed in the following:

$$H = \frac{p^2}{2m} + \frac{1}{2}m\omega_0^2 q^2$$

$$<\frac{1}{2}m\omega_0^2 q^2> = \frac{1}{2}k_B T$$

$$\omega_0^2 = \frac{k}{m}$$

$$k = \frac{k_B T}{<q^2>} \tag{5.1}$$

Using Fourier relations one can have the following:

$$\int |q(t)|^2 \, dt = \int |q(\omega)|^2 \, d\omega = P$$

$$k = \frac{k_B T}{P}$$

The thermal noise spectrum of the cantilever can be obtained by sampling the amplified photodiode difference signal at high speed, and Fourier transforming the results [5.28]. Further analysis of the detailed deflections of the cantilevers suggested that the amplitude of thermal noise could be affected by temperature, the spring constant, and measurement configurations [5.29, 5.30]. It was further suggested that the thermal noise method is particularly suitable for soft cantilevers [5.30].

The other method is by adding a known mass M to the cantilever, and measuring the shift of resonance frequency, as illustrated in the following

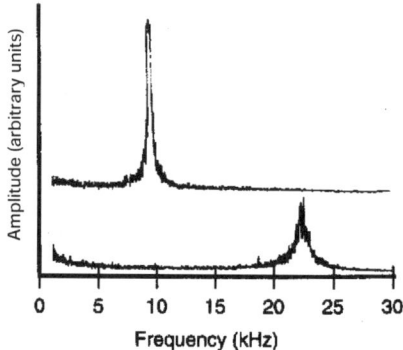

Fig. 5.3. The resonance spectra of an AFM cantilever **a** before and **b** after adding a 44-ng end mass (extracted from [5.31])

expressions and the example given in Fig. 5.3 [5.31]:

$$\omega = \sqrt{\frac{k}{M + m^*}}$$

$$M = \frac{k}{\omega^2} - m^* \quad , \quad m^* = \frac{k}{\omega_0^2}$$

$$k = \frac{M}{\frac{1}{\omega^2} - \frac{1}{\omega_0^2}} \tag{5.2}$$

An alternative method is to use a pre-calibrated cantilever as reference. This calibration method is performed by pressing the cantilever to a pre-calibrated glass fiber while recording the deflection. The calibration of glass fiber can be obtained by attaching a small load [5.32].

5.2 AFM Operating in Contact Mode

5.2.1 Contact Mode

In the contact mode of AFM operations, the interaction force is represented by the deflection of the cantilever. The magnitude of the interaction is derived from the displacement and the spring constant of the cantilever. It should be noted that the cantilever deflection is generated by forces both perpendicular to the surface as well as in the lateral direction.

The effects of atomic structures of a tip on AFM images have been studied theoretically using both single atom and multiple-atom tips [5.33]. The simulated images showed that single atom tips can generate anomalous patterns and are not appropriate for assessing experimental results. In the case of multiple-atom tips (Fig. 5.4), the applied load, tip geometry and bond length

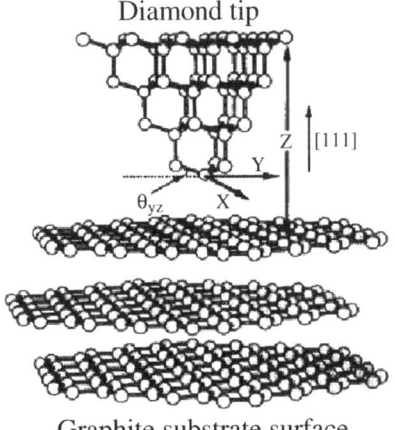

Fig. 5.4. Cluster model of a diamond (111) tip consisting of 54 atoms over a graphite surface (extracted from [5.33])

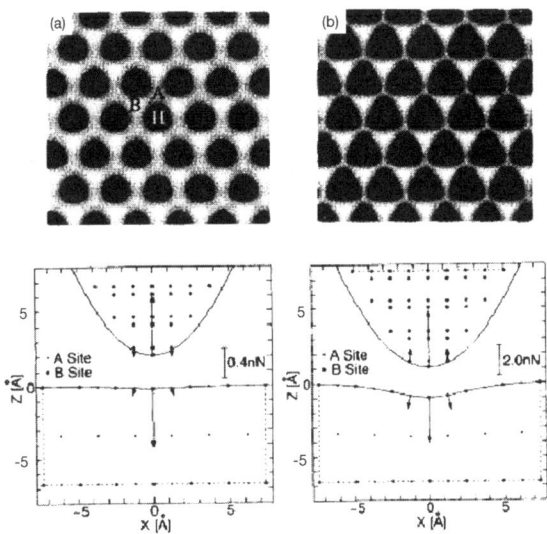

Fig. 5.5. Calculated topography image using a 54-atom cluster tip at constant repulsive force of **a** 0.5 nN and **b** 5.0 nN (extracted from [5.33])

can indeed affect the observed lattice patterns (Fig. 5.5). The distribution of the interacting force in the vicinity of the tip apex is also affected by the tip apex structures.

Besides these atomically flat surfaces, using an AFM, it has been possible to image organic molecules on molecular crystal surfaces such as amino acids [5.34] and nitronyl nitroxide [5.35]. Surfaces in air are typically covered with a layer of liquid contaminants, and when the probe comes near a sample, there is a capillary force that pulls the tip toward the sample (with a typical value of 10^{-7} N). Operating the AFM within water can reduce this undesirable and often destructive force, and allows one to better control the force between the tip and the sample.

Topography imaging can also resolve molecular domains among different hydrocarbon chain lengths [5.36, 5.37]. For example, phase separation of mixed fatty acids (palmitic acid $C_{15}H_{31}COOH$ and lignoceric acid $C_{23}H_{47}COOH$) prepared by the Langmuir-Blodgett method was characterized by both topographic and frictional imaging [5.37]. The height difference equivalent to the difference of molecular chain length in topographic images can be used to identify the composition of the domains.

Nanometer-scale resolution AFM studies of multicomponent polymer surfaces can enable detailed analysis of properties such as phase separation behavior, and surface wetting/dewetting. The distinction of surface compositions can be rationalized by arguments based on bulk compositions and surface free energy [5.38]. As an example, direct AFM observations of thin films of immiscible blends of polystyrene (PS) and poly(methyl methacrylate) (PMMA) revealed the surface phase separation behavior of the polymer film. The continuous domains displaying high contrast are attributed to PMMA as a result of either chain conformation or chain aggregation structures being frozen at the air–polymer interface. Several types of phase separation characteristics can be resolved at various film thicknesses, and can be interpreted within the framework of thickness dependence of both the Flory-Higgins interaction parameter and the degree of polymer entanglement.

Microscope studies of organic–inorganic composite materials using AFM also provided interesting insight into nucleation and growth processes. An example can be found in the study of biomineralization processes [5.39–5.41]. The crystal structure of calcium carbonate was shown to be subtly affected by the inclusion of proteins during growth processes [5.42–5.44]. Such biocomposite materials have demonstrated superb toughness and strength [5.45, 5.46].

AFM has found extensive applications in biological systems [5.47]. The AFM has been used to image various cell surfaces, (such as red blood cells [5.48]), and DNA [5.49]. A major challenge for investigators in this field is to develop appropriate sample preparation techniques that will allow optimal details to be imaged. The sample must be rigid enough not to be damaged or distorted by the applied force. As mentioned above, imaging with the sample in liquid eliminates meniscus forces that can pull the stylus destructively into the sample, thus making it possible to use smaller forces, usually in the 10^{-9} N range or less.

With technical advances in AFM, especially the optical beam deflection techniques, the AFM has been used successfully to study electrode surfaces under potential control in a fluid electrolyte. AFM has also been used successfully to image Au(111) electrode surfaces showing atomic periodicity while the electrode was under potential control in a fluid electrolyte [5.50]. AFM images in electrolytes taken at +0.7 V prior to Cu deposition showed large areas exhibiting Au(111) structures. By sweeping the potential, the Cu can be stripped down to an under-potential-deposited monolayer and finally returned to a bare Au(111) surface. The images revealed that the underpotential-deposited monolayer has a correlation with different electrolytes. Specifically, for a perchloric acid electrolyte the Cu atoms are in a closepacked lattice with a spacing of 0.29 nm. For a sulfate electrolyte, they are in a more open lattice with a spacing of 0.49 nm. As the deposited Cu layer grew thicker, the Cu atoms converged to a (111)-oriented layer with a lattice spacing of 0.26 nm for both electrolytes. Images were also obtained of an atomically resolved Cu monolayer in one region, and an atomically resolved Au substrate in another region in which a 30° rotation of the Cu monolayer lattice from the Au lattice is clearly visible. The observation of a complete adsorption and desorption cycle for a metal onto another metal surface with such high resolution demonstrates that the AFM is a valuable technique in studying other electrochemical processes in situ.

Studies of individual biomolecules such as DNAs under ambient conditions provide many examples of contact-mode AFM applications. AFM has displayed great potential in the study of the structure of surface-bound DNAs and DNA–protein complexes. Full-length DNA molecules have been studied extensively by AFM [5.51–5.53]. The difficulty in imaging large DNA molecules comes from the fact that the longer the strand, the stronger the tendency to form entanglements. One way to solve this problem is to align the long DNA strands in a single direction. Several approaches have been proposed to straighten large DNA segments, such as by surface tension, and hydrodynamic force [5.54–5.59]. Figure 5.6 presents some examples of aligned DNA molecules on mica surfaces prepared by flowing gas [5.59].

It has been known from experiments that 3-aminopropyltriethoxysilane (APTES)-covered mica surfaces can effectively immobilize DNA molecules under water to allow high-resolution AFM studies [5.52]. The measured width of DNA molecules by AFM is typically in the range 10–40 nm, compared with the structural width of around 2 nm. The broadening effect can come from several factors, including tip broadening, and the presence of a hydration layer around the DNA. It is worth pointing out that freshly cleaved surfaces of highly oriented pyrolytic graphite (HOPG) can have intriguing features closely resembling those of polymeric molecules including DNAs observed by STM [5.60, 5.61]. HOPG is therefore not the ideal substrate for such studies.

Direct observations of site-specific interaction between DNA and enzymes can be achieved with AFM. Allison et al. [5.62] demonstrated the mapping of EcoRI endonuclease attached to single cosmid DNA [5.62, 5.63]. The AFM images provided consistent information on the binding sites of EcoRI, with the known values of the separations between these sites (Fig. 5.7). In another study, protein binding was found to cause local conformation change of DNA segments, and resulted in sharp bends in stretched DNAs [5.64].

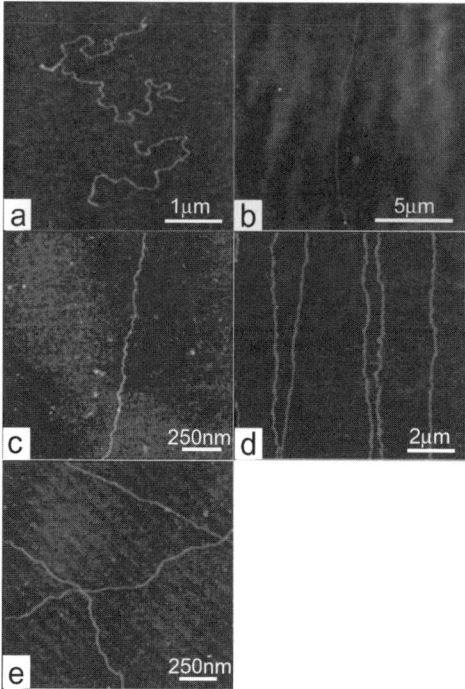

Fig. 5.6. a–e Various appearances of stretched DNA molecules on mica surfaces observed by AFM (extracted from [5.59])

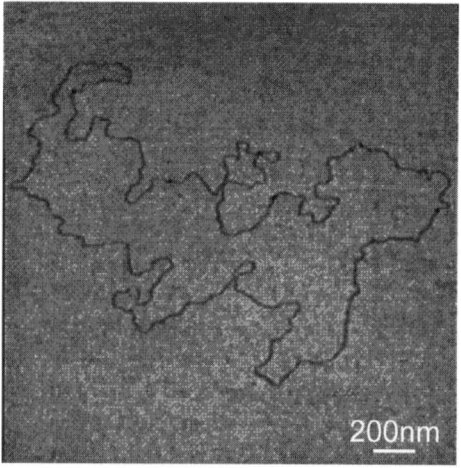

Fig. 5.7. Site-specific binding of DNA/EcoRI observed by AFM (extracted from [5.62])

The strand-exchange reaction of nucleoprotein and double-stranded DNA was observed with AFM [5.65]. The reaction involves a RecA-coated single-stranded DNA as a probe and plasmid DNA. The probe targets a specific site on the double-stranded DNA consisting of 36 base pairs. The attachment of the probe protein to DNA was identified with high precision using AFM.

AFM was also used in analyzing the proteolysis process of collagen I molecules by *Clostridium histolyticum* collagenases in a phosphate-buffered saline (PBS) solution [5.66]. The digestion of individual collagen molecules can be identified. By monitoring the number of collagen I molecules in real time, the hydrolysis rate constant can be obtained, and the reaction time for single collagen digestion was estimated in the range of 15 s to several minutes. With the advantage of introducing light illumination while sensing local normal force interactions simultaneously, the photo-induced motion of protein-bacteriorhopsin (bR) within membranes can be followed [5.67].

5.2.2 Friction Force Microscopy

The friction force microscope (FFM) is an extension of the AFM that allows one to measure the bending and twisting of the cantilever in the imaging process simultaneously [5.68, 5.69]. It measures topography just like the AFM, but additional sensors are used to measure the frictional forces between the sample and the tip from the same scan.

It is instructive to examine briefly the nature of the cantilever response to frictional and surface forces. In the latter case, the direction of the acting force is perpendicular to the surface of the sample, and results in a vertical bending (z-axis direction) of the free end of the anchored cantilever. By contrast, in the lateral or frictional regime, where the cantilever is presumed in contact with the surface, upon scanning the cantilever undergoes a torsion (twist) motion about its long axis (in the x-y plane). Both motions are orthogonal to each other. It is orthogonality that enables the simultaneous, yet independent, acquisition of topographic images and frictional data.

It should be noted that the torsion force constant depends critically on the tip length. Additionally, the tip should be centered at the free end of the cantilever. Both factors can be readily controlled with the use of microfabricated cantilevers and tips.

Friction force microscopy is able to reveal structures of molecular crystal surfaces, $4, 4', 7, 7'$-tetrachlorothioindigo (TCTI), with high resolution, as shown in Fig. 5.8 [5.70]. Frictional images were also able to distinguish the orientational difference in self-assembled monolayers (SAM) of n-alkanethiolate in Fig. 5.9 [5.71]. This difference between alkane chains with even or odd numbers of methylene units is due to the maximization of surface free energy. In this case, the SAM hexadecanethiol was prepared by microcontact printing followed by the deposition of heptadecanethiol from solution.

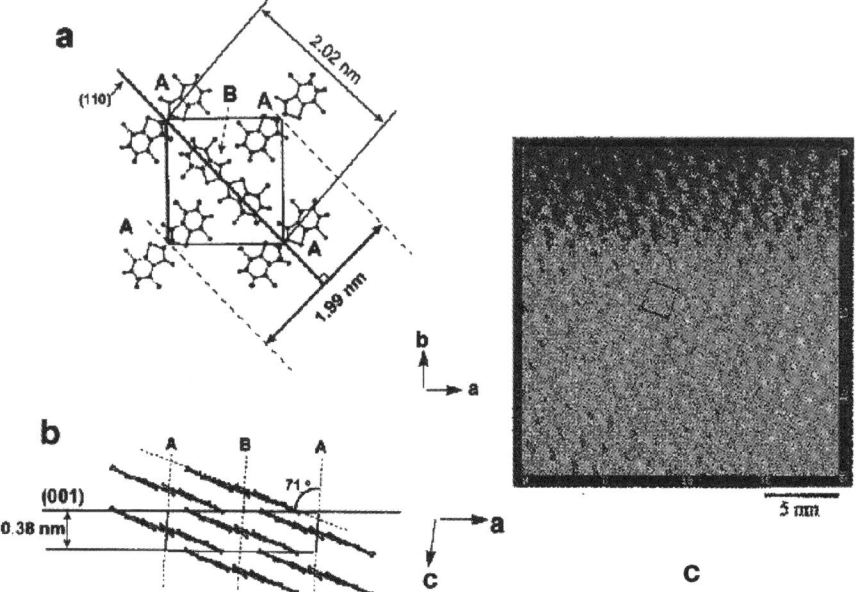

Fig. 5.8. Molecular arrangement of TCTI crystal in **a** a–b plane and **b** projection in the (001) direction **c** Frictional force image of the molecular lattice structure of the (001) surface (extracted from [5.70])

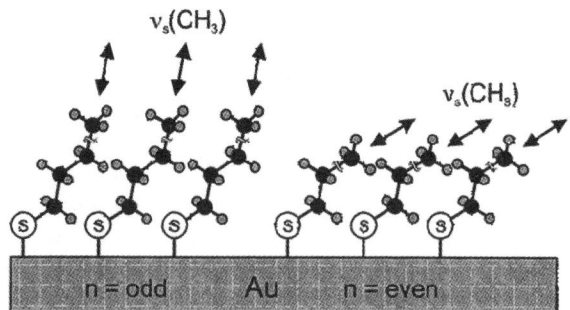

Fig. 5.9. Schematic illustration of molecule orientations of alkanethiolate assembled on gold surfaces associated with even (*left*) and odd (*right*) numbers of methylene units (extracted from [5.71])

5.3 AFM Operating in Oscillatory Modes

5.3.1 Tapping Mode

An AFM tip operating in tapping mode is driven by an oscillating piezoelectric actuator at its resonant frequency [5.72]. In the tapping mode, high

5.3 AFM Operating in Oscillatory Modes

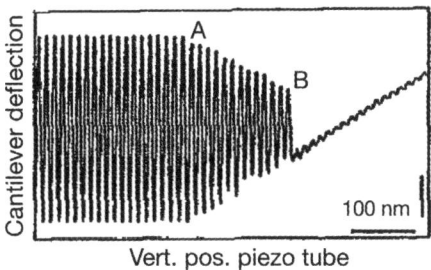

Fig. 5.10. Example of amplitude variation as a tapping tip approaches the sample surface (extracted from [5.73])

aspect ratio tips with small radii of curvature are used, and the vibrating tip contacts the sample surface many times during collection of each data point. The cantilever oscillation is damped when the tip contacts the water layer and the sample surface, but the larger vibration amplitude gives the cantilever sufficient energy to overcome the surface tension of the adsorbed water layer. The advantages of tapping mode are that the impact of adhesion force in ambient conditions is reduced due to the high restoring force of the stiff cantilever. In addition, the energy dissipation due to lateral force is minimal because of the greatly reduced contact time between the tip and sample surface. The force imparted onto the sample surface by the cantilever can be very small because small shifts in the vibration amplitude can be detected. The tapping mode allows delicate samples to be imaged with normal forces on the order of fractions of a nN, and shear forces that are essentially zero. The applied force is significantly lower than the force applied by the contact AFM. The lateral resolution in the tapping mode compared favorably to the contact AFM because high vibration frequencies allow the tip to contact the sample surface many times before it translates laterally by one tip diameter. Therefore, the tip shape defines the lateral resolution in the tapping mode, as it does with the contact AFM. In practice, this mode can be used to get a topographic image of any sample, regardless of its conductivity or mechanical composition. The amplitude of the cantilever oscillation is monitored as the tip approaches the sample surface (Fig. 5.10), and used as a feedback parameter [5.72, 5.73].

For an AFM tip in harmonic oscillation with period T and amplitude A_0, the average force of the tip in contact with the surface during time τ is [5.73]:

$$\begin{aligned} F_{\text{avg}} &= kA_0 \left[\frac{2\pi\tau}{T} \left(\sin\left(\frac{2\pi\tau}{T}\right) - \cos\left(\frac{2\pi\tau}{T}\right) \right) \right] \\ &\approx kA_0 \frac{4\pi^2}{3} \left(\frac{\tau}{T}\right)^2 \end{aligned} \tag{5.3}$$

The oscillation behavior of the cantilever in air is observed to be symmetric. Under liquid conditions, the transition from free oscillation to pivoting movement is similar to the force-distance curves in contact-mode AFM.

There is fast-growing interest in using tapping force microscopy to study surface properties, especially of soft materials. This approach, though in essence a modified contact-mode AFM, minimizes effects due to friction or other lateral forces. The key factors in the measurements are the amplitude and phase variations, compared with free resonant vibrations. Much effort has been invested in the quantitative description of these variations [5.74–5.76]. Such effort is not a trivial exercise, as it involves both the interface and bulk properties of the sample materials.

5.3.2 Phase Imaging

Another operation mode of tapping AFM is phase imaging, in which the phase variation of the oscillating cantilever relative to the excitation signal is used as the imaging parameter. At the same time, the amplitude is kept constant. The variation of the phase of oscillatory motion has been shown to be related to several factors, such as tip–sample interaction, energy dissipation of viscous interaction with the surrounding medium, surface topography, and inelastic interaction between the tip and sample [5.76–5.84]. Phase imaging has been extensively applied in the study of heterogeneous surface structures to resolve different compositions at nanometer scale [5.85–5.89].

Marcus et al. [5.90] recently considered the effect of in-plane tip motion on the magnitude of the phase shift, particularly for the case in which the tip has a finite tilt angle. The additional energy dissipation due to in-plane forces (friction and shear inelastic deformation) will be reflected in anisotropic features in the phase image.

5.3.3 Operations Under Liquids

The oscillation frequency spectrum for AFM under liquid conditions has been shown to be the convolution of the fluid drive spectrum and the thermal noise spectrum of the cantilever, and is independent of the geometry of the cantilever [5.28]. The frequency spectrum recorded in liquids often displays multiple peaks around the resonant peak position, as a result of acoustically coupled excitation of the liquid (Fig. 5.11). The detection frequency in tapping-mode operation should be chosen as one of these peak frequencies.

A different approach for tip oscillation is achieved with magnetic excitation of coated cantilevers, also termed magnetic a/c mode (MAC mode) [5.91]. The method uses cantilevers that are coated with magnetic materials and magnetized prior to experiments. During experiments, a magnetic solenoid driven by an AC current is used to induce oscillation of the cantilever. This method has improved frequency spectra under liquid conditions

5.3 AFM Operating in Oscillatory Modes   121

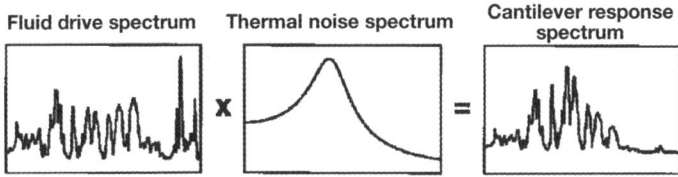

Fig. 5.11. Schematic of cantilever oscillation spectrum in liquid conditions. Multiple peaks can be seen in the vicinity of resonance frequency (extracted from [5.28])

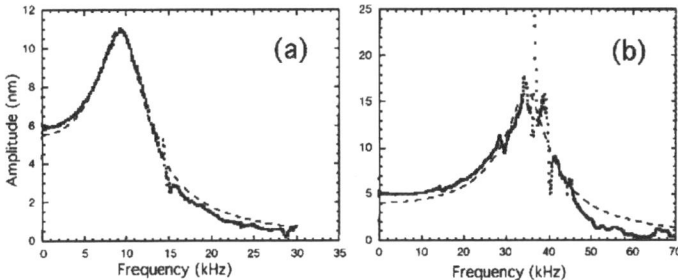

Fig. 5.12. Oscillation spectrum for a magnetically stimulated cantilever in water. The *dashed line* is the simulated result using a harmonic oscillator model. The cantilever has a length of **a** 140 μm and **b** 85 μm (extracted from [5.91])

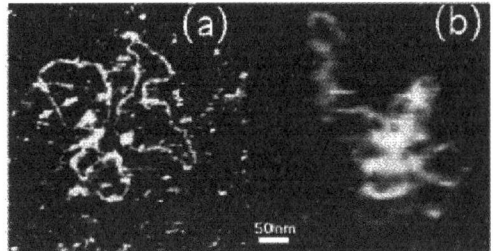

Fig. 5.13. Topography image of a plasmid DNA in solution using **a** magnetic a/c mode and **b** contact mode (extracted from [5.91])

(Fig. 5.12). Better signal–noise ratios have been demonstrated in imaging of DNA molecules, and organic SAMs under liquid conditions (Fig. 5.13) [5.92].

Controlled cantilever vibration can also be obtained by passing an AC current through the cantilever in the presence of an external magnetic field [5.93]. The cantilever is attached with a current circuitry. The current frequency can be set at close to the resonance frequency of the cantilever while operating in the tapping mode (Fig. 5.14).

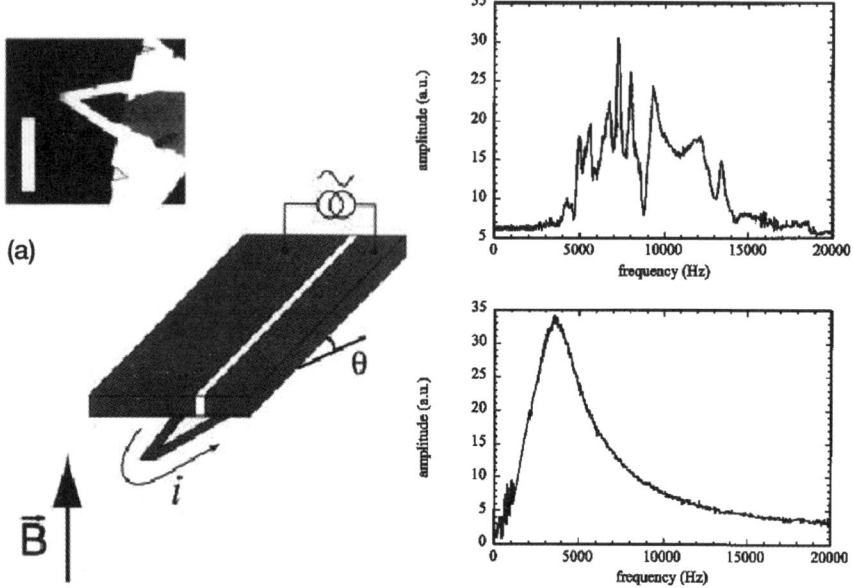

Fig. 5.14. a Schematic of a magnetic field-coupled cantilever with an ac current control mechanism. **b** and **c** are the oscillation spectra in water using acoustic excitation- and magnetic field-controlled methods, respectively (extracted from [5.93])

5.3.4 Non-Contact Mode

Non-contact AFM (NC-AFM), or frequency modulation AFM, uses the frequency shift as the imaging parameter [5.76, 5.94–5.96]. During NC-AFM imaging acquisition, the cantilever is kept at constant amplitude, and the tip–sample separation is adjusted to keep the frequency shift at the set value. The frequency shift depends on the force and interaction potential, and also the operation conditions. The frequency shift can be expressed in a general form [5.76]:

$$\Delta f(d, k, A_0, f_0) = -\frac{1}{2\pi} \frac{f_0}{kA_0} \int_0^{2\pi} F_{\text{ts}}[d + A_0 + A_0 \cos \varphi] \cos \varphi \, d\varphi \quad (5.4)$$

where k is the spring constant of the cantilever, A_0 the oscillation amplitude, f_0 the resonance frequency, and d the closest distance between the tip and sample.

The imaging mechanism in non-contact AFM is associated with the covalent interaction between the semiconductor surface and tip (as for the atomic resolution obtained on Si(111) [5.97–5.99] and InP(110) [5.100] surfaces), the short-range electrostatic interaction between ionic crystal surfaces and the tip (as in the study of alkali halides and oxides [5.101]), and short-range dispersion forces between noble gas atoms and the tip (as in the observation of the atomic structure of Xe films adsorbed on graphite surfaces [5.102]). A recent low-temperature NC-AFM study also revealed for the first time the full lattice of graphite [5.103]. Such resolution capability could be highly desirable in studying the surfaces of soft materials.

The resolution capability of NC-AFM has been manifested in the imaging of the unit cell structure of a tungsten tip apex using graphite surface atoms as the probe [5.104]. The charge density of $2p_z$ orbitals (which points perpendicular to the surface plane) of C atoms in the graphite lattice can serve as a local probe. The fine subatomic features of the tungsten cell can be resolved only in the higher harmonics of the oscillating force frequencies. The characteristic geometry of the observed images within a tungsten unit cell was ascribed to the total electron charge density distribution of the outmost tungsten atom within the AFM tip, approximated by a W(001) surface with bcc crystal structure. The tungsten atom actually measured can be located at different symmetry sites within the bcc cell unit, resulting in differences in the observed charge distribution symmetries. The highest spatial resolution by this approach is 77 pm. These data provide an example of reverse imaging in AFM operations, where the atomic protrusions of the atomically flat surface can actually be used as an imaging "tip".

Three basic forces can be mapped across a sample surface in the non-contact mode by detecting the deflection of the cantilever under the influence of the desired force. These consist of electrostatic, magnetic, and van der Waals forces, in the order of complexity of the interaction. The first deals with monopoles, the second with dipoles, and the third requires a quantum mechanical treatment. These forces and their derivatives can be as small as 10^{-12} N and 10^{-14} N m^{-1}, respectively, which is much smaller than those encountered with other AFM techniques. Much effort has gone into investigating magnetic force in force microscopy. This kind of microscopy has developed into an independent field as magnetic force microscopy (MFM), and is discussed in the following section. Non-contact AFM has also been applied to distinguish organic molecules such as thymine and adenine [5.105]. The contrast variation can be associated with the electrostatic coupling with the tip, such as for the example of fluorine-substituted carboxylate surfactants [5.106].

5.4 Magnetic Force Microscopy (MFM)

5.4.1 Basic Imaging Mechanism

The magnetic force microscope is essentially a kind of force microscope operated in the non-contact mode, except that the silicon tip is replaced by a magnetic tip (made of ferromagnetic metals or metal composites) that is magnetized along its length [5.106, 5.107]. Tips coated with ferromagnetic thin films were investigated successfully by a number of groups and have become commercially available. Magnetic thin film tips have a significantly reduced tip stray field, compared to bulk-wire tips. Another advantage is that their magnetic properties can be controlled by choosing an appropriate coating material. Thus, it is possible to measure the selected components of the sample field by coating the tips with high-coercivity films and magnetizing these appropriately in an external field.

The magnetic forces measured in the MFM are purely magnetostatic. They arise from the magnetic dipoles in the tip interacting with local magnetization in the sample. When the cantilever, which is oscillating at its resonant frequency, is brought near a magnetic sample surface (typically between 10 and 200 nm), the tip encounters a magnetic force gradient. At the same time, the effective spring constant and the resonance frequency are also shifted. By driving the cantilever above or below the resonant frequency, the oscillation amplitude varies as the resonance shifts. In the amplitude detection mode, an image of the magnetic field gradient is obtained by recording the oscillation amplitude as the tip is scanned over the sample.

The imaging mechanism of MFM can be initiated by considering an oscillating cantilever described by the following equation:

$$\frac{\partial^2 d(t)}{\partial t^2} + \frac{\omega_0}{Q}\frac{\partial d(t)}{\partial t} + \omega_0^2(d(t) - d_0) = A_0\omega_0 \cos(\omega_D t) \tag{5.5}$$

where $d(t)$ is the tip–sample separation, d_0 the tip–sample separation at zero oscillation, ω_0 the resonant frequency of the cantilever, ω_D the driving frequency, and Q quality factor. ω_0 and Q are defined as:

$$\omega_0 = \sqrt{\frac{c}{m}}$$

$$Q = \frac{m\omega_0}{2\gamma}$$

where c is the spring constant of the cantilever and γ the damping factor associated with the medium surrounding the cantilever.

The oscillation amplitude of the cantilever can be explicitly expressed in Lorentzian form [5.108]:

$$A = \frac{A_0(\omega_0/\omega_D)}{[1 + Q^2(\omega_D/\omega_0 - \omega_0/\omega_D)^2]^{1/2}} \tag{5.6}$$

The phase shift between the actual oscillation and the driving signal is:

$$\alpha = \arctan \frac{2\gamma\omega_D}{\omega_0^2 - \omega_D^2}$$

In the presence of a force gradient $(\partial F/\partial z)$ (i.e., non-uniform force distribution), the effective spring constant will be changed into:

$$c_{\text{eff}} = c - \frac{\partial F}{\partial z} = c - F' \tag{5.7}$$

The corresponding resonance frequency will be changed into:

$$\omega_0' = \left(\frac{c_{\text{eff}}}{m}\right)^{1/2} = \left[\frac{c-F}{m}\right]^{1/2}$$

$$\Delta\omega = \omega_0' - \omega_0 \approx -\frac{\omega_0 F'}{2c}$$

Similarly, the phase shift will also be affected by the same force gradient.

The driving frequency is usually selected at the position of largest slope in the amplitude spectrum in order to obtain high sensitivity.

$$\omega_D \simeq \omega_0'\left(1 \pm \frac{1}{\sqrt{8Q}}\right)$$

The change in oscillation amplitude is expressed as:

$$\Delta A = \left(\frac{2A_0 Q}{3\sqrt{3}c}\right) F'$$

It was pointed out that the time interval required for the transition between different oscillation states under the influence of the force gradient is restricted by:

$$\tau = \frac{2Q}{\omega_0}$$

Therefore, a high Q-factor is not desirable for such detections. This restriction can be severe enough to hamper the application of the method to vacuum conditions [5.109]. A different detection approach can be achieved by the frequency modulation AFM mechanism that keeps the cantilever (with high Q-factor) at resonance, and the frequency variation induced by the force gradient is measured [5.94].

In MFM studies, the force gradient is originated from the interaction between the magnetic tip and the stray field of the sample surface. The tip can be simplified as a magnetic point monopole with moment q and dipole with moment m. The stray magnetic field above the sample surface can be expressed as:

$$H(r) = \int_{\text{sample}} \frac{3\bm{n} \cdot [\bm{n} \cdot \bm{m_s}] - \bm{m_s}}{r^3} dV'$$

where $\bm{m_s}$ is the magnetization of the sample. The magnetic force experienced by the tip can be written as:

$$F_{\text{mag}} = \mu_0 (q + m \cdot \nabla) H$$

5.4.2 Examples of MFM Studies of Molecular Structures

Much of the interest in MFM is due to the fact that the technique is the only magnetic imaging technique providing high lateral resolution (in the range of 10–100 nm) with essentially no special sample preparation requirements. This microscopy method enables us to determine the surface uniformity and strength of the magnetic media, giving insight into both the performance and quality of the magnetic storage media. High-quality images can be taken even when the magnetic material is covered with a thin coating, an important feature when imaging many technologically important samples. A resolution of 10 nm has been shown on rapidly quenched FeNdB thin films by MFM [5.110].

An example can be seen in the observation of the static changes of domain configuration in $(YGdBi)_3(GaFe)_5O_{12}$ garnet using MFM in an external magnetic field [5.111]. Sequential variations of the magnetic domain configuration in the garnet were observed, and can be associated with the strength of the applied field. It has been demonstrated that the undisturbed magnetic domain structure can be discerned with a soft magnetic tip. The stray field emanating from a hard tip could magnetize the garnet and alter the widths of the domain region, an effect that can be minimized by choosing an appropriate tip–sample separation.

In the short period since the first MFM image was obtained in late 1980s, the field has grown rapidly. MFM is already a powerful tool for magnetic imaging, with many scientific and practical applications [5.109]. Future advances in MFM include improved tips, matching of tip–sample characteristics, and determination of field strengths and domain structure from force gradient images.

Attention has been given to improving the MFM probe performance by reducing the effective magnetic apex dimension. The electron beam deposition method produces tips consisting of magnetic particles (fabricated from CoCr coating) protected by carbon overlayers [5.112]. The use of multiwalled carbon nanotubes coated with magnetic material and carbon coatings presents another alternative of fabricating MFM tips. Carbon nanotubes capped with Fe alloy or Ni_3C particles were shown to be potential candidates for MFM tips [5.113].

5.4.3 Imaging Single Molecule Magnets

Molecules with paramagnetic transition metal centers have been selected as magnetic materials. These species could provide potentials for data storage strategies at single molecule level. A family of molecules exemplified by Mn_{12} ($[Mn_{12}O_{12}(O_2CMe)_{16}(H_2O)_4] \times 4H_2O \times 2MeCOOH$) and Fe_8 ($[Fe_8O_2(OH)_{12}(tacn)_6]+8$) has been studied [5.114–5.116]. Due to the weak intermolecular interactions between magnetic centers, these species are widely used as single molecular magnets (SMM). Monodispersed Mn_{12} molecules in polycarbonate matrix have also been observed by MFM (Fig. 5.15). The polymer film was treated by CH_2Cl_2 vapor to ensure high solubility of Mn_{12} molecules [5.117].

A molecular magnet of high-spin polyphenoxyl radicals has been successfully synthesized [5.118]. The polyradical molecule has a star-shaped struc-

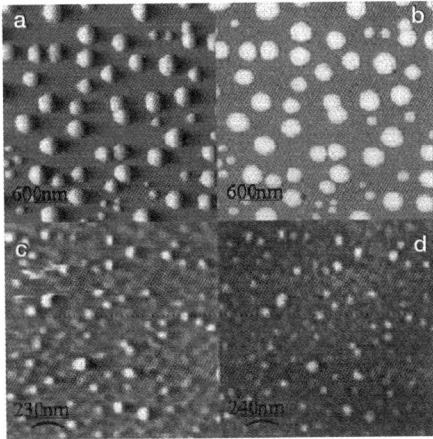

Fig. 5.15. Images of Mn_{12} complexes prepared with **a, b** CH_2Cl_2/hexane (1:1) and **c, d** CH_2Cl_2. **a** and **c** are topography images. **b** and **d** are MFM images recorded after retracting the tip 30 nm (extracted from [5.117])

ture and displays ferromagnetic behavior, as measured by SQUID magnetometer. The molecules are stable in ambient conditions, and the dispersed species embedded in a polymer matrix of polystyrene has been studied by MFM [5.118]. The individually dispersed molecules can be detected by a magnetic tip held about 20 nm above the sample surface. Magnetic imaging can be performed for time periods as long as nearly one day before the high-spin polyradical molecules become magnetively inactive.

The complementary development of MFM in the study of single spin centers may be worth noting. The recent advance of magnetic resonance force microscopy (MRFM) has set the record of detecting single electron spins [5.119]. With a force sensitivity of 10^{-18} N, the effect of the interacting force between the resonating SmCo magnetic tip (150 nm wide) and the electron spins (generated by gamma ray irradiation of the vitreous silica) can be detected as a perturbation in the resonance frequency. The frequency shift due to a single spin interaction was estimated as on the order of a few mHz. Since the magnetic interaction is long range, it was proposed that the detectable spins could be embedded as far as 100 nm under the surface.

5.5 Force Spectrum and Surface Mapping

5.5.1 Force Spectrum and Imaging

The core of AFM studies is the understanding of probe–surface interactions. Though available knowledge provides general ideas of this interaction [5.3, 5.4, 5.120], the microscopic behavior is far from having been thoroughly clarified. The key factors in determining these forces, such as molecular arrangements, various functional end groups, and adsorption properties have all been investigated extensively. Adhesions associated with small molecules such as water and organic species are important not only for the understanding of the adsorbed films, but also for the enhancement of resolution power of AFM, since the hydration force is considered as an essential factor in improving the resolution of SPM in ambient and solvent conditions.

By recording the interacting force between the probe and sample surface while changing their separations, one can obtain much information about the nature of the probe–surface interaction. This so-called force spectrum has been proven to be an effective method to measure probe–sample interactions and intermolecular forces [5.121–5.123]. By analyzing the characteristics of the force spectra, it is possible to retrieve information concerning the nature of the interacting force, such as van der Waals, electrostatic, and elastic forces. This information could have far-reaching implications.

There has been a fast-growing interest in using the force spectrum method to probe characteristic interactions between biological species, as well as a large variety of organic molecules and polymers. Examples can be found in the exploration of the binding strength of cell adhesion proteoglycans from marine sponges, which yielded a value of 400 pN, leading to the contention that a single pair of molecules could be composed by as many as 1,600 cells [5.124]. Using the measured force and binding energies, the effective rupture length of avidin-biotin was estimated [5.122]. The technique has also been applied to investigate the bonding strength of the base pairs of nucleic acids [5.121] and ligand–receptor pairs [5.123]. Force spectrum measurements have also been used to study local elastic properties such as in polysaccharide chains [5.125, 5.126]. Adhesion force imaging on intercellular adhesion molecule-covered mica surfaces provides a possible method to obtain specific recognition on samples such as cell surfaces [5.127].

It should be noted that surface energy is not the sole factor important in adhesion measurements, because other factors such as contact geometry and deformations are also crucial in making the technique a quantitative one. The Hertzian model for spherical and cylindrical contacts provides a good simplification [5.128, 5.129], which also points to the necessity to have a regularly shaped probe. This method is successfully used to study the adhesion mapping of various films, including self-assembled films, polymeric films, and other surface adsorbates, and is also an important factor in surface frictional analysis [5.130, 5.131]. The precision force measurements have also prompted explorations of novel sensory systems that can detect interactions of single molecules, an aspect of particular interest to biomaterial studies.

5.5.2 Chemical Force Microscopy

It has been demonstrated that by decorating the AFM cantilevers with molecules of specific functionalities, additional resolution capabilities could

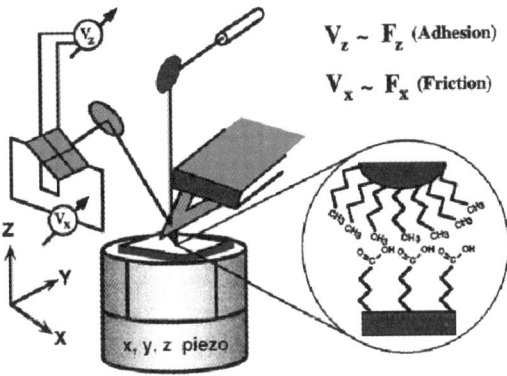

Fig. 5.16. Schematic of force measurements using a chemically modified cantilever probe (extracted from [5.3])

be obtained. Such efforts extend the functions of regular contact-mode AFM and frictional force microscopy, leading to many interesting studies on surface compositional analysis at nanometer scale [5.3].

The AFM cantilever can be modified by covalently attaching a monolayer of functional molecules (Fig. 5.16). The exposed probe surface in this case consists of chemically uniform groups. This interaction between the chemically decorated probes and the sample surface can be highly sensitive to local chemical compositions and environments, and could be used to differentiate surface regions of different chemical nature but with nearly identical topographical features. Applications of this method can be seen in the study of, for example, bonding strength between different functional groups, and titration effects. This method is also successfully used in surface frictional analysis [5.130, 5.131]

Carbon nanotubes have been extensively pursued as scanning probes [5.132–5.138]. The nanometer-diameter nanotubes possess novel mechanical properties, and can be functionalized at the ends [5.139–5.141]. The high aspect ratio of the CNT enables improved resolution of fine surface features.

6 Intermolecular and Intramolecular Interactions

6.1 Techniques for Studying Intermolecular and Intramolecular Interactions

Throughout the discussions in the preceding chapter, the center of the mechanism for topography mapping at nanometer-scale resolution using AFM is the proximal interaction between the AFM cantilever and the sample surface. The detected force is the summation of forces from different origins, such as van der Waals, electrostatic, and magnetic forces. Besides topography imaging, the investigation of intermolecular forces can improve our understanding of microscopic phenomena, such as adhesion, friction, and lubrication. In addition to the force measurements using AFM introduced in the preceding chapter, several other detective techniques, such as biomembrane force probe (BFP) [6.1–6.5], surface force apparatus (SFA) [6.6–6.8], and optical tweezers [6.9–6.21] have been used to measure intermolecular forces. As will be presented below, although these techniques can be applied to investigate intermolecular interactions, the capability of high spatial and force resolutions is critical to measure discrete molecule–molecule interactions. The surface force apparatus is used mainly to study the force interactions between large (centimeter scale) areas, is rarely applied for single molecules, and thus will not be discussed in this chapter. The BFP and optical tweezers have less spring constant than in the case for AFM, thereby providing a complementary range of force detections. Since the mechanisms of AFM have been given in the preceding chapter, we will in this chapter begin with a brief introduction of techniques other than AFM applied to single molecule force detection.

6.1.1 Biomembrane Force Probe (BFP)

The deformation of a capsule membrane has been adapted for studying intermolecular interactions under biological conditions. The osmotically pressurized capsule (such as a human red blood cell) is immobilized at the open end of a micropipet by sucking pressure (Fig. 6.1). A microbead is typically used at the opposite end of the capsule for measuring intermolecular interactions [6.2,6.3]. The tension of the membrane surface can be expressed as [6.4]:

$$\tau_m = \frac{P \cdot R_p}{2(1 - \frac{R_p}{R_0})} \tag{6.1}$$

6 Intermolecular and Intramolecular Interactions

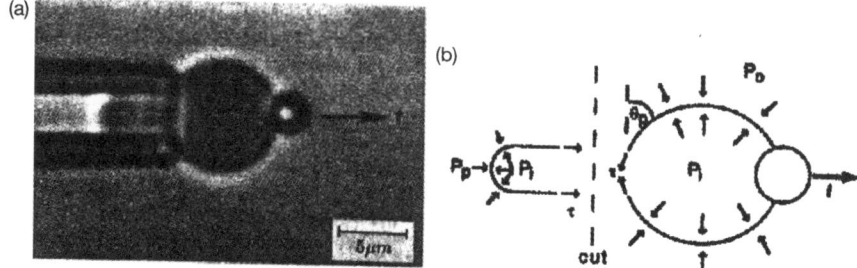

Fig. 6.1. a Schematic of biomembrane force measurement setup (extracted from [6.2]) **b** Force distribution of the measurement study (extracted from [6.3])

where R_p and R_0 are the radi of the micropipet and capsule, respectively. The stiffness can be approximated as:

$$k_t \approx \frac{2\pi}{\tau_m} \left[\ln\left(\frac{2R_0}{R_p}\right) + \ln\left(\frac{R_0}{r_b}\right) \right] \qquad (6.2)$$

where r_b is the circular radius of the contact between the microbead and the capsule. The typical range for the stiffness is 45 µN m^{-1} to 10 mN m^{-1}. The interacting force can therefore be obtained by measuring the deformation of the contacting membrane. By optimizing the displacement detection transducer, a force range of 0.01 to 10^3 pN can be achieved with this technique.

6.1.2 Optical Tweezers

Optical tweezers are constructed from a heavily focused laser beam. A dielectric particle experiences strongly enhanced interaction in the vicinity of the focal regions, with the dimension of diffraction limit. The interacting forces originating from scattering, field gradient, and radiometric effects are the contributing factors in stabilizing the particles. Particles with diameters much larger than the wavelength (Mie regime) or much smaller than the optical wavelength (Rayleigh regime) can be effectively trapped in TEM$_{00}$ mode.

For metallic and dielectric particles with diameters larger than irradiation wavelengths, the estimate of the force exerted by the laser beam can be obtained by the geometric optic method [6.9–6.11] and electromagnetic wave formalism [6.12–6.14]. The theoretical analysis pointed to an optical potential well that causes a radial inward force on the high-index particles. A multiple-beam configuration can therefore be an effective method to trap particles with micron or submicron diameters in their free standing state, and is widely adopted in optical tweezers experiments. In addition, trapping of biological structures has also been successfully demonstrated [6.15–6.19].

For Rayleigh particles (radius r much less than the wavelength), the stabilizing force within the medium (with index n_b) can be expressed in the

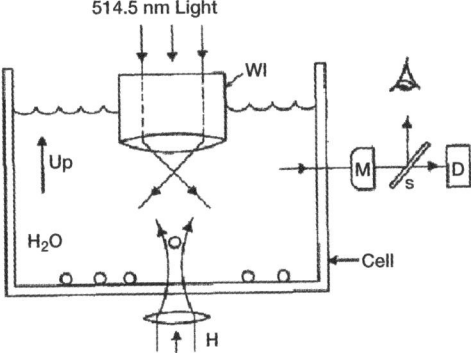

Fig. 6.2. Schematic of particle trapping in optical tweezers setup (extracted from [6.20])

following [6.20]:

$$F_{\text{scat}} = \frac{I_0}{c} \frac{128\pi^5 r^6}{3\lambda^4} \left(\frac{m^2-1}{m^2+2}\right) n_{\text{b}} \qquad (6.3)$$

$$F_{\text{grad}} = -\frac{n_{\text{b}}}{2}\alpha \nabla E^2 = -\frac{n_{\text{b}}^3 r^3}{2}\left(\frac{m^2-1}{m^2+2}\right)\nabla E^2 \qquad (6.4)$$

where I_0 is the beam intensity with wavelength λ, and m the effective index of the particle (for example, latex (polystyrene) or silica). Figure 6.2 shows schematically the particle trapping by a single laser beam [6.20].

The stability of the trapped particle is measured by the ratio of the back-scattered axial gradient force to the forward-scattering force. Under the approximation of a Gaussian beam with focal spot width W_0 and the position for maximum axial intensity gradient at $z = \pi w_0^2/\sqrt{3}\lambda$, the condition for the ratio of stabilization can be as:

$$R = \frac{F_{\text{grad}}}{F_{\text{scat}}} = \frac{3\sqrt{3}}{64\pi^5} \frac{n_{\text{b}}^2}{\left(\frac{m^2-1}{m^2+2}\right)} \frac{\lambda^5}{r^3 w_0^2} \geq 1 \qquad (6.5)$$

The above condition applies to particles with radius range $r \leq 0.1\lambda$. The other requirement for stabilization is that the thermal diffusion of the particle should be within the limit $\exp(-U/kT) \ll 1$, where $U = n_{\text{b}}\alpha E^2/2$ is the potential of the gradient force. The typical lasers used are Nd:YAG, argon-ion, or diode lasers. The gradient force is the dominant factor and is proportional to the polarizability of the particles. It was also shown that metallic Rayleigh particles (such as gold) can also be effectively trapped [6.21].

Other force detection techniques, such as the magnetic bead method using ferromagnetic or paramagnetic beads, have also be adapted to study intermolecular forces [6.5]. A force on the order of piconewton has been obtained using a paramagnetic microbead.

6.2 Static Force Measurements of Single Molecules

A straightforward force measurement can be performed using the above techniques to measure the interaction between the matching surfaces. The experimental realization of single molecule resolution may be obtained by controlled attachment of species to the detection surfaces.

6.2.1 Single Bond Interaction

Force spectrum studies can be performed by chemically attaching molecules at the tip and substrate surface, either by direct attachment or via a tether molecule, as described briefly at the end of the preceding chapter. By controlling the tip–sample separation, and recording the interacting force simultaneously, the magnitude of interactions can be obtained. This is a direct approach to gain insight into intermolecular and intramolecular interactions. Measurements of rupture force are reflected in intermolecular interactions, ligand–receptor chemical bonding, etc. The elasticity of polymeric and biomolecules presents opportunities to individually assess the mechanical response of these molecules, which is closely related to their internal structures. By studying the force spectra, one can correlate elastic behavior to the conformational transitions. Findings such as the enforced unfolding of polymeric molecules have opened a new field in dynamic force spectroscopy.

Three main methods can be used to derive individual bond-rupture forces. The first, or so-called force quantum method, involves a histogram of pull-off force differences between pairs for large numbers of measurements. The histogram exhibits a peak progression, the spacing of which can be regarded as the force required to break a single interaction. Hoh et al. [6.22] detected a quantized magnitude distribution in adhesive forces (Fig. 6.3), and the quantum of the force was assigned to a single hydrogen bond. This method requires a large number of force curves, and this may cause sample damage due to repetitive contact with the tip. The second approach is related to Johnson-Kendall-Roberts (JKR) theory [6.23], which can also be used to determine single-bond rupture forces if surface energies and tip radii are known [6.24].

The third approach utilizes Poisson statistics, a statistical method adopted by Beebe et al. [6.25–6.27] to deduce the single-bond force between the tip and substrate, modified with self-assembled monolayers terminating in various functional groups. The validity of this method has been demonstrated in the recent literature [6.25–6.28]. Moreover, van der Vegte et al. [6.24] have proved that this approach and the JKR theory are equivalent in deriving single-bond forces. The Poisson statistics method requires no assumptions about the surface energies or contact areas between the tip and sample surface. It is also not limited by the force resolution of the instrument, and the number of measurements required to derive the individual bond forces is significantly lower than that required by the former two methods.

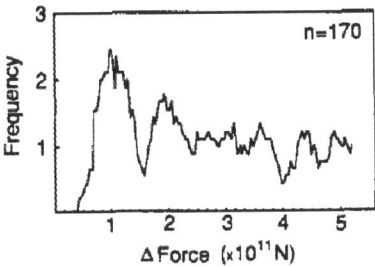

Fig. 6.3. Histogram of the measured adhesion force between an AFM tip and glass surface under water. The force quantum of 12 pN was attributed to the hydrogen bonds established during contact (extracted from [6.22])

There are two assumptions in this approach [6.25]. The first is that the total adhesion force (F) develops as the sum of any possible discrete chemical bond and any possible non-covalent bond interaction, such as the hydrogen bond and van der Waals force. The second assumption is that these bonds form randomly and all have similar force values (F_i). With the assumption of integer numbers of interacting bonds at the pull-off point, the Poisson distribution can be used for the estimation

$$P(n) = \frac{(n_{\text{av}})^n}{n!} \exp(-n_{\text{av}})$$

and the standard deviation

$$\sigma_s^2 = n_{\text{av}}$$

where n_{av} is the average number of interacting bonds. The total adhesion force F_{av} can be expressed as:

$$F_{\text{av}} = n_{\text{av}} F_i + F_0 \tag{6.6}$$

where F_i is the single-bond force, and F_0 the non-specific interaction. The standard deviation of the adhesion force is:

$$\sigma_F^2 = (\sigma_n F_i)^2 = n_{\text{av}} F_i^2 = F_{\text{av}} F_i - F_i F_0$$

Therefore, the plot of σ_F^2 versus F_{av} yields the F_i as the slope, and $F_i F_0$ as the intercept.

Several SAM systems terminated with different functional groups have been studied by this method. A single-bond force of 181 ± 35 pN was obtained for one OH–OH group pair in water medium [6.25]. Figure 6.4 presents an example of single-bond interaction between amino-terminated self-assembled organosilane monolayers based on atomic force microscopy [6.28]. The plot shows the force variance versus the mean force. Fitting these points with a linear regression, results in a slope of 200.0 ± 43.6 pN (ascribed to the single-bond force) and a small intercept. The nominal intercept implies that any

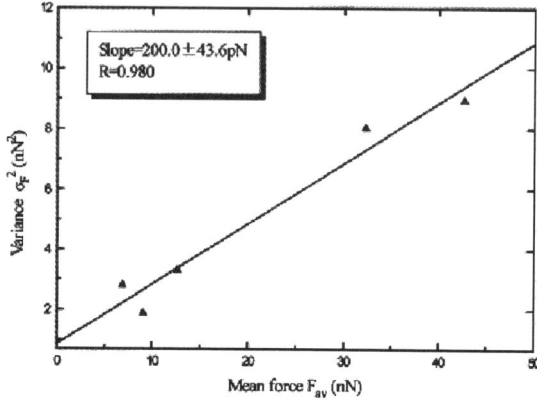

Fig. 6.4. Linear dependence of variance on the mean force between NH_2–NH_2 groups yielding the slope as the single bond interaction force (extracted from [6.28])

non-specific forces in our measurements can be neglected. According to the description in Poisson statistics analysis, the interaction between one NH_2–NH_2 pair is regarded as an individual bond, and Poisson statistics are used to describe the distribution of a discrete number of interacting NH_2–NH_2 pairs during a set of force measurements. Therefore, the so-called single-bond force extracted from the Poisson statistical analysis is not a single interaction but rather an average value of overall interaction between a single group pair. The "overall" interaction is a sum of several forces, including any covalent chemical bonds and non-covalent interactions such as hydrogen bond and van der Waals forces. For the NH_2–NH_2 system in this example, there is no formation of chemical bonds, so the extracted single-bond force is equal to the summation of the hydrogen bond and van der Waals forces.

For contact-mode AFM, the total amount of force exerted on a sample comes from two sources – the cantilever spring force, and any adhesion force between the tip and sample. These two factors represent a total tracking force, which is the sum of capillary force, molecular interactions (e.g., hydrogen bond, van der Waals interaction), and applied load. To investigate the single-bond force between the tip and silicon substrate covered with SAMs, the measurements are preferably carried out in water, rather than in air to eliminate capillary force.

Direct measurement of hydrogen bonding between DNA base pairs was tested using molecular-decorated AFM tips and Au(111) surfaces covered with nucleotides [6.29]. The nucleotide molecules were assembled on Au(111) surfaces with an "edge-on" orientation that would allow the formation of hydrogen bonds between complementary base pairs decorated on the tip and sample surface, respectively. A quantized statistical distribution in the adenine (tip) and thymine interactions led to the result of 54 pN per A–T pair.

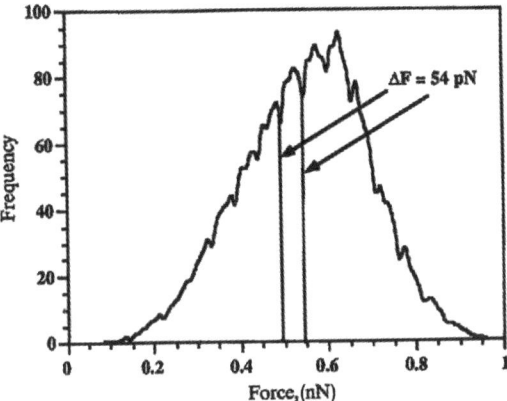

Fig. 6.5. Histogram of A–T interactions with a force quantum of about 54 pN, possibly due to the single A–T pair interaction (extracted from [6.30])

Sequence-specific interaction between the complementary oligonucleotides ACTG$_5$ and CAGT$_5$ was obtained with the force spectrum method [6.30]. The effect of interchain interactions on the extension spectrum of homopolymer [poly(I)] was analyzed using the freely jointed chain model. Unzipping double helix structure of single molecule DNA was realized by attaching the DNA molecule between the glass substrate and a polystyrene bead [6.31]. A glass microneedle was connected to the microbead and used in this case as a force sensor in the range of piconewton. The unzipping of the helical structure revealed a sawtooth pattern that is consistent with statistical mechanics simulations (Fig. 6.5). The results also indicated that molecular stick-slip motion during opening is an equilibrium process.

Direct measurement of Si–Si bond strength was achieved by using a noncontact AFM. The data revealed differences among adatoms in the Si(111) 7 × 7 unit cell interacting with the Si tip [6.32].

Single-bond strength can also be measured from the detachment of longchain molecules from support surfaces. One such study was performed by stretching polysaccharide (amylose) molecules covalently anchored to amino group-decorated silicon oxide surfaces [6.33]. During the stretching process, multiple force peaks appeared prior to the final rupture of the molecular bridge between the tip and sample. Such peaks can be attributed to the detachment of the flat-lying segments of the amylose molecule from the silicon surface, likely through S–C bonds. Thus, the rupture force for the Si–C bond was inferred to be 2.0 ± 0.3 nN. In a similar way, the detachment of amylose molecules anchored between an aminothiol-modified gold surface and the AFM tip showed a rupture force of 1.4 ± 0.3 nN. From molecular dynamics simulations on the pulling process involving a single thiolate molecule adsorbed on a gold surface, the sulfur–gold chemical bond was found to be not preferentially broken at first. The rupture force of Au–Au bond (or abstrac-

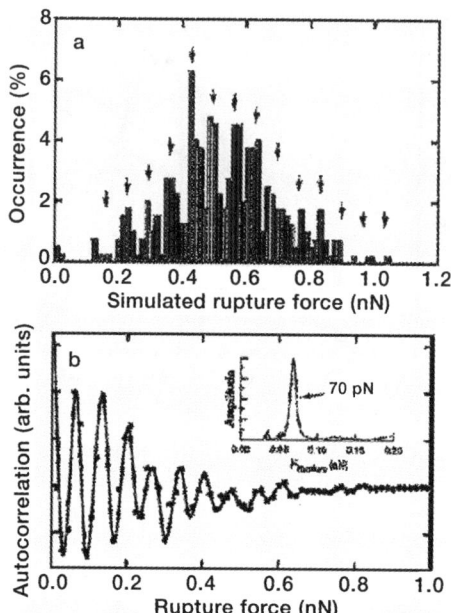

Fig. 6.6. a Histogram and b autocorrelation function of the adhesion force of the TMPD–TCNQ complex (extracted from [6.36])

tion of Au atoms) was estimated as 1.2 nN [6.34], which is consistent with the above-mentioned experimental value.

A considerably lower force value for thiol–Au complex abstraction was reported, by adopting the configuration of chemical force microscopy under solvent conditions [6.35]. The rupture force of thiol terminal groups from Au-coated tips has been measured with a force quantum of about 0.1 nN in the histogram of rupture force. The reduced force magnitude was argued to be associated with the weakening of intermetallic bonding due to chemisorption of thiol groups. In addition, solvent conditions were observed to significantly affect the measured rupture force. A factor of 7 difference in average rupture forces was identified for Au/S-acetate and Au/O-acetate contacts under ethanol, whereas no appreciable difference was obtained under water. This difference has been ascribed to the variation of interfacial energy under solvent conditions. These studies suggest that multiple factors are involved in the atomistic rupturing process.

The interaction between single pairs of the charge-transfer complex of N,N,N'N'-tetramethyl-phenylenediamine (TMPD, electron donor), and 7,7,8,8-tetracyanoquinodimethane (TCNQ, electron acceptor) was determined as 70 ± 15 pN from the discrete-shaped histograms and autocorrelation function (Fig. 6.6), leading to a value of 4–5 kJ mol^{-1} for the single-bond energy [6.36].

With a hybrid molecular simulation method, the adhesion and friction characteristics between the chemically modified tip and sample have been analyzed, showing good agreement with experimental results [6.37]. The atomistic molecular dynamics simulations on the interaction between an alkylthiolate-decorated tip and substrate surface favored the Johnson-Kendall-Roberts model, compared with the Derjaguin-Muller-Toporov model and Hertz model [6.38].

6.2.2 Single Pair Ligand–Receptor Interactions

Applications of force spectroscopy measurements in molecular recognition by studying the binding process of a ligand molecule to a receptor protein have been actively pursued [6.39]. Such studies could help reveal specific interactions accompaning molecular recognition processes. The elastic behavior of polymers and biomolecules is a reflection of intramolecular interactions, and the structural transformations induced by external force. Applications to study the folding and unfolding processes of proteins are illuminating.

It was demonstrated that in order to measure the rupture force between antigen and antibody molecules, a flexible spacer between the sensor molecule and the tip is desirable [6.40]. This is because antigen–antibody interactions are highly site-specific, and a certain mobility of the pairing molecules would be beneficial to establish such interactions. An example is shown in Fig. 6.7, between an antigen (human serum albumin, HSA) and antibody (affinity-purified polyclonal anti-HSA antibody). The sensor molecule of HSA antibody was anchored to the AFM tip through polyethylene glycol (PEG) derivative (8 nm long). The antigen molecules were adsorbed to a mica surface through the same spacer molecule. There are two binding sites for the antibody, and the single spectrum peak obtained indicated that only one of these two possible binding sites is effective in the binding–unbinding process. The measured antigen–antibody unbinding force is 244 ± 22 pN. It was also observed that the asymmetry in coupling of the antigen (antibody) molecules to the spacer molecules does not affect the measured unbinding force.

Site-specific antigen–antibody interactions were shown to be helpful in developing new topographic mapping methods. In a study of lysozyme

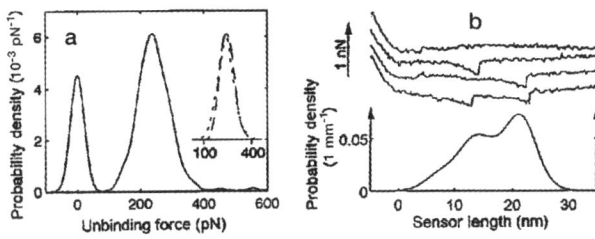

Fig. 6.7. Distribution of the measured **a** unbinding force and **b** sensor length of single antigen–antibody pairs (extracted from [6.40])

molecules adsorbed on mica surface, a half-antibody-modified tip (cf. flexible tether molecule) was used. The dynamic force microscope was operated with the magnetically driven oscillating cantilever under buffer solution. A well-resolved distribution of lysozyme molecules was obtained using the sensor-decorated tip [6.41]. By contrast, the characteristic features of the antigen molecules could not be observed with an unmodified tip. It was pointed out that the measured width of the lysozyme molecules was broadened by the length of the tether molecule and the width of the sensor molecule. A force quantum of 60 pN was observed between biotinylated bovin serum albumin (BBSA) and polyclonal, biotin-directed IgG antibodies [6.42]. In this case, the biotin was connected to bovin serum albumin (BSA) through a 2.24-nm spacer molecule to reduce the influence of conformational change, steric hindrance and orientation. The interacting force between a single biotin–avidin pair was determined as 160 ± 20 pN from histograms and the autocorrelation function of adhesion force measurements [6.43]. The tip was modified by BSA based on non-specific adsorption followed by attachment of avidin.

The rupture force between wild-type and mutant antiboby single-chain Fv fragments (scFv) with the antigen fluorescein (connected to the tip by means of a tether molecule of polyethylene glycol) was obtained and displayed a discernible difference (50 ± 4 to 40 ± 3 pN) [6.44].

The interaction between cell adhesion molecules was studied under biological conditions using the model system of cell adhesion proteoglycan (AP) of the marine sponge *Microciona prolifera*, based on the AFM force spectrum method [6.45]. An average interacting force of 125 pN was determined at 10 mM Ca^{2+}, with a maximum force around 400 pN for individual AP–AP pairs, and was ascribed to the binding between three and 10 pairs of AP arms.

Carbon nanotubes (CNT) decorated with carboxy groups were shown to be capable of resolving CH_3- and COOH- terminated regions based on phase contrast in tapping AFM operation [6.46]. The COOH- functionalized CNTs could also be attached to 5-(biotinamido)pentylamine and used to interact with streptavitin-covered surfaces (Fig. 6.8). A characteristic force quantum of about 200 pN was identified as the biotin–streptavitin interaction [6.46].

Theoretical analysis on mechanisms of the rupturing process of ligand–receptor complexes under external force has been carried out, the results suggesting a multiple unbinding step behavior, and that the environment can play an important role in this respect [6.47].

The measured adhesive force for single pairs of ligands (biotin, iminobiotin, or desthiobiotin) and receptors (avidin or streptavidin) was found to correlate linearly with the corresponding enthalpy change of the unbinding of the complexes but was independent of the free energy [6.48]. This observation suggests that the unbinding process is adiabatic, and the entropy change becomes relevant after unbinding. In addition, adhesion force mapping of intercellular adhesion molecule-covered mica surfaces was shown as a possible method to obtain specific recognitions on surfaces such as cells [6.49].

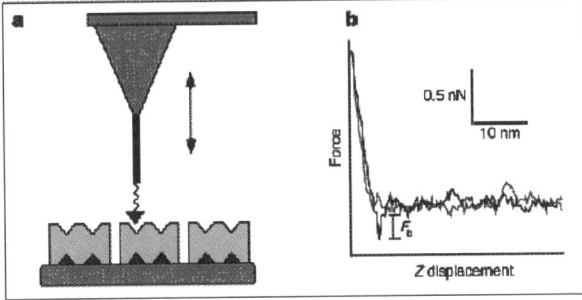

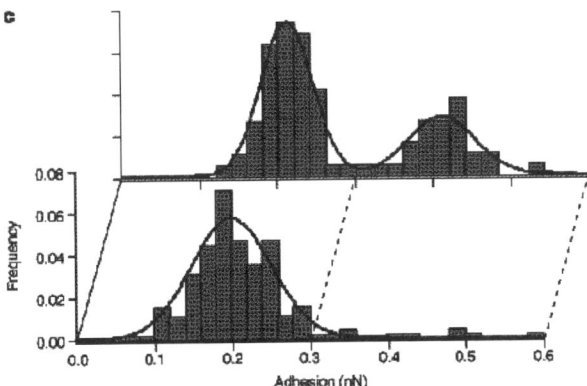

Fig. 6.8. a , b Unbinding force measurement using a biotin-decorated CNT tip and streptavidin. The histograms in **c** show the unbinding individual pairs of biotin–streptavitin with two different tips (extracted from [6.46])

6.2.3 Guest–Host Interactions

Direct measurement of guest–host interactions in β-cyclodextrin/ferrocene complexes has been performed [6.50]. The cyclodextrin molecule has a cyclic ring-like structure and these were assembled on Au(111) surfaces through thiol groups in substituent segments. The cavities of the cyclodextrin molecules can have specific interactions with apolar species, such as ferrocene, adamantane, and 1-anilinonaphthalene-8 sulfonic acid (1,8-ANS), to form guest–host complexes in aqueous environments. In this case, the 6-ferrocenyl-hexanethiol was coadsorbed with 2-hydroxyl-ethanethiol to the gold-coated AFM probe. The concentration of the mixed SAM was calibrated to expose about 14–28 ferrocene terminals at the apex of the tip of radius 100 nm. The measured force-distance spectra show multiple peaks corresponding to consecutive breaking of the guest–host complexes. The distribution of the rupture force is given in Fig. 6.9. By using the autocorrelation function de-

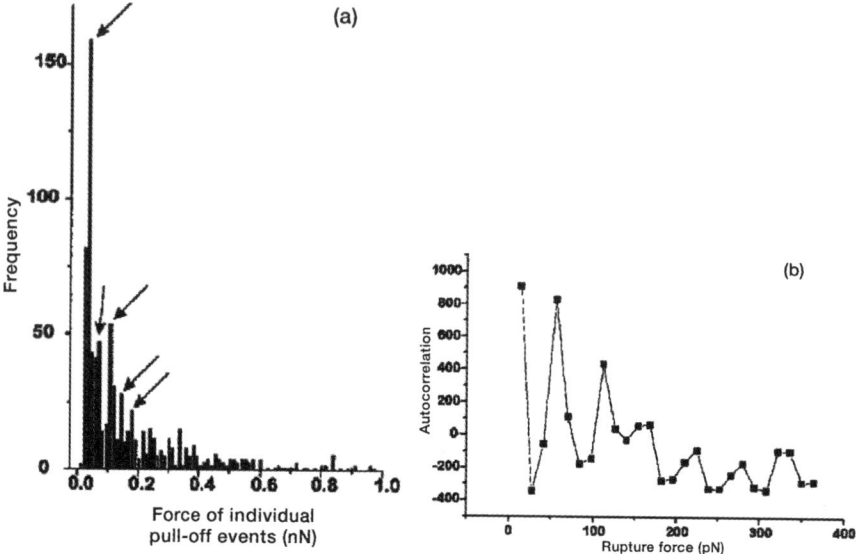

Fig. 6.9. a Histogram of unbinding force for a hydroxyl-ferrocene-modified tip and SAM of β-CD surface. **b** Corresponding autocorrelation function (extracted from [6.50])

fined as

$$G(x) = \frac{1}{N} \sum_{k=1}^{N} \Delta f(k) \cdot \Delta f(k+x) \qquad (6.7)$$

and

$$\Delta f(k) = f(k) - \frac{1}{N} \sum_{j=1}^{N} f(j) \qquad (6.8)$$

the maxima of the force correlation function appear at multiples of 56 ± 10 pN, indicative of single guest–host pair interactions. The force distribution has a characteristic multiple-peak feature possibly due to the finite tip curvature or the granular nature of metal coating that leads to multiple pull-off events.

As a control experiment, another guest agent (1,8-ANS) was introduced into the solution, leading to the disappearance of most of the discrete features in the force spectrum. This is clear evidence that the cavities of β-cyclodextrin have been filled by the guest agent (1,8-ANS), leaving no accessible sites for the ferrocenes attached to the AFM probe. The discrete force distribution characteristics was seen to assume after the blocking agent was washed off.

6.2.4 Desorption of Single Molecules at Interfaces

The desorption forces of poly(4-vinylpyridine) (PVP) on surfaces modified with hydroxy or amino groups were measured [6.51]. The characteristic saw-

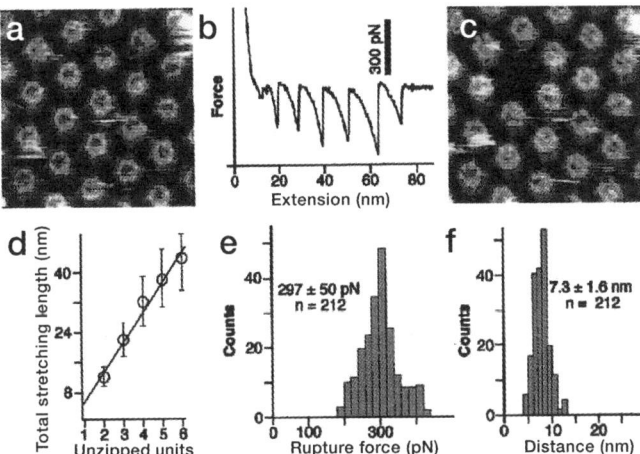

Fig. 6.10. a AFM image of the inner surface of an HPI layer and **b** the force spectrum. The six peaks in the spectrum represent sequential removal of individual protomers. **c** The image after removal of an HPI protein molecule. **d** and **e** are the statistics of the total stretching length and rupture force required for removing individual protomers. **f** is the distribution of the stretching distance between events of single protomer removals (extracted from [6.52])

tooth-like rupture behavior of the single molecule stretching operation was correlated with the desorption of the polymer segments. The statistically obtained desorption force of individual polymer segments is 180 ± 20 pN on amino-terminated surfaces. This value was slightly higher on hydroxy-terminated surfaces than that of the amino-terminated surfaces, due to the higher hydrogen bond interaction strength between pyridine and hydroxy groups compared to that of amino groups. The magnitude of the desorption force was also shown to be dependent on the solvent, by affecting the hydrogen bond interaction between pyridine and terminal groups (hydroxy or amino) on the surface.

An example of detaching biomolecules from mica surfaces can be seen in a study of hexagonally packed intermediate (HPI) layers consisting of *Deinococcus radiodurans*. Each HPI protein contains six units of protomers, and the protein has two different surfaces. The outer surface, the close-packed HPI layer, is hydrophilic and adsorbs directly to the mica surface whereas the inner surface is hydrophobic and exposed to the environment. The imaging of these two protein surfaces can be observed with different imaging forces [6.52]. In addition, individual protomers can be removed from the HPI layer during force spectra measurements (Fig. 6.10). The force extension spectrum indicates that the protomers were removed sequentially. The required force for the removal of single protomers was statistically measured at around 300 pN, with a characteristic saw-tooth pattern in the force-distance spectrum. The stretching distance for removing an individual unit of protomer is 7.3 ± 1.6 nm, which implies a flexible polypeptide linkage between the protomers.

The above examples provide a qualitative measure of the adhesion strength for polymers and biomolecules at functionalized interfaces, and hopefully could develop into a quantitative characterization technique, particularly in benign environments such as under ambient and aqueous conditions.

6.3 Intramolecular Interactions of Single Molecules

The mechanical properties of single polymeric molecules and biomolecules can provide vital insight into intramolecular interactions. There are a number of factors, such as structural conformation and environmental conditions, which can appreciably affect the characteristics of intramolecular interactions. Force measurements at single molecule level present a unique opportunity to look into the physical and chemical nature of such interactions.

6.3.1 Elasticity of DNA Molecules

The forced extension characteristics of double-stranded DNA have been demonstrated using optical fiber as a force sensor [6.53], optical tweezers [6.54], and the receding meniscus method [6.55]. A coherent transformation represented by the base-pair spacing change has been identified at an external force of around 70 pN. In addition, the stretching of single-stranded DNA (ssDNA) resulted in an estimated persistence length of 7.5 Å [6.54]. Two typical regimes have been reported to describe the mechanical properties of DNA molecules. The entropic regime is associated thermal energy, and the elastic regime is due to base-pair interactions. In addition, for double-stranded DNAs, an overstreching transition will occur at high loading force.

The stretching of poly(dG-dC) and poly(dA-dT) molecules revealed a sequence-dependent behavior [6.56]. The transitions of B-type to overstretched configuration (S), and eventually melting are evident in the forced unzipping process. The typical force for melting transition of λ-DNA has been measured at around 150 pN. The stretching forces required for these transitions are lower in the case of poly(dA-dT), compared to poly(dG-dC). Note that in Fig. 6.11a the B–S transition is at 65 pN and the melting transition at 300 pN, whereas in Fig. 6.11b the B–S transition is around 35 pN and no distinct melting transition can be seen. Based on the hairpin formation during zipping/unzipping cycles, the unbinding forces for G–C and A–T pairs are estimated as 20 ± 3 and 9 ± 3 pN, respectively.

A combined setup of optical tweezers and AFM has been demonstrated for studying DNA stretching (Fig. 6.12) [6.57]. A microbead (latex spheres) was first attached with single DNA molecules, manipulated by the optical tweezers toward, and grafted onto the AFM cantilever by laser heating. This setup can extend the measurement range of unfolding force from that of

6.3 Intramolecular Interactions of Single Molecules 145

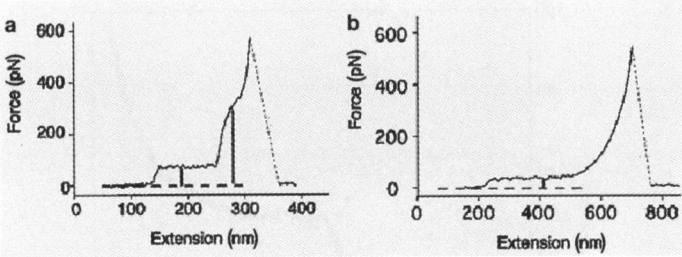

Fig. 6.11. The force spectra for DNA molecules with sequence of **a** poly(dG-dC) and **b** poly(dA-dT) (extracted from [6.56])

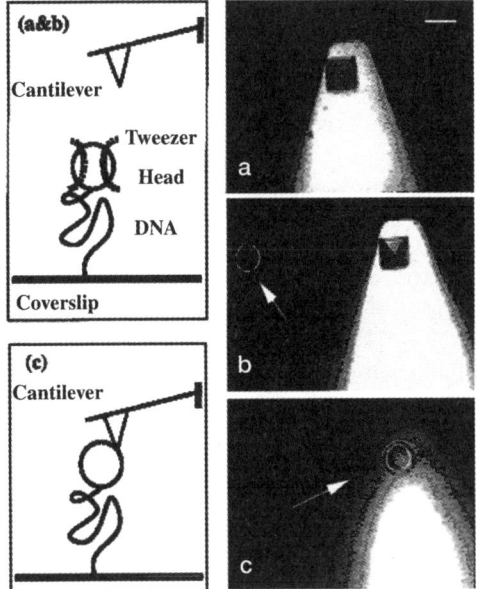

Fig. 6.12. Schematics of the process of combined stretching of DNA molecules using optical tweezers and AFM (extracted from [6.57])

optical tweezers (0–15 pN) to that of AFM cantilevers (100s pN and higher) (Fig. 6.13).

Enhanced force spectra resolution can be achieved by analyzing the thermal fluctuation of beads that are connected to single DNA molecules. The analysis was performed using dual beam optical tweezers. The time-dependent autocorrelation and cross-correlation function of thermal fluctuation of bead positions enabled quantitative determination of the longitudinal and transverse spring constants and friction coefficients of DNA molecules at various extension ratios (74% – 92%) [6.58]. The friction coefficients were found independent on extension ratios, with a ratio of 2.28 between transverse

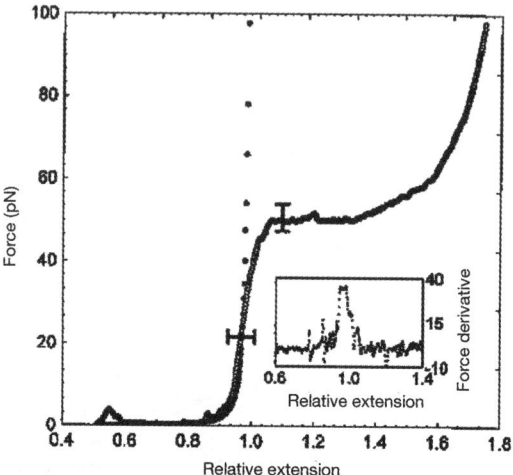

Fig. 6.13. Force spectrum of DNA extension measured by AFM; the *filled circles* represent fitted results using the WLC model (*dotted line*). There are two stages of the extension spectrum representing the entropic (lower force load) and elastic (higher load) regimes. The inset presents the force derivative versus relative extension showing the position of the transition between the entropic regime to the elastic regime (extracted from [6.57])

and longitudinal coefficients. The spring constants are also highly anisotropic, and show appreciable dependence on extensions, especially for the longitudinal spring constant that varies from 0.2 pN μm^{-1} at 74% extension to 4.7 pN μm^{-1} at 92% extension.

Simulation studies based on entropic considerations pointed out that the freely jointed chain (FJC) model can fit the dsDNA extension spectra at low force, whereas the warm-like-chain model (WLC) can generate consistent fitting in most of the experimental range (Fig. 6.14) [6.59]. Stretching-induced conformational changes have also been simulated by using the molecular dynamics method [6.60]. It was suggested that most of the hydrogen-bonded base pairs remained intact while a sudden change in length occurs at a threshold force that corresponds to the B-state to S-state transformation [6.60].

The stretching of single-stranded DNA was studied with the magnetic bead method and optical tweezers [6.54, 6.61]. Theoretical simulation of the extension spectra using the FJC model suggests effects from the intrinsic elasticity of the segments [6.54, 6.62], pairing interactions with hairpin structures [6.63], and electrostatic self-avoiding interactions due to the phosphate backbone [6.64].

The tension applied to DNA molecules was shown to affect the replication activity of T7 DNA polymerase (DNAp). The study was performed with a DNA molecule that had both double-stranded (dsDNA) and single-stranded (ssDNA) sections [6.65]. The DNA molecule was extended between

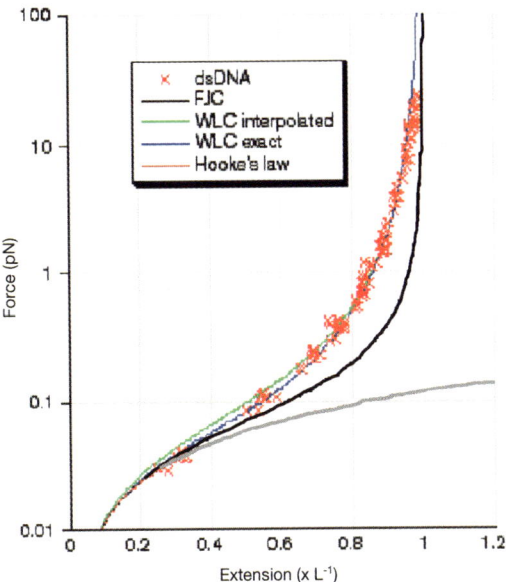

Fig. 6.14. Comparison of simulation results using different models with the extension spectrum of dsDNA (extracted from [6.59])

microbeads held by a pipette and optical tweezers, respectively. The activity of T7 DNA polymerase was observed as a change in overall length under constant tension. Due to the different length for ssDNA and dsDNA under the same tension, the variation of the measured overall length was directly related to the progress of DNAp. A burst behavior was observed during the progressing of DNAp. It was found that the replication rate depended on the tension applied to the DNA molecule. In addition, activities of DNAp were found to switch between exnucleolysis and polymerization at around 34 pN. The results suggest that single molecule techniques can be applied to investigate the mechanochemistry of DNA polymerase processes.

The presence of cations is known to induce DNA condensations, or transition from extended to coiled and globule states. The force spectrum of the DNA molecules under the influence of spermidine, a trivalent cation, measured by dual optical tweezers revealed a characteristic stick release length of about 300 nm [6.66]. This was attributed to the spermidine-induced supercoiling that was relaxed under the pulling force.

The uncoiling process of DNA molecules under the influence of type II DNA topoisomerases has been investigated with the magnetic bead method [6.67]. The DNA molecules are stretched and twisted to form supercoiled structures. The addition of topoisomerases II in the absence of ATP serves to stabilize the coils. The supercoiled DNA will uncoil in a stepwise manner when ATP is added. As a result, the overall length of the DNA will increase by

148   6 Intermolecular and Intramolecular Interactions

steps equivalent to two supercoils. The turnover rate of the uncoiling process was found to depend monotonically on the ATP concentration. However, a decrease in turnover rate was observed with increasing tension applied to the DNA. The latter effect was attributed to an inhibitory effect of enzymes on the force.

In another study, it was found that both positively and negatively supercoiled DNAs display similar extension spectra [6.68]. For high degrees of supercoiling, the positively supercoiled DNA becomes more difficult to stretch than the negatively supercoiled DNA.

6.3.2 Folding and Refolding of Single Protein Molecules

In addition to stretching DNA molecules, the force spectrum method has been actively pursued as a direct approach to study the folding and unfolding processes of protein molecules.

Folding/unfolding processes of the extracellular matrix protein of human tenascin-C have been studied with an AFM in force spectrum operation mode. The force-distance spectra displayed equally spaced peaks in the stretching process, corresponding to FN-III domains stretched to 90% of their maximum length (based on the assumption that each amino acid contributes 0.38 nm to the overall contour length) (Fig. 6.15) [6.69]. The average force that each FN-III domain could sustain is about 137 pN. Pulling rate was also found to affect the measured unfolding force. The details of the extension spectrum can be nicely fitted by the warm-like-chain (WLC) model. The relaxation, or refolding, of the protein tenascin-C was monotonic. Similar results for the unfolding process have been observed for the recombinant proteins TNfnALL and TNfnA-D. A more complicated refolding behavior was identified for the TNfnALL protein, in which the FN-III domains do not renature at the same rate [6.69].

The unfolding process of the polymeric I27 protein has been analyzed with the force spectrum method. The protein consists of eight tandem repeats of single domains, manifested as a clearly saw-tooth pattern in the force-extension spectrum [6.70]. It was found that the unfolding is triggered by a 2.5-Å extension according to Monte Carlo simulation, and the refolding proceeds exponentially with time. Titin molecules were unfolded by using optical tweezers, and a characteristic stepwise pattern was observed in the force-extension spectrum as a result of the relaxation of constituent domains [6.71]. Force spectrum studies on single protein molecules of titin illustrate the details of the unfolding process [6.72]. The restoring force displayed appreciable step-like patterns. The modulation of stretching force was attributed to the unfolding of immunoglobin (Ig)-like domains. It should be noted that the unfolding process always begins with the weakest domain. By fitting the extension feature corresponding to individual Ig domains with the WLC model, a persistence length of 0.4 nm and contour length of 28–29 nm were obtained.

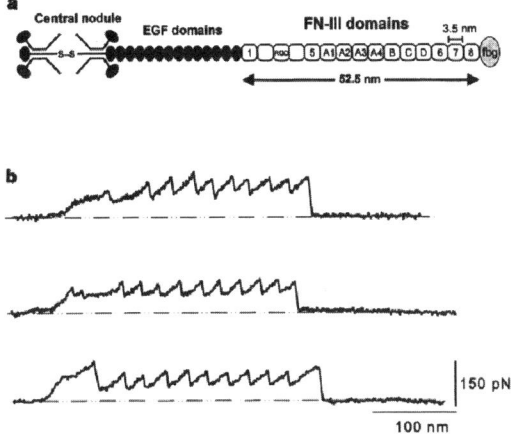

Fig. 6.15. (a) Molecule structure of a subunit of native tenascin hexabrachion. (b) Typical force-extension spectrum (extracted from [6.69])

Force spectrum studies on folding–unfolding processes of polymeric T4 lysozyme molecules resulted in a similar behavior as that described above for other protein molecules (Fig. 6.16) [6.73]. The persistence length obtained from the WLC model fitting is 0.65 ± 0.25 nm. In addition, a lower unfolding force and higher refolding efficiency, compared to titin molecules, have been observed. This could be associated with the structural difference between T4 lysozyme (α-helical) and titin (predominantly β-sheet structures).

Unfolding of individual membrane protein molecules of bacteriorhodopsin (BR) was achieved by attaching the molecules to the AFM tip at the cytoplastic COOH terminus. The measured force spectra revealed that the helices contained in the protein can be associated with different unfolding characteristics in the force range of 100–200 pN [6.74].

Bustamante et al. [6.75] attached titin filaments to latex beads and used optical tweezers to study the elastic behavior. It was found that the force-extension curve can be better described with the warm-like-chain model, yielding a persistence length of 20 Å. Significant hysteresis in stretch-release cycles was observed when the force load rate is higher than the presumed rate for unfolding and refolding of the titin molecule.

6.3.3 Stretching Other Biomolecules

The modulus of the peptide cysteine$_3$–lysine$_{30}$–cysteine (C_3–K_{30}–C) in buffer solution was obtained by AFM (Fig. 6.17) [6.76]. The peptide molecules were connected to gold film on a mica support and a gold-coated tip. The connection was established utilizing the thiol group of the cysteine terminal. The AFM was operated in force modulation mode by magnetically controlled cantilever, through a magnetic particle attached to the back of the cantilever. The

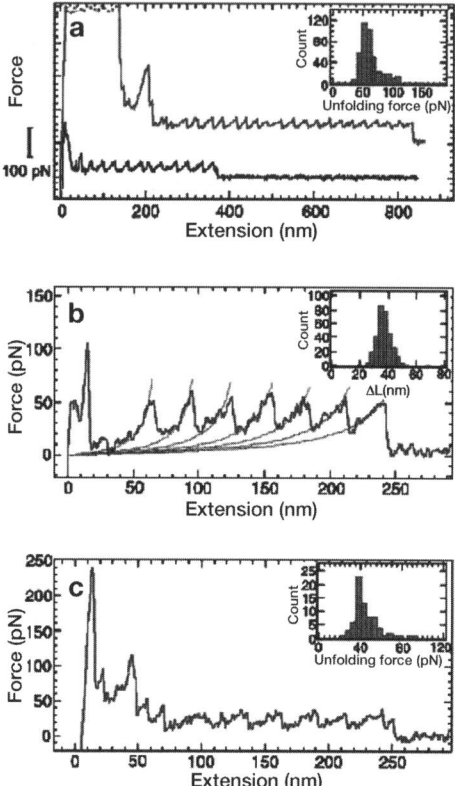

Fig. 6.16. Force spectra of single T4 lysozyme molecules. **a** In 10x PBS solution (1,260 mM NaCl, 72 mM Na_2HPO_4, 30 mM NaH_2PO_4, adjusted to pH 7.0 with HCl). **b** Fitting results using the WLC model for the stretching part of the force spectrum; **c** Force spectrum in a solution of 1 M GuHCl (extracted from [6.73])

longitudinal Young's modulus for the α-helix (1.2 ± 0.3 GPa) and elongated backbone (50 ± 15 GPa) were obtained from the force-distance spectra [6.76].

A similar experiment was performed to open up single DNA molecules attached with a DNA-binding protein BsoBI or XhoI [6.77]. In this work, the DNA molecules were attached between a glass substrate and an optically trapped microbead. The pulling was performed by moving the glass substrate while the force was recorded from the microbead movements. The presence of the single protein molecule was manifested in a surge of measured force, as the unzipping event proceeded to the protein binding site. The disruption force associated with protein–DNA binding was found to depend on the pulling rate, and also the binding sites, which could enable the study of site-specific interactions between DNA and proteins.

The stretching of long-chain molecules has the advantage of minimizing non-specific interactions between the tip and sample surface because these are

6.3 Intramolecular Interactions of Single Molecules   151

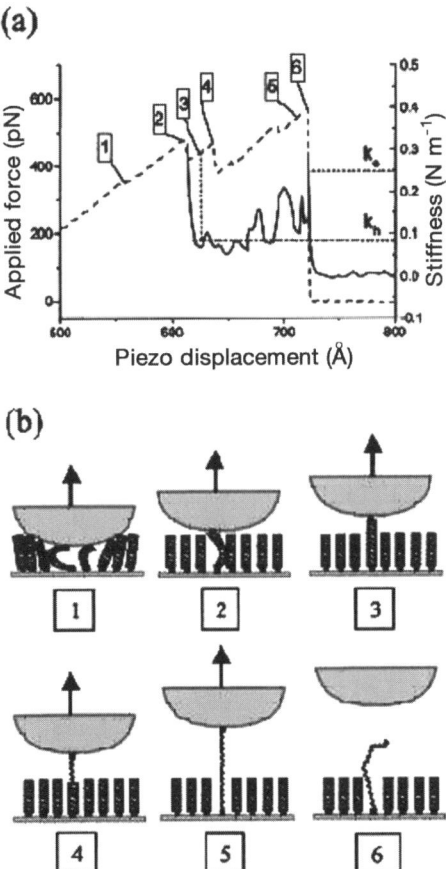

Fig. 6.17. a Force spectrum (*dashed line*) and corresponding stiffness (*solid line*) of the $C_3K_{30}C$ peptide. b Schematics of various stages of the pulling process (extracted from [6.76])

kept apart by a molecular bridge. Connecting the tip to the target molecule is typically by pressing the tip to the adsorbate-covered surface with a force of a few nN for a few seconds [6.78]. The nature of the connection should be covalent in order to sustain a high stretching force, usually on the order of nN.

For molecules of short length, the interference of non-specific interaction may be pronounced. Improvements have been demonstrated by introducing tether molecules (molecular spacers) or working under vacuum or liquid conditions. In the case of molecular spacers, the measured rupture force can be modified as a result of bonding configuration, and the force resolution may also be reduced [6.79, 6.80].

6.3.4 Polysaccharides

The deformation of polysacchrides has been investigated since the early stage of force spectrum experiments. Dextran filaments were bound to gold surfaces by means of epoxy-alkanethiols, and were activated with carboxymethyl group per glucose unit. As a result, the dextrans could be connected to streptavidin. The force spectrum reveals a two-stage behavior, namely, an entropic elastic regime at low stretching force with a 6-Å Kuhn length, and a structural transition at higher force. The structural transition at high force can be characterized by the twist of bond angles [6.81].

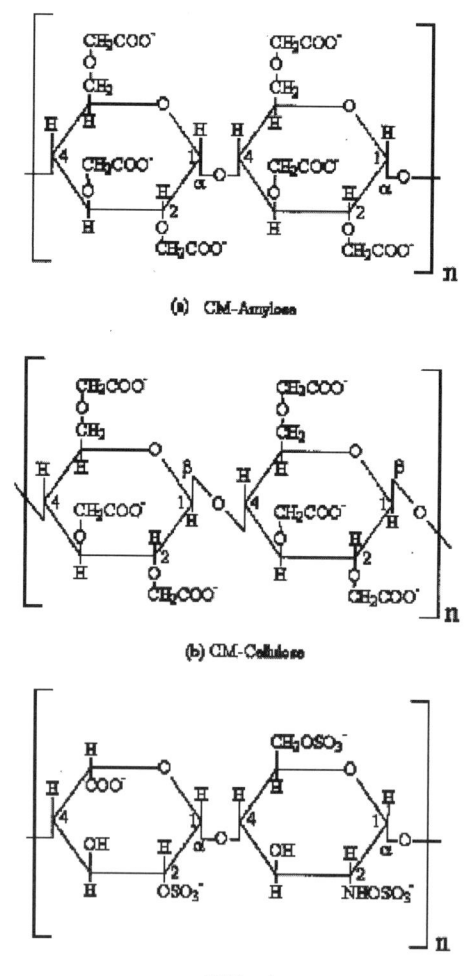

Fig. 6.18. Molecular structures of **a** carboxymethyl amylose (CM-amylose); **b** carboxymethyl cellose (CM-cellose); **c** heparin (extracted from [6.82])

6.3 Intramolecular Interactions of Single Molecules

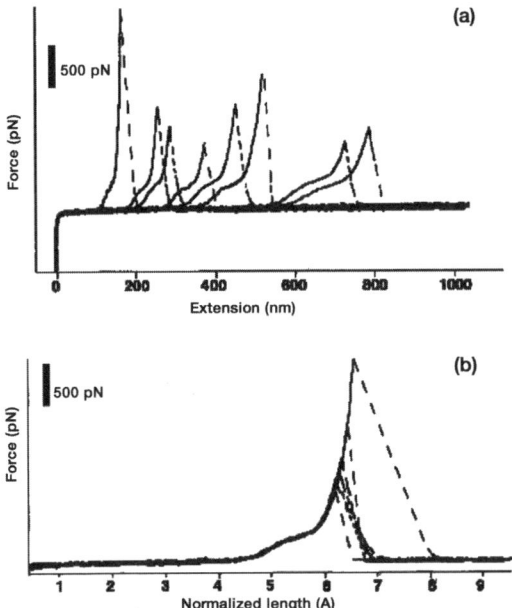

Fig. 6.19. Force spectrum of CM-amylose **a** with different contour lengths and **b** the overlapped spectrum (extracted from [6.82])

The force spectrum of CM-amylose resulted in a persistence length of 0.54 nm based upon WLC model simulation [6.82]. The first plateau in the force-extension spectrum is typical for entropy-driven elastic regimes (Figs. 6.18, 6.19). Further elongation (reflected in increased stiffness) can be ascribed to the conformational transition of chair-twist type, and could be a characteristic feature for the compounds.

A study of carrageenan polysaccharides of indicated that force spectrum characteristics are sensitive to the environment [6.83]. Whereas native carrageenan displays typical extension spectra that can be satisfactorily described by the freely jointed chain (FJC) model, the addition of salt (0.1 M NaI) led to the appearance of additional features assigned to coil-to-helix transition. The two-state Monte Carlo simulation model was proposed to account for the extension behavior of both the polysaccharide dextran and the titin protein [6.84].

6.3.5 Other Polymers

A polymer containing azobenzene units and polypeptide backbone was shown to display substantially different force spectrum characteristics when excited to the extended state or relaxed state. A switching behavior was observed in association with the transition between *cis* and *trans* states of azobenzene [6.85]. An optical-driven weight-lifting molecular machine was proposed, based on the mechanical switching behavior of the polymer.

A polyelectrolyte molecule of polyvinylamine (PVA) was subjected to force spectrum analysis in varying salt concentrations (5–100 mM NaCl) [6.86]. The persistence length for single polyelectrolyte chains was derived by fitting the force-extension curves with the warm-like-chain (WLC) model. The results indicate that the electrostatic effect due to the ionic environment is not significant for molecules under high mechanical stress. This is in contrast to the general knowledge for polymers in mechanically relaxed state. The detachment force of the polymer from silica surfaces showed good correlation with salt concentrations, as a result of a Debye screening effect and charge density of the polymer.

The elastic properties of poly(ethylene-glycol) (PEG) molecules under various solvent conditions showed that a possible deformation of the helical suprastructure could evolve in water, a phenomenon that was not observable in hexadecane solvent [6.87]. The molecular dynamics (MD) simulation provided supportive information on the localized helical structure in association with water molecule bridges between PEG monomers [6.88].

Stretching-induced conformation changes in single pectin molecules (1-4-linked α-D-galactouronic acid polymer) were studied by the force spectrum method. A two-step chair inversion reaction of pectin monomers was identified and supported by *ab initio* calculations [6.89].

6.4 Dynamic Force Measurements of Single Molecules

6.4.1 Pulling Rate Effect on Force Spectrum Measurements

It has been observed in several detailed studies that the stretching force is exponentially dependent on the pulling rate. The unfolding process of polymeric molecules can be considered as a reaction process with a free energy barrier. The external force applied to the molecule will deform the energy barrier by a time-dependent potential [6.90–6.94]:

$$\Delta E(x,t) = f \cdot \Delta x = f \cdot vt$$

where f is the applied external force, Δx the distance travelled along the direction of the applied force, and the v the pulling rate. The energy barrier height is superimposed by the above potential, and the dissociation or unfolding will proceed by thermal excitation as the barrier is reduced. The rate of the dissociation event (k_{off}) can be expressed as:

$$k_{\text{off}} \propto \exp[-(E_{\text{b}} - \Delta E)/k_{\text{B}}T] = k_0 \exp(f/f_0)$$

The force-scale parameter F_0 is defined as $f_0 = k_B T/\Delta x$ and the pre-exponential factor $k_0 = \exp(-E_{\text{b}}/k_{\text{B}}T)$.

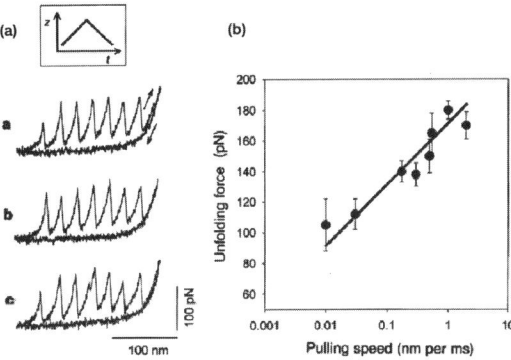

Fig. 6.20. Schematic of the dependence of unfolding force on the pulling rate (extracted from [6.69])

As a result, the most probable dissociation or unfolding force will depend logarithmically on the pulling rate [6.93]:

$$F(v) = F_0 \ln\left(\frac{kv}{F_0 k_0}\right)$$

where k is the stiffness of the force sensor. Note that the slope of $F(v)$ versus v provides a measure of the energy barrier, i.e., through k_0. In principle, the pulling rate (defined as $k \cdot v$) dependence can provide insight into the spatial distribution of energy landscape relevant to the unfolding process. Such logarithmic dependence has been confirmed experimentally in unfolding studies of titin immunoglobutin domains [6.72], DNA molecules [6.95], and extracellular matrix protein tenascin (Fig. 6.20) [6.69]. The energy landscape obtained by dynamic force spectra is affected by the spring constant [6.91]. The dependence was also shown to be affected by the number of base pairs and temperature in an unbinding experiment of DNA molecules [6.96].

The observation of more than one slope in the $F(v)$ versus v relationship reveals the possibility of intermediate state in the dissociation pathway [6.94].

The methodology could help study the intermediate state for streptavidin (avidin)–biotin complexes [6.94, 6.97]. Effort was placed in exploiting rupture force measurements with the thermodynamic properties of molecular dissociation of the biotin/streptavidin complex [6.98]. It was suggested that a pulling rate much higher than the currently adopted range will be desirable to retrieve such information.

6.4.2 Pulling Rate Effect on Rupture Force Measurements

The dynamic force spectroscopy of streptavidin–biotin and avidin–biotin was obtained using a biomembrane force probe (BFP) and red blood cells as force transducer [6.94]. The force constant is in the range of 0.1–3 pN nm^{-1} for the

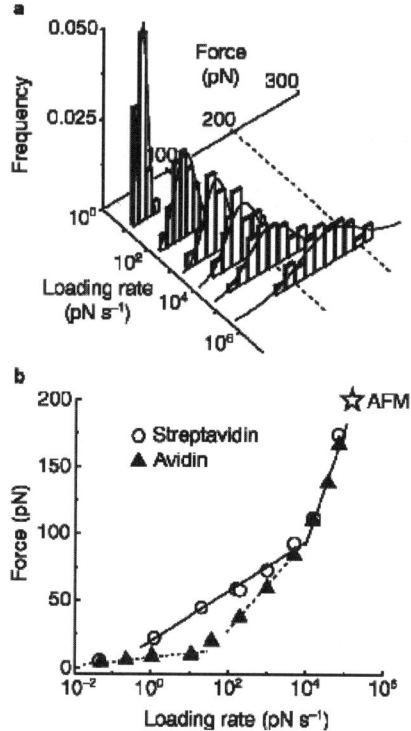

Fig. 6.21. a Histogram of the biotin–streptavidin interacting force on the pulling rate. **b** The averaged force measured by the biotin tip versus pulling rate (extracted from [6.94])

force range 0.5–1,000 pN (Fig. 6.21). The pulling rate can be adjusted in the range 1–20,000 nm s^{-1}. The observed dependence of the measured intermolecular force on the pulling rate (Fig. 6.21) and the instantaneous interaction energy provide insight into the energy barrier landscape of the streptavidin–biotin complex, with indications of transition states [6.99]. Three stages of the unbinding process were identified, these being initial displacement of the biotin from the hydrogen bonds, water bridges and non-polar interaction in the binding pocket, and a sudden displacement and prominent jump in rupture force.

The analysis of the pulling rate effect on the bond dissociation barrier led to the prediction of spontaneous bond dissociation at sufficiently low pulling rate, as a result of thermal activation to overcome the barrier [6.100]. This effect was confirmed in the dynamic force spectrum measurements of the single rabbit immunoglobulin G (IgG) and single molecules of protein A using biomembrane force probe method (Fig. 6.22). An appreciable transition of median rupture force was observed as the pulling rate was reduced, leading to a spontaneous dissociation at low pulling rate.

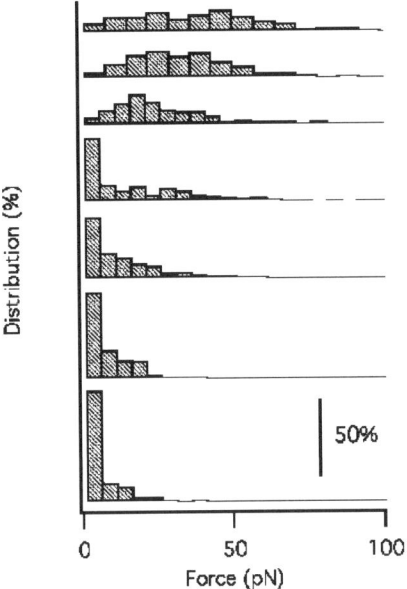

Fig. 6.22. Histograms of unbinding force between IgG and the receptor molecule at loading rates of 630, 280, 75, 27, 11, 5.5 and 2.8 pN s^{-1} (from *top* to *bottom*) (extracted from [6.100])

The unbinding force between complementary oligonucleotides was also found to depend logarithmically on the pulling rate [6.95]. A single energy barrier model can be invoked to accommodate the single-slope behavior. The parameters of the energy barrier, i.e., thermal-off rate (or dissociation rate), and the separation between the maximum and minimum of the barrier, were also found to depend on the number of base pairs.

The dependence of rupture force on pulling rate was shown to be associated with the dissociation rate of the ligand–receptor complex through the following relationship:

$$k_{\text{off}}(F) = k_{\text{off}}^0 \exp(-F(t) \cdot s_{\text{pot}}/k_B T) \tag{6.9}$$

where s_{pot} is the average width of the potential barrier.

The interaction between P-selectin and its counter-ligand P-selectin glycoprotein ligand-1 (PSGL-1) was studied for its relevance to the selectin-mediated cell–cell interactions of the immune system. A rupture force of up to 165 pN was observed from the force spectrum experiments [6.101]. The simulation of the force-distance spectrum using a modified freely jointed chain (FJC) model yielded a spring constant of 5.3 pN nm^{-1} and persistence length of 0.53 nm for the P-selectin/PSGL-1 complex. The Monte Carlo simulation was shown to nicely describe the rupture process under external force by simplifying the process as random events. The simulation yielded an off-rate

k_{off}^0 of 0.022 s^{-1} under no external force, and 15 s^{-1} under external force applied by AFM.

Comparative assessments were carried out on the dissociation of antibody–antigen complexes studied by dynamic force spectrum and fluorescence methods in solutions. The rupture force obtained by the dynamic force method is rather well correlated with the off-rates from the solution method [6.102].

With a glass microsphere attached to the AFM cantilever, and subsequently modified with biotin, the interaction between biotin and streptavidin was estimated at around 0.3 nN and found to depend on the spring constant of the cantilever [6.103].

6.4.3 Force Measurements Relevant to Movements of Biomolecules

The movement of kinesin enzyme molecules along microtubules was followed by means of optical tweezers [6.17, 6.18, 6.104]. The results suggest that silica beads with low kinesin density (single kinesin molecule, or less) tend to detach from the microtubule briefly during the mechanochemical cycle [6.104].

An infrared optical tweezers was used to detect the force of mitochondria moving along microtubules under in vivo environmental conditions. Data were derived from the laser power needed to stop and release the trapped particle [6.105]. A force magnitude of 2.6×10^{-7} dyne was estimated for mitochondria with one motor can run at a speed of 10 µm per second.

By using optical tweezers, the isometric force of a single kinesin enzyme molecule on a biotinylated microtubule was estimated at about 1.9 pN in the presence of uncleotide GTP or ATP [6.106].

With an optical interferometer, the movement of kinesin-decorated silica beads along microtubules under the control of an optical tweezers was directly measured [6.107]. A discrete moving behavior was identified, with a step size of 8 nm. In addition, it was found that the kinesin can transport against loads up to 5 pN, contrasting with the value of 1.9 pN shown above. This can at least partly be explained by a difference in manipulation conditions between the two studies.

Force measurements at single molecule level have become a vital addition to the high topography resolution capability of AFM. Further improvements of force spectrum resolution may lead to even more interesting topics in the study of intermolecular interactions.

General aspects dealing with the specificity of single molecule force measurements relate to environmental fluctuations such as solvent, surface heterogeneity, and experimental errors. As the the detection of single molecules remains a highly challenging arena, endeavors to solve such problems will surely continue to improve the quality of the results.

7 Electrical Conductivity of Single Molecules

7.1 Introduction

This chapter focuses on another important aspect of molecular properties, i.e., electron transportation behavior. Such studies are enabled by, and closely related to the molecular level exploration of adsorption and assembling processes presented in the preceding chapters. Promoted by conceptual advances in molecular electronics and other fields, the design and synthesis of electrical conductive molecular structures have drawn extensive interest [7.1]. This effort, together with the development of novel nanolithography techniques, is aimed at obtaining structures with electronic functions in the length scale below 100 nm. Chemical approaches in this scale regime are unique owing to the diversity of molecular functions. We wish to provide in this chapter a brief summary on recent progress in this fast evolving field.

7.1.1 One-Dimensional Molecular Conductance Structures

The consideration for transportation processes in molecular wires can be described in the form of homogeneous carrier transport kinetics in which conductance can be expressed in terms of the rate constant k_{et}:

$$-\frac{d\{D\}}{dt} = k_{et}\{D\}$$

where k_{et} represents the efficiency of electron transfer. Under the approximation of non-adiabatic intramolecular carrier transfer, a superexchange mechanism was proposed whereby electrons tunnel through a barrier in a coherent way and do not reside within the barrier. The expression may be simplified in an exponential form [7.2]:

$$k_{et} = k_{tunnel} = k_0 \exp(-\beta d)$$

with

$$\beta = -\frac{2}{a} \ln\left(\frac{H_{BB}}{\Delta E_{DB}}\right)$$

where k_0 is the pre-exponential factor, β the attenuation factor, H_{BB} the internal coupling energy between the bridge units, a the bridge unit length, and

$\Delta E_{\rm DB}$ the energy difference between the intermediate state and the ground state. The formalism was developed and tested in solutions and applied successfully in solid junction studies [7.3, 7.4].

In the case of the carrier residing on the bridge molecule with an effective lifetime and transport by hopping between bridge sites (total number of bridge unit: N), transport is of Ohmic type with a rate expressed as [7.5]:

$$k_{\rm et} = k_{\rm hopping} \propto \frac{1}{N}$$

The rate also depends on the energy gap, as shown in the following:

$$k_{\rm et} = k_{\rm hopping} \propto \frac{1}{N} \exp\left(-\frac{\Delta E_{\rm DB}}{RT}\right)$$

Seeing that, actual transport may occur through both pathways, the rate expression should be:

$$k_{\rm et} = k_{\rm tunneling} + k_{\rm hopping}$$

The contributions from the tunneling and hopping mechanisms can be differentiated by their temperature dependence.

Under the situation of $\Delta E_{\rm DB} = 0$, i.e., the energy level of intermediate state coincides with the Fermi level, the transport is essentially a resonant process. Since there is little energy loss due to the bridge molecules, the carrier transport is dominated by the contact behavior.

The second approach is to directly consider the resistance/conductance of the carrier transport process as first proposed by Landauer [7.6]:

$$g(V) = \frac{\partial I(V)}{\partial V} = \frac{2e^2}{h} \sum_{ij} T_{ij}(E, V)$$

where $g(V)$ is the conductance, and T_{ij} the transmission probability of electrons across the junction. $g_0 = 2e^2/h = (12.9\ {\rm k}\Omega)^{-1}$ is the quantum unit of conductance. This quantized conductance behavior has been identified in a number of studies.

The scattering occurs both at the molecule–electrode interface and the molecules. The scattering due to the presence of molecules is largely considered as an elastic process [7.7–7.10].

The above expression again can be simplified into exponential dependence:

$$g = g_0 \exp(-\beta d_{\rm DA})$$

or Ohmic behavior

$$g \propto \frac{1}{N} \exp\left(-\frac{\Delta E_{\rm DB}}{RT}\right)$$

The correlation between the Landauer mechanism with the kinetics model can be established by considering the coherent tunneling process:

$$g = e^2 k_{\rm et} \rho_i(E_{\rm F})$$

where k_{et} is defined as in the kinetics model, and $\rho_i(E_F)$ is the density of states of the initial electrode.

Under the approximation of strong chemisorption, the relationship can be expressed as:

$$g \sim \frac{8e^2}{\pi^2 \Gamma_D \Gamma_A} \left(\frac{k_{\text{et}}}{DOS}\right)$$

Further simplification on the contact with electrodes can lead to another expression:

$$g = \frac{e^2}{2\pi^2} \left(\frac{k_{\text{et}}}{DOS}\right) \left(\frac{H_{LB}^2 H_{BR}^2}{H_{DB}^2 H_{BA}^2}\right)(\rho_L \rho_R)$$

where

$$DOS = \gamma_M(\text{FCWD})$$

$$\gamma_M = \pi k_B T \rho_M(E_F)$$

$$\text{FCWD} = \left[\frac{1}{4\pi\lambda k_B T}\right]^{1/2} \exp\left[-\frac{\lambda}{4k_B T}\right]$$

and $\rho_M(E_F)$ is the effective density of states near the Fermi level of the metal electrode, FCWD the Frank-Condon weighted density of vibronic states that controls the electron transfer when D and A sites are coupled to the vibronic modes, and λ a nuclear reorganization parameter. The numerical calculation of β can be performed using recursion relation [7.11] and band structure methods [7.12].

The third widely used approach of analyzing tunneling characteristics is based on formalisms for M-I-M junctions [7.13]:

$$j = \frac{2b}{A(\Delta x)^2 e\varphi_r^{1/2}} \{\varphi_r \exp(-A\Delta x \sqrt{\varphi_r}) - \varphi_r + eV)\exp[-A\Delta x(\varphi_r + eV)^{1/2}]\}$$

$$b = e^2(2m\varphi_r)^{1/2}/h^2 \qquad (7.1)$$

$$A = 2(2m)^{1/2}/\hbar = 1.025 \; (eV)^{-1/2} \; \text{Å}^{-1}$$

where j is current density, and Δx and φ_r the effective junction width and barrier height, respectively. These parameters assume different magnitudes in the following regimes:

(1) Electrode 1 positive, $0 \leq V \leq \varphi_1/e$,

$$\Delta x = d$$
$$\varphi_r = \frac{2}{3}\frac{\varphi_2^{3/2} - (\varphi_1 - eV)}{\varphi_2 - \varphi_1 + eV}$$

(2) Electrode 1 positive, $V > \varphi_1/e$,

$$\Delta x = d\varphi_2/(\varphi_2 - \varphi_1 + eV)$$
$$\varphi_r = 4\varphi_2/9$$

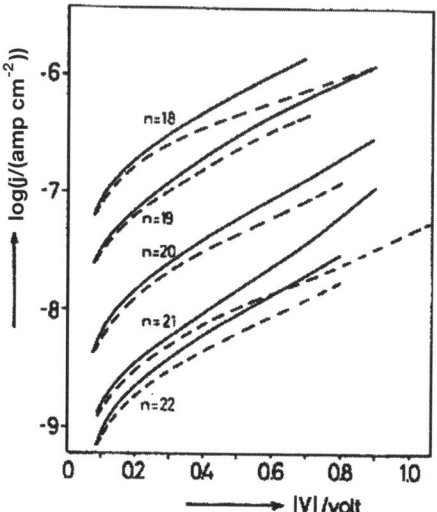

Fig. 7.1. Current-voltage characteristics for fatty acid salts of different chain length. *Solid* and *dashed curves* correspond to opposite tunneling bias (extracted from [7.13])

(3) Electrode 2 positive, $0 \le V \le \varphi_2/e$,

$$\Delta x = d$$
$$\varphi_r = \frac{2}{3} \frac{\varphi_1^{3/2} - (\varphi_2 - eV)}{\varphi_1 - \varphi_2 + eV}$$

(4) Electrode 2 positive, $V > \varphi_2/e$,

$$\Delta x = \frac{d\varphi_1}{\varphi_1 - \varphi_2 + eV}$$
$$\varphi_r = 4\varphi_1/9$$

where φ_1 and φ_2 are the work functions for the electrodes, and d the separation between the electrodes. Figure 7.1 is an example of the current-voltage characteristics for fatty acid salts of different chain lengths [7.13].

So far, a wide range of species have been subjected to direct conductance measurements at single molecule level. The development of, amongst others, self-assembling and Langmuir-Blodgett techniques, self-assembled monolayers (SAM), scanning probe microscopy (SPM), and nanolithography techniques has enabled direct experimental characterization of single molecules, greatly enhance the feasibility of constructing molecular devices based on carrier transportation processes. Currently, a wide range of molecules being systematically analyzed for electrical conductivity [7.14, 7.15], including linear saturated alkyl derivatives, conjugated molecules (generally represented

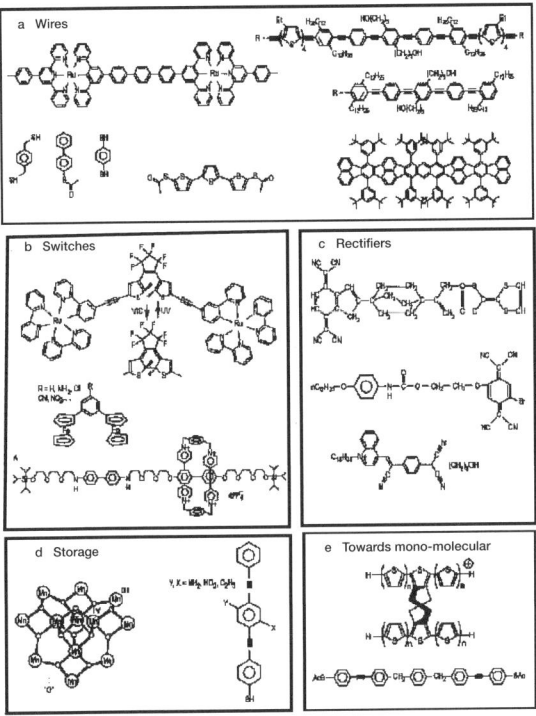

Fig. 7.2. Brief summary of the molecular structures selected for electrical property studies (extracted from [7.15])

by the family of conductive polymers), and carbon nanotubes. Figure 7.2 is a brief summary of the types of molecular structures selected for electrical property studies.

Along with work on transporting electrons or holes using small-dimensional molecular nanostructures, there have also been studies dealing with ionic transportations, which is important for both biological and chemical processes [7.16–7.19]. An example of ionic channels is one-dimensional columnar structures. Low generation of dendrimers is also known to form columnar structures via non-bonding interactions [7.20–7.22].

7.1.2 Methods for Measuring Molecular Conductivity

Techniques for preparing tunneling junctions include:

(1) Evaporation processes [7.13]
(2) Break-point method [7.23, 7.24]
(3) Mercury column method [7.25, 7.26]
(4) Nanowires method [7.27]
(5) Nanolithographically defined pores [7.28]

(6) Capillary molecular junction [7.29]
(7) Scanning probe microscopy (SPM) [7.30–7.38]
(8) Cross-wire method [7.39]
(9) Metallic nanoparticle-based contact [7.40]
(10) Junctions prepared by using the electromigration effect [7.41]

Figure 7.3 illustrates several methods for molecular conductivity measurements. The results obtained from these junctions mainly represent ensemble averages, and recent studies have demonstrated the feasibility of single molecule junctions. For example, a mechanically controllable break junction was used to form a statically stable gold–sulfur–aryl–sulfur–gold system with molecules of benzene-1, 4-dithiol self-assembled onto the two facing gold electrodes [7.23]. Such junctions allow direct observation of charge transport

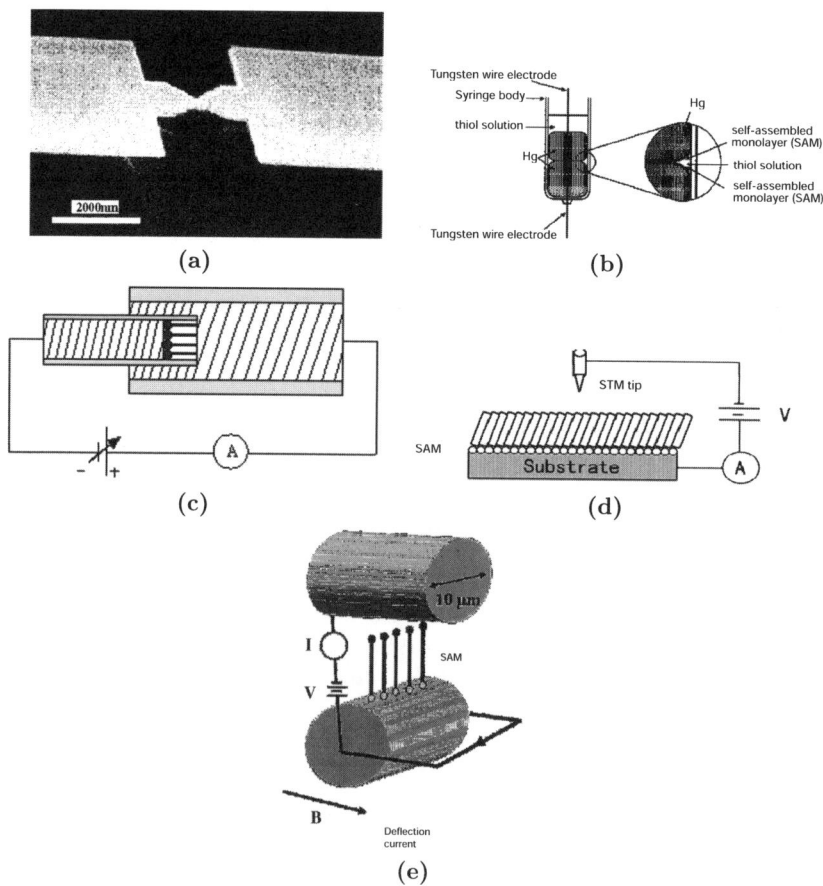

Fig. 7.3. Methods for molecular conductivity measurements. **a** Break point junction [7.24] **b** Mercury column [7.25] **c** Capillary junction [7.29] **d** SPM **e** Cross wire method [7.39]

through the molecules. The obtained conductance-voltage curve provides a quantitative measure of the conductance of a junction containing a single molecule, which is a fundamental step in the emerging field of molecular-scale electronics.

In the mercury column junction method [7.25, 7.26], two Hg drops a few millimeters in diameter are initially modified with thiol SAM and then brought into contact within a microsyringe. Capacitance measurements of the as-formed junctions can provide a reliable assessment of the quality of the junction. The liquid drop of Hg provides a homogeneous surface that promotes the formation of highly uniform SAMs. Such junctions can be operated in electrolytes under electrochemical conditions. The SAMs on the Hg drops can be different, and redox molecules may also be included between the interface of two SAMs in such junctions. One should be aware of possible solvent effects, especially when using non-polar solvents that may be trapped inside the SAM during the approaching process.

In cross-wire junction configuration, the separation between the current-carrying wires is adjusted by an external magnetic field. The effective tunneling cross section is proportional to the diameter of the wires [7.39]. Using the cross-wire junction method, it was shown that bond-length alternation has significant effects on the conductance of SAMs of oligo(phenylene ethynylene) (OPE) and oligo(phenylene vinylene) (OPV), as a result of gap separation between HOMO–LUMO orbitals. In addition, the σ-bonded alkanes have much lower conductance than the π-conjugated molecules. The SPM method appears to have better capability to address individual molecules embedded in SAMs.

7.2 Electrical Conductivity of Molecular Monolayers

7.2.1 Linear Alkane Derivatives

Studies on aliphatic molecules (exemplified by alkanethiols) have focused on constructing devices directly based on single molecule conductivity. This work has spanned a time period of nearly three decades since an early study on stearic acetate using the mercury junction method [7.13], recently complemented by several groups [7.25, 7.26]. The SPM method has also contributed to research in this field [7.30–7.38]. The results have revealed the effects of, for example, the tunneling barrier, and terminal groups, and have helped examine the electron transfer mechanism.

The reported results revealed that values for the electron decay attenuation factor β for a series of alkanethiolate on gold and mercury surfaces are in the range 0.8–1.5 and 0.4–0.6 Å^{-1} for phenyl group [7.25a, 7.25c, 7.27, 7.31, 7.37, 7.38] and nitro-based molecules [7.37]. The difference between the attenuation factor of alkanethiolates and that of molecules with phenyls and electroactive groups can be qualitatively correlated with the nature of the electron transportation process in these specimens [7.37]. For alkanethiolate, the molecular orbitals are typically far away from the Fermi level, and the tunneling is probably dominated by the superexchange mechanism. The attenuation factor is expectedly large for this process. By contrast, for molecules with phenyl and other electroactive groups, the molecular orbitals of the functional groups may be positioned closer to the Fermi level. As a result, resonant tunneling may play a role in the transportation process in this case, and the attenuation factor should be smaller.

An electron charging effect was observed for nitro-based oligo(phenylene ethynylene) molecules [7.37]. The evidence is based on the additional anodic peaks in I-V characteristics after an initial scan to negative bias (tip voltage). In addition, the attenuation factor for nitro-based molecules was found to vary significantly with the applied bias voltage [7.37]. Such variations in attenuation factor may be considered within a broader range of controlling factors, including the properties of substitution groups, molecular backbone, interactions between these, and tip pressures [7.37, 7.42]. In another study of theoretical calculations, the observed conductance on benzene-1,4-dithiol was found to be due mainly to the resonant tunneling effect, with little charging effect, because the electrons do not reside on the the benzene group long enough [7.43].

In more recent studies, dielectric characteristics of capacitors, electrical breakdown voltage (BDV) and electron transport properties have been studied using Hg–SAM–metal (Hg, Ag, Au, and Cu) systems as nanometer-thick organic dielectrics [7.25, 7.26]. It was demonstrated that alkanethiols on atomically flat surfaces of mercury form monolayers of extremely low conductivity, close to that of bulk polyethylene. Notably, the magnitude of BDV was found to depend on the metal electrodes and correlates with the organizational parameters of the SAM on the metal electrodes. In addition, the BDV depends on the chain length of alkanethiol forming the SAM for the same metal surface. A survey of SAMs with different chemical structures shows that the BDV correlates overall with the thickness of the densely packed hydrocarbon portion of the SAM: aliphatic and aromatic SAMs of the same thickness show similar BDVs. Using the same approach, it was found that current density decreased with increasing distance between the electrodes, and increased roughly linearly with the area of contact between SAM(1) and SAM(2) in the Ag–SAM(1) SAM(2)–Hg junctions.

7.2.2 Conjugated Molecules

Conjugated molecules are known to have higher carrier mobility than do aliphatic molecules, due to the nature of delocalized electronic states. There are a large number of one-dimensional molecules that have been studied [7.44, 7.45], and can be considered as potential candidates for conductive wires.

The resistance of a number of differently sized molecules (monomers, oligomers) with aromatic components was measured. The resistance of double-ended aryl dithiols (xylyldithiol) was estimated as 18 ± 12 MΩ [7.46], and of benzene-1,4-dithiol as about 22 MΩ [7.23]. One should note that resistance measurements using metal–molecule–metal junctions can be greatly affected by contact properties [7.47, 7.48].

Oligomers with linearly conjugated double and triple bonds are also typical candidates for conductive molecules. The effective conjugation length for poly(triacetylenes) was determined as being in the range 7–10 monomer units, based on optical and non-linear optical measurements, UV/Vis and Raman spectroscopies [7.49, 7.50]. There have been extensive studies on using phenylene-ethynylene oligomer as molecular wires, substituted with p-active groups [7.51]. In addition, molecules with ferrocene as terminal groups were also studied for oligo(phenylethynyl) [7.52] and oligo(phenylvinylene) [7.53]. The ferrocene was connected to the gold substrate by means of spacer molecules of different lengths. The attenuation factor was obtained from the dependence of the standard electron-transfer rate constant on the length of the spacer molecule, by using ac voltammetry. In this study, the attenuation factors for aliphatic spacers and conjugated spacers were determined as 0.90 and 0.36 Å^{-1}, respectively [7.52]. The results are consistent with those reported for other methods, as introduced in the prceding section.

7.2.3 Rectification Molecular Conductance

The concept of electronic rectification of molecules was first explored with a conjectured donor–spacer–acceptor model (D-S-A) in 1974, and has been extensively tested in molecular wire structures [7.54]. Even though the original molecular structure, the donor (TTF) and acceptor (TCNQ) separated by a σ part, has yet to be realized, these stimulating studies can be considered within a broader scope of research on intramolecular electron transfer processes.

Several linear molecular structures of D-S-A have been investigated as conductive wires. These molecules provide ideal examples to test electron transfer formalisms. A number of wire structures manifested appreciable rectification behavior, and could be utilized for transistors.

A well-documented molecular structure with rectification behavior is g-hexadecylquinolinium tricyanoquinomethanide ($C_{16}H_{33}Q$-3CNQ) (Fig. 7.4) [7.55–7.57]. Characteristic rectification behavior was also obtained from the LB film of a donor-π-acceptor molecule, thioacetic acid S-(10-4-[2-cyano-2-(4-dicyanomethylene-cyclohexa-2,5-dienylidene)-4H-quinolin-1-yl]decyl ester [7.58]. A more complex rectification behavior was reported for this molecule [7.59]. Several other types of molecules have been reported with observable rectification properties, including zwitterions [7.60–7.62], hemicyanine derivatives [7.63–7.65], and chevron-shaped dye [7.66, 7.67].

Rectification effect was observed for planar binuclear phthalocyanine Co(I)Pc-Co(III)Pc. The molecular monolayer was self-assembled on a graphite surface with a titled orientation. The Co(I)Pc is the donor part, and Co(III)Pc the acceptor part [7.68].

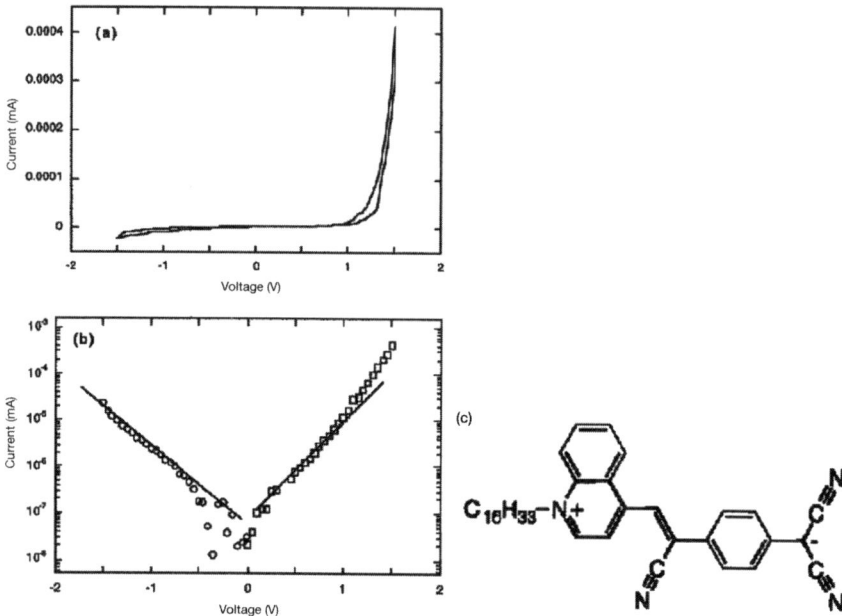

Fig. 7.4. Schematic and I-V curves of $C_{16}H_{33}$Q-3CNQ. **a** Linear plot and **b** Logarithmic plot **c** Structure schematic of $C_{16}H_{33}$Q-3CNQ (extracted from [7.55])

Rectification effect was also observed for the asymmetrically substituted phthalocyanine molecules (NtBuPc) [7.69]. In this molecule, the donor t-butyl and acceptor NO_2 are separated by the π-conjugated molecule phthalocyanine.

The mechanisms of the observed rectification behavior are currently under extensive study. In the case of $C_{16}H_{33}$Q-3CNQ [7.70], the rectification was proposed as electric field-induced excitation of the ground state and electron transfer across the molecule–electrode interfaces. Asymmetric molecular geometry was also suggested to contribute to the rectification effect [7.71, 7.72]. Considering that the Q-3CNQ part is close to one electrode and the alkyl part is close to the other electrode, the asymmetric electrostatic field distribution along the molecule may induce an asymmetric transport behavior.

Another approach using similarly asymmetric molecules with only one conjugated group (thiophene, phenyl) also led to rectification effect, attributed to the resonant tunneling of the molecular front orbitals of the conjugated group [7.73].

In another study, asymmetric distribution of electronic density of states of copper phthalocyanine (CuPc) molecules was considered the cause for rectification behavior [7.74]. The CuPc molecules were adsorbed on acidified graphite surfaces in a tilted orientation and displayed an asymmetry ratio as high as 40 in I-V characteristics, which is much higher than the typical value for clean graphite surfaces. The presence of peak features in the occupied density of state distribution around the Fermi level has been argued to be the most likely explanation for the observed large asymmetry in *I-V* characteristics, excluding factors such as the barrier height effect, Schottky-Mott-type interface between molecule and substrate, electrical break-down, electrical dipole effect, and asymmetric geometry of tip and sample.

7.2.4 Switching Behavior of Molecular Conductance

Nanopore experiments [7.75, 7.76] have indicated that the nitroamino compound has two distinct conduction states (conducting and non-conducting), the so-called negative differential resistance (NDR) effect. The switching behavior is associated with the ensemble of approximately 1,000 molecules in the nanopore junction region. The observed NDR phenomena has been attributed to the combined effect of electron charging and resonant tunneling of substituents [7.76]. It was also suggested that the charged state of dinitro compounds, 2'-nitro-4-ethynylphenyl-4'ethynylphenyl-5'-nitro-1-benzene thiolate, can be considered as a prototype of molecular switch [7.77].

Another study on the complex of 3-nitrobenzal malononitril and 1,4-phenylenediamine (NBMN-pDA), p-nitrobenzonitrile (PNBN) demonstrated distinctly different conduction states that could be controlled by the electric field under the STM tip [7.77]. The tip-induced conductance change appears in localized spots with diameters as small as 0.6 nm, which is very attractive for high-density data storage applications. Possible effects due to localized polymerization and electric field-induced molecular structural change have been discussed.

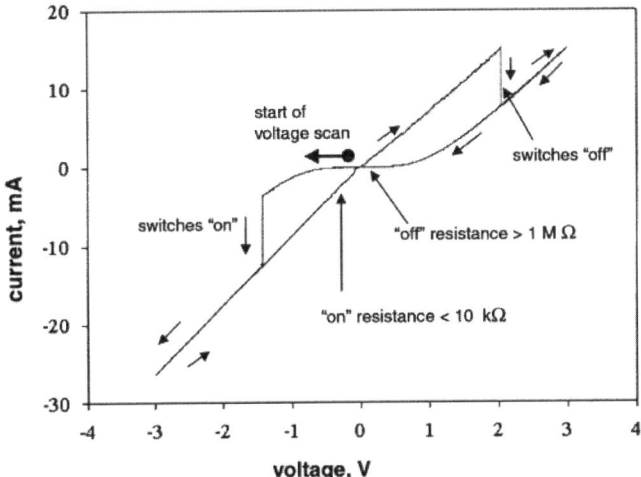

Fig. 7.5. Bistable conductance of nitroazobenzene junction with thickness of approximately 47 Å and contact area of about 0.8 mm^2 (extracted from [7.78])

170   7 Electrical Conductivity of Single Molecules

Switching behavior was also reported for a multilayered nitroazobenzene assembly formed between a pyrolyzed photoresistant film (a graphitic material similar to glassy carbon) and a mercury electrode, as shown in Fig. 7.5 [7.78]. The nitroazobenzene molecules are connected by conjugated C–C bonds to the carbon substrate through electrochemical reaction of diazonium salt. The low conductance state was seen to switch into a high conductance state at bias voltage less than -1.5 V (graphite relative to mercury electrode), and the low conductance state can be regained at voltages higher than 2 V. This was attributed to a possible electron-induced redox process in which the phenyl ring is rotated and forms a planar quinoid structure, similarly to the effect observed by STM [7.79]. Similar bistable conductance behavior was also observed for monolayer structures [7.80].

7.3 Single Molecule Conductance

The determination of molecular conductivity at single molecule level is a challenging experimental task. The use of the SPM technique has greatly enhanced the capability of studying transportation properties at single molecule level. Reported results on single molecules not only reconfirm those obtained for monolayer molecules, but also provide new insights into the nature of electron conduction across molecular junctions, such as possible coupling effects in electron transportation along molecules.

7.3.1 Molecule–Electrode Contact Effect

Among many of the important issues concerning electron transfer mechanisms [7.47, 7.60, 7.81] via organic molecules, the effect of electrical contact between the molecule and electrode surface is one of the most extensively studied aspects [7.82]. From experimental results focused on thiolated conjugated molecules (with aliphatic or aromatic moieties), with strong chemical bonds to metal electrodes, much information has been gained on the conductance of single molecules. The effect of terminal atoms and electrode metal combinations upon the measured conductance has been assessed theoretically [7.47], and the conductance was conjectured substantially higher when the molecular wire is terminated by selenium rather than sulfur or oxygen on gold or silver electrodes (Fig. 7.6). In addition, the analysis suggests that Ag electrodes lead to reduced conductance compared to that of Au electrodes. Direct comparisons of CN- and S-groups on metals (Ni, Cu, Pd, Ag, or Au) have also made with quantum chemistry simulations [7.81].

First-principle calculations of benzene-1,4-dithiol with metallic contacts revealed that the atomic connection between the molecule and the electrode is critical to the overall magnitude of the measured conductance [7.61]. The usage of a single Au atom as contact can reduce the conductance by 2 orders of magnitude. The argument is based on bonding between the s states of Au and p states of sulfur. The symmetry promoted coupling between the surface

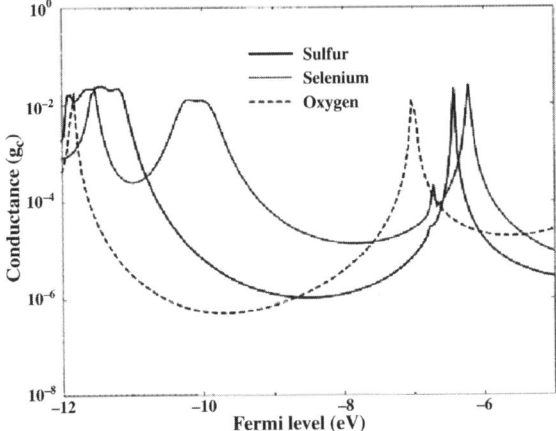

Fig. 7.6. Calculated conductance of the molecules terminated by selenium, sulfur or oxygen atoms (extracted from [7.47])

normal-oriented p states and the s state. As a result, π-bonding interaction will be reduced. By contrast, an aluminum atom at the contact point can enhance the conductance by contributing p states to the coupling. It should be noted that the calculations showed similar I-V characteristic shapes for different metal contacts, indicative of the dominance of molecular electronic structures in determinig the line shape of tunneling characteristics.

With the aid of atomistic manipulation, a width-adjustable junction made of Au atoms was demonstrated on NiAl(110) surfaces [7.83]. Each electrode consisted of 1–6 Au atoms. The single copper(II) phthalocyanine (CuPc) molecules were brought into the junctions made of different Au chain electrode lengths. The interaction between the positively charged Au atoms on NiAl(110) surfaces and the benzene group of CuPc helps stabilize the assembly of the Au atom chain/CuPc/Au atom chain. The corresponding tunneling characteristics suggested the degeneracy effect of molecular LUMO orbitals of $(e_g\pi^*)$. The number of Au atoms within each electrode does indeed cause additional splitting of the molecular orbitals. This result provides a direct atomistic visualization of the metal–molecule–metal contact, and evidence of the electronic coupling between the contact atom and the molecule.

The enhanced resistance of thiol molecules was speculated for non-bonded molecules, compared with chemically bonded species, as revealed in earlier SPM studies on thiol molecules assembled on Au surfaces [7.36]. Fluctuating conductance of single thiol molecules immobilized on Au(111) surfaces through coadsorbed SAM was observed and interpreted as a switching between "on" and "off" states [7.62]. The rate of occurence of such fluctuations was found to follow an exponential distribution, associated with the bonding characteristics of the thiol terminal group with the Au substrate. The "on" and "off" states are likely due to the stabilized bonding and de-stabilized

172   7 Electrical Conductivity of Single Molecules

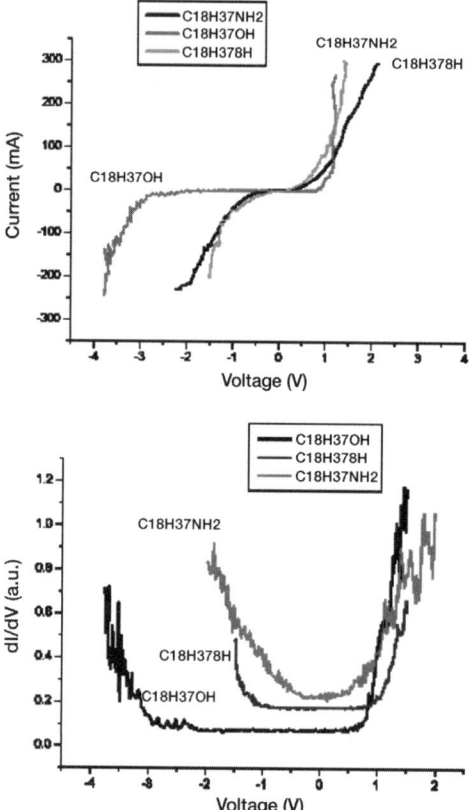

Fig. 7.7. Electrical conductance of the tunnel junctions using tin as electrode **a** Measured I-V curves for octadecane derivatives. **b** Measured dI/dV versus V characteristics showing the gap regions for octadecane derivatives. A pronounced offset of Fermi level can be identified for $C_{18}H_{37}OH$

bonding, respectively. From the variation of the rate distribution at 25 and 60 °C, an activation energy of about 0.1 eV for bond destabilization was estimated.

Recent studies were able to obtain the tunneling characteristics of alkane derivatives formed on indium oxide and tin oxide surfaces (Fig. 7.7) [7.29], indicating possible effects of functional groups on the tunneling behavior, as theoretically predicated. In another study, temperature-induced resistance enhancement was observed for alkyls assembled on oxide surfaces. The effect was ascribed to the possible breaking of hydrogen bonds between the terminal groups of SAM and the oxide surface [7.84].

Along with achievements in atomic manipulation and nanofabrications, considerable interest, both technologically and theoretically, has been focused on the quantized conductance behavior of nanowires or nanocontacts produced by STM. The experimental results have provided direct evidence of the dimensional restriction on the electron density of states. A zero- or one-dimensional contact is defined as two electrodes connected by a single atom or a linear chain of atoms, respectively. A zero-dimensional contact can be readily formed by the attachment of an individual atom to the tip [7.85, 7.86], or the sample surface, whereas a one-dimensional contact can be formed by mechanical deformation of the STM tip from its contact state [7.87, 7.88]. Closely related studies on the deformation of the nanocontact [7.89, 7.90] and electron scattering [7.91, 7.92] have helped to clarify the observed quantized conductance.

It should be noted that other types of quantizations in tunneling conductance are also very interesting. The study of the single electron tunneling process has been very active for several decades, and continued with refreshing new results. As early as 1965, Giaever et al. [7.93] discovered the Coulomb blockade effect unique to the single electron tunneling process. This effect is based on the electronic charging energy associated to finite capacitance of the tunneling junction. The tunneling electron needs to possess a certain energy to overcome this charging energy barrier. Qualitatively, the Coulumb blockade effect is reflected in step-like features in I-V characteristics, or periodic peak structures in dI/dV versus V curves [7.93–7.95]. Early experiments were performed with microparticles of metal oxide embedded in the tunneling junction. As a result, the tunneling junction can be approximated by a double junction model. Recent studies have revealed similar characteristics in tunneling junctions in a wide range of nanostructures, including various nanoparticles, and molecules [7.96].

Resistance of chemically bonded octanedithiol has been determined as being 900 ± 50 MΩ, whereas non-bonded molecules may have resistance at least 4 orders of magnitude higher [7.36]. This value has been revised to 51 ± 5 MΩ in a recent measurement using molecular junctions formed between gold contacts under STM [7.97]. In this study, the contact between the tip and sample surface was continuously varied while monitoring the conductance. The final stage prior to the complete breaking apart of the tip–sample contact consists of discrete breaking events of single chains of Au atoms. The difference in these results may come from the finite contact resistance in association with the AFM probe used in [7.36]. In the same study, the single molecule conductance of 4,4'-bipyridine was determined as 1.3 ± 0.1 MΩ [7.97]. For molecules containing phenylene groups, the values are higher (for example, the resistance of benzene-1,4-dithiol is about 22 MΩ [7.23]).

The evolving behavior of conductance for single atom contact between electrodes (Pb, Al, Nb, Au) revealed a dependence on the chemical effect of the contact atom [7.98]. By analyzing the step-like conductance variation function, and the accompanying I-V characteristics, a qualitative correlation was proposed between the number of conduction channels and the number of valence electrons of the contact atom.

7.3.2 Conductance of Single Organic Molecules

The magnetic field-induced splitting of zero-bias conductance peaks at low temperatures was also observed in the single molecule junction of a divanadium molecule (V_2, [(N,N′,N″-trimethyl-1,4,7-triazacyclononane)$_2$-V$_2$(CN)$_4$(μ-C$_4$V$_4$]) [7.99]. The inclusion of vanadium atoms in the molecule resulted in three different charge and spin states, namely, neutral V_2^0 has spin $S=0$, V_2^- is a spin-quadruplet ($S=3/2$), and V_2^+ a spin-doublet ($S=1/2$). According to the mechanism of Kondo effect, the zero-bias conductance peak follows the expression:

$$G_k(T) = \frac{G_0}{(1 + (2^{1/s} - 1)\frac{T^2}{T_K^2})^s} \quad (7.2)$$

G_0 is a constant. T_K (Kondo temperature) and s are determined from the coupling strength of the tunneling electron with the electrode, and the energy level of the scattering center.

Single molecules containing Co ions and polypyridyl ligand molecules revealed step-like features in I-V characteristics, which is reminiscent of the Coulomb blockade effect for the single electron tunneling process [7.100]. The conductance peak magnitude at zero bias also showed logarithmic dependence on temperature and splitting under an external magnetic field, as a result of the Kondo scattering effect associated with Co ions.

Conductance of single H_2 molecules was measured using a mechanical break point junction at 4.2 K [7.101]. The measured conductance has a peak value at $G_0 = 2e^2/h$, which is strong evidence that the tunneling occurred mainly through a single channel. The differential conductance for the Pt/H$_2$/Pt junction revealed a resonance feature at 63.5 meV, identified as the vibrational energy for H$_2$ molecules (Fig. 7.8).

Single C_{60} molecules located within a break junction have been shown to display discrete conductance gaps under gate voltage controls at 1.5 K (Fig. 7.9) [7.102]. The observed 5-meV gap value was ascribed to the excitation of mechanical oscillation of C_{60} molecules within the junction. The vibrational quantum was determined from the basic vibration frequency of a C_{60} molecule interacting with a metal substrate via van der Waals potential (which is approximatly 1.2 THz). Conformation change of fullerene molecules can lead to substantially different response in electronic behavior, such as for C_{60} under deformation by an STM tip [7.103], or by controlling the gate potential in a three-terminal geometry [7.104]. A discernible negative differential resistance (NDR) effect was observed in a junction formed by two C_{60} molecules, due to the local density distribution at the Fermi level of C_{60} molecules (Fig. 7.10) [7.105].

An effective method was developed to immobilize single molecules of conjugated phenylene ethynylene oligomer with nitroamino groups using alkanethiolate assemblies as the host matrices [7.106]. Switching behavior in conductance is reflected in the change of apparent height of the oligomers in STM

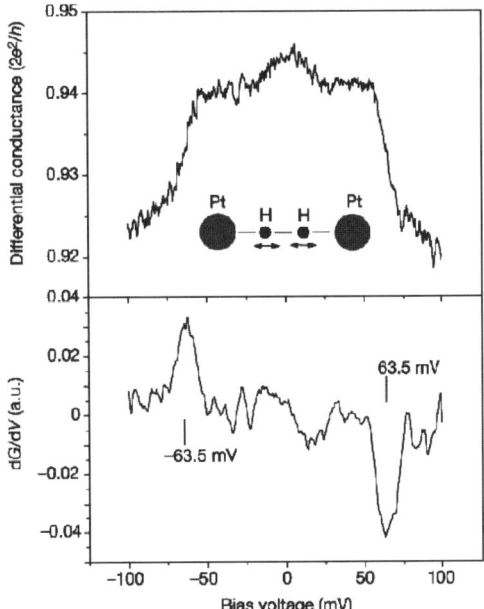

Fig. 7.8. Differential conductance for the Pt/H$_2$/Pt junction showing a resonance feature at 63.5 meV (extracted from [7.101])

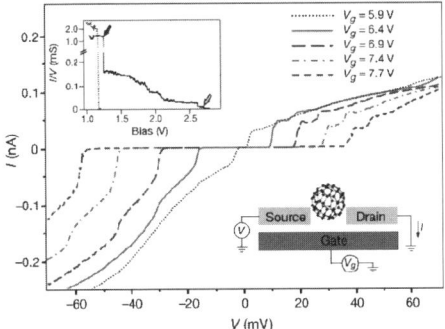

Fig. 7.9. *I-V* characteristics of single C$_{60}$ molecules at various gate voltages. A change in the apparent gap width in multiples of 5 mV can be identified (extracted from [7.102])

images. This change in conductance is due to the electron-induced conformation change of the nitroamino groups, similarly to that observed when using the nanopore method [7.75]. It was found that tunneling conditions (bias voltage and current) do not affect the switching efficiency, and the switching efficiency is dependent on the defect density and film quality of the host matrices.

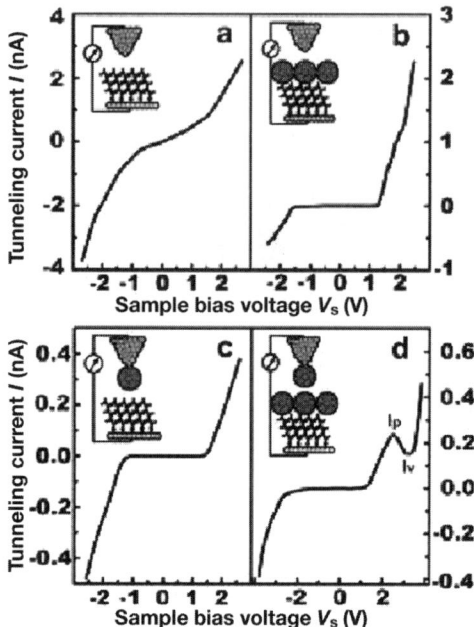

Fig. 7.10. Negative differential resistance of C$_{60}$ (extracted from [7.105])

7.3.3 Conductance of Single Nanotubes and Nanowires

Carbon nanotubes (CNTs) can be approximated as graphitic sheets with hexagonal lattices that are wrapped up into seamless cylinders (Fig. 7.11) [7.107, 7.108]. Both single-walled carbon nanotubes (SWCNT) and multi-walled carbon nanotubes (MWCNT) have been examined for electron and hole transportation. CNT-based transistors have been extensively studied and demonstrated many excellent properties. Transistor arrays and logic gates using CNTs show promising application potential (Figs. 7.12 and 7.13) [7.109].

Substitutional doping of boron and nitrogen in multiwalled carbon nanotubes has been shown to change the electric conductive behavior from semiconducting to metallic [7.110]. The inclusion of alkali and halogen in single-walled carbon nanotubes enhances electrical conductivity [7.111, 7.112]. Hybrid carbon nanotubes have been shown to possess excellent rectification behavior [7.113]. The conductance of SWCNTs was found to be affected by the chemical environment, such as the presence of NH$_3$ and NO$_2$ [7.114], which could be applicable to chemical sensing.

The conductance of boron-doped silicon nanowires (SiNW) was also shown to be sensitive to the chemical environment, as a result of surface charge variations. Such effects could be explored in developing novel chemical and biological sensors [7.115].

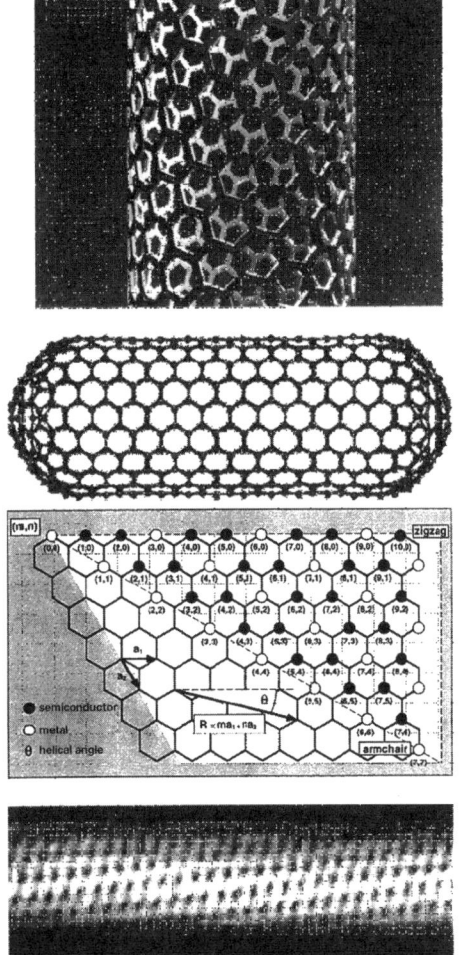

Fig. 7.11. Structural models of carbon nanotubes (extracted from [7.107])

7.3.4 DNA Molecules

Attempts to use DNA as an electronic conductive wire have been widely pursued in the past decade. This is rooted in the well-defined, one-dimensional helical structure of DNA that is composed of uniformly spaced base pairs rich in p-carriers. The distance range for electron transfer was proposed to be as long as 40 Å, with the aid of covalently linked donor and acceptor intercalators [7.116]. A number of mechanisms have been investigated to elucidate charge transfer pathways under various conditions [7.117, 7.118].

On-going, extensive exploration will help to clarify the nature of charge migration processes for DNA, in terms of, for example, intercalators, and se-

178   7 Electrical Conductivity of Single Molecules

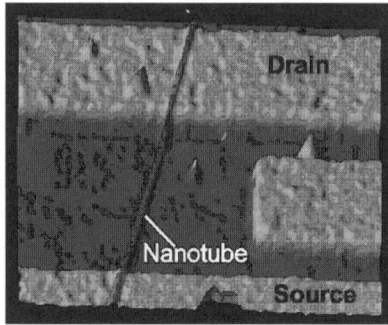

Fig. 7.12. Conductivity measurements of a single carbon nanotube between source and drain electrodes (extracted from [7.108])

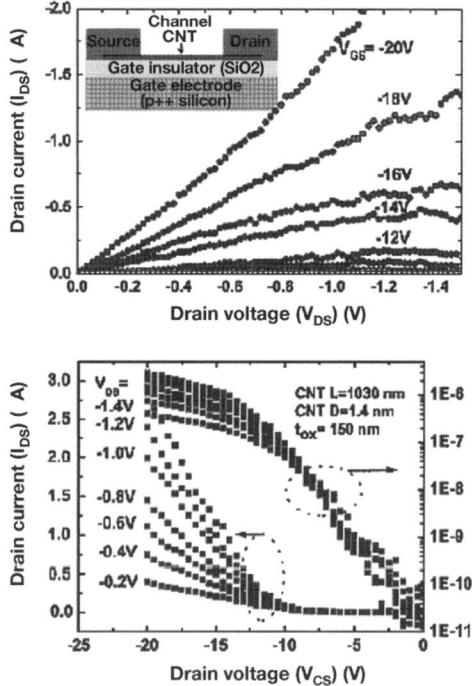

Fig. 7.13. *I-V* characteristics of a carbon nanotube under the influence of gate voltage (extracted from [7.108])

quence. [7.115–7.121]. Conductance measurements on poly(dA)-poly(dT) and poly(dG)-poly(dC) revealed similar behavior that can be interpreted on the basis of the polaron hopping mechanism [7.122]. The gated *I-V* characteristics further suggest that the conduction in poly(dA)-poly(dT) is reminiscent of n-type semiconductors, whereas poly(dG)-poly(dC) behaves more like p-type semiconductors. Other models proposed for interpreting DNA-related conductance include ionic conduction [7.123], polaron transportation [7.124], soliton [7.125], and electron or hole transportation [7.126].

The high resistivity of λ-DNA molecules was confirmed by first-principle calculations on the electronic structures. The results also suggest that the low-energy electron induces significant change in conductivity as a result of contamination [7.127].

7.3.5 Single Molecule Devices

It has been an inspiring field to design and build molecular devices based on a wide range of chemical and physical properties of various molecules. Molecular electronic elements, potentially leading to devices at single molecule level, require synthesizing molecules with electrical functionalities and connecting these to external electrodes. Earlier explorations proposed that a single molecule with a donor–spacer–acceptor structure could develop rectification properties and behave as a diode [7.1, 7.54]. Such components could have wide applications in fast switches, oscillators and frequency-locking circuits. Since then, a variety of molecules of such types have been synthesized, and appropriate detection techniques have been developed.

The molecular wire structure was expanded to branched structures of three-terminal and four-terminal systems intended for molecular transistor designs (Fig. 7.14) [7.128].

Monolayers of rotaxanes [7.129] and [2]-cantenane displayed bistable behavior [7.130]. The behavior led to the cross-bar design of current-driven solid state molecular logic circuits and addressable memory [7.130–7.134]. It was further pointed out that the molecular domains, rather than individual molecules, are responsible for the observed switching behavior [7.134].

It has been proposed that intramolecular electron transportation could be used to develop logic gate functions [7.135]. Based on elastic scattering quantum chemistry analysis, multiple-arm molecular structures, incorporating donor and acceptor groups, could develop similar logic functions such as "OR" and "AND" (Figs. 7.15 and 7.16) [7.136, 7.137].

Several circuit structures with logic functions of inverter, "NOR", SRAM, and ring oscillators were demonstrated based on individual carbon nanotubes

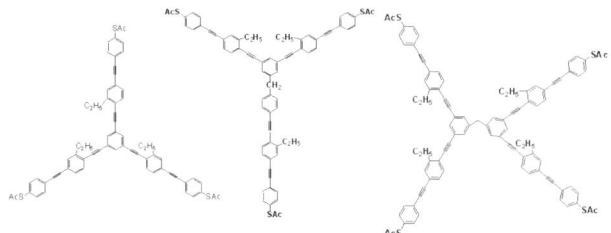

Fig. 7.14. Branched molecular wire structure (extracted from [7.128b])

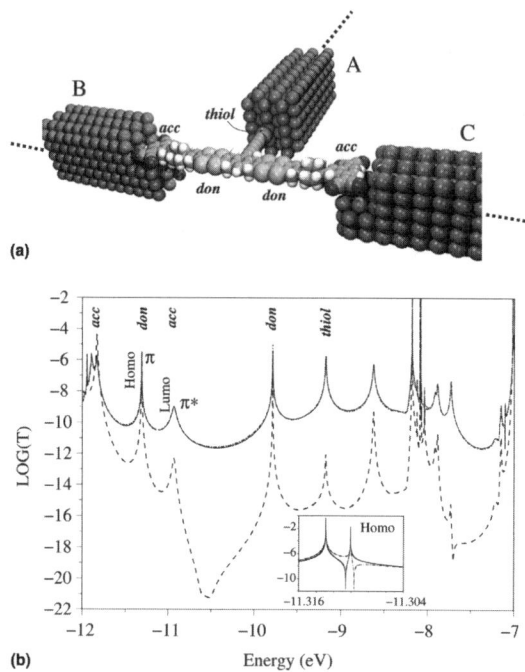

Fig. 7.15. Proposed intramolecular circuit structure (extracted from [7.136])

[7.138]. The structures were built on silicon oxide surfaces by electron beam lithography. An aluminum film was used as gate electrode. The Fermi level of the nanotube can be tuned between the valence band to gap, and conduction band by adjusting the gate voltage, covering the whole doping regime of p-type to n-type.

Intramolecular logic gates have been demonstrated based on single CNT doped at separate sections [7.139]. By using nanowires of p-type silicon and n-type gallium nitride, circuitries with various logic functions can be achieved [7.140]. The p–n diode behavior was derived from the crossed single nanowires. The basic computation processes have also been demonstrated based on the nanowire-derived logic gates.

The nature of molecular conductivity has been gaining increasing attention. It has evolved as a multidisciplinary topic that sees effort from synthetic chemistry, physical chemistry, solid state physics, and so on. It is evident that even though conceptual aspects have been explored for about three decades, we are still at an early stage in our attempts to fully understand the mechanisms governing the process of electric conduction across molecules. More rigorous studies are needed to further advance our knowledge in this important field.

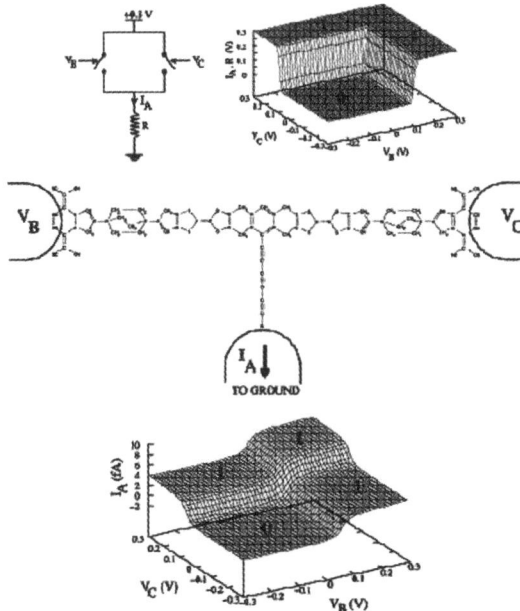

Fig. 7.16. Logic function of the intermolecular circuit (extracted from [7.136])

8 Single Molecule Fluorescence Imaging and Spectroscopy: Far-Field Studies

8.1 Introduction

8.1.1 Fluorescence of Molecules

The preceding chapters have focused on single molecule studies based exclusively on electrical and mechanical properties. Since optical characteristics are vital to the understanding of single molecule properties, we focus the following chapters on the optical properties of single molecules. These data represent fundamental aspects in our understanding of single molecules. This section provides a brief introduction of the principles of molecular fluorescence, and the main factors that affect the fluorescence of molecules.

The fluorescence of molecules is generated by the transition from an excited state to a ground state. The excitation is achieved by absorbing exciting photons, typically on the order of 10^{-15} second. The electron spin in the singlet excited state is opposite to that in the ground state. Therefore, the transition from the singlet excited state to the ground state does not require spin adjustment. The excited electronic state consists of vibrational energy levels with much smaller energy spacings. Once in the excited singlet state, the molecule will relax to the vibrational and rotational equilibrated state in approx 10^{-12} second, followed by a transition back to the singlet ground state (typically in 10^{-9} second), as illustrated in Fig. 8.1 [8.1, 8.2].

Conversion from the singlet excited state to the first triplet state is a slow process (known as intersystem crossing, ISC, that accompanies the fluorescence-generating transition between singlet states. The electron spin in the triplet state is parallel to that in the singlet ground state. Therefore, a transition from the triplet state to the ground state is forbidden. Such emission is known as phosphorescent, and the lifetime is typically in the range of milliseconds to seconds.

In brief, the fluorescence emission of molecules can be described as a four-step process [8.3]:

(1) Electronic excitation from the ground state to an excited singlet state
(2) Internal relaxation in the excited state
(3) Radiative decay from the excited state to the ground state
(4) Internal relaxation in the ground state

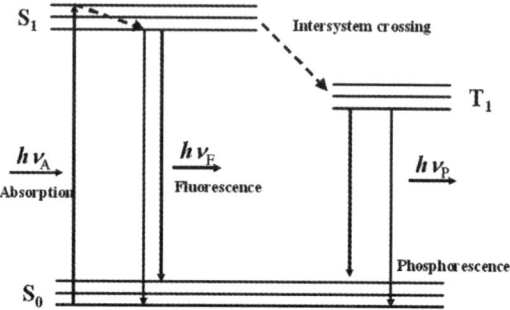

Fig. 8.1. Energy diagram for fluorescence and phosphorescence processes

For both biological and organic species, the fluorophores can sustain only limited irradiation before irreversible structural changes occur that lead to complete loss of fluorescence. There are several photo-induced processes known to cause inactivation of a fluorophore [8.3–8.5], including:

- Photooxidation
- Photoionization
- Photodissociation
- Photoisomerization
- Photochemical transformation of fluorophores by multiphoton process.

In addition, the intersystem crossing (ISC) between the excited singlet state to the triplet state could also represent a dark state in fluorescence. A typical dye molecule could emit 10^5–10^6 photons before photobleaching. Considering the detection efficiency of about 1–5%, there would be 1,000 to 50,000 photons available for single fluorophore molecule spectroscopy. With an improved collection speed of about 500,000 datapoints per second, one could follow the real-time trajectories of fluorescence intensity of single fluorophores.

In the photooxidation process, the fluorophore in the triplet state reacts with ground-state molecular oxygen through a triplet–triplet annihilation process, generating singlet oxygen and ground-state fluorophores. The ground-state fluorophores can be further irreversibly oxidized by the singlet oxygen.

Another mechanism of photobleaching is multiphoton absorption of the long-lifetime triplet state, in which the triplet state becomes highly reactive and irreversible photochemistry may occur. The photon antibunching effect can be explained by the fact that the photon emissions of single fluorophores should be separated by the recovery time of the ground state (on the order of nanoseconds). The effect can be measured by using a dual-beam setup known as the Hanbury-Brown-Twiss correlator [8.6], and characterized by a dip in the autocorrelation function of the fluorescence intensity [8.7, 8.8]. The effect has been demonstrated at low temperature [8.9], room temperature [8.10, 8.11], as well as in solutions [8.12].

A generally adopted method to reduce the effect of photobleaching is by adding reduction agents to remove oxygen and hydroxyl radicals. Typical agents include mercaptoethanol, or a mixture of catalase and glucose oxidase.

The fluorescence behavior is also known to depend on excitation power. Three characteristic regimes can be typically observed [8.3]. The absorption time is the dominant factor at low excitation power, and the fluorescence intensity is correlated linearly with the excitation intensity. At intermediate excitation intensity, the lifetimes for absorption and excited state are comparable. As a result, saturation of excited states will occur, the fluorescence showing only weak dependence on the excitation intensity. At even higher laser intensity, the excited state lifetime is dominant and the fluorescence intensity is independent of the excitation power. The experimental temperature may affect the saturation behavior of single molecules by coupling with the local phonon mode, such as for the broadening of linewidth at increased temperatures for single pentacence molecules in para-terphenyl [8.13].

A typical experimental spectrofluorometer setup for studying fluorescence includes a light source such as a xenon arc lamp or mercury lamp, polarizers, excitation and emission monochromators, and photomultiplier tubes (PMTs).

8.1.2 General Considerations for Experimental Setup

One of the criteria to observe single molecule fluorescence is the excitation and collection efficiencies. In the confocal experiment configuration, the detection volume (for both solid and liquid media) should be appropriate to contain single molecules at experimental concentrations. A typical detection volume is on the order of 1 fl or 1 μm^3. The excitation beam needs to be highly focused to achieve such small detection volumes.

By contrast, a completely different set of experimental conditions are required for the wide-field experiments. Larger detection volumes have been used with wide-field epi-illumination or total-internal-reflection (TIR) combined with charge-coupled device (CCD) detection. This approach can detect large numbers of mobile molecules simultaneously.

Photon emission during transition from the excited singlet state to ground state is characterized by the radiative rate k_r. The non-radiative rate k_{nr} describes the energy dissipation due to collisional quenching. The fluorescence lifetime is then determined by [8.14]:

$$\tau = (k_r + k_{nr})^{-1}$$

Fluorescence excitation can be achieved mainly by four approaches:

(1) Wide-field epifluorescence illumination
(2) Total internal excitation
(3) Confocal excitation
(4) Near-field excitation.

Common to both confocal and wide-field configurations, minimizing background interference is an important condition for obtaining high-quality results. Main contributing factors to the background noise include [8.15]:

(1) Dark count of photodetector
(2) Autofluorescence of optics
(3) Impurities in the matrix or solvent
(4) Fluctuation of the detected signal
(5) Scattering from the matrix or solvent.

Several types of detectors have been used effectively for single molecule studies:

(1) Back-thinned charge-coupled device (CCD) with quantum efficiency (QE) up to 90%
(2) Intensified CCD, with higher frame rate but reduced QE ($\leq 40\%$)
(3) Electron-multiplying CCD with single molecule sensitivity.

The signal-to-noise ratio for single molecule fluorescence measurements is influenced by a number of factors [8.16]:

$$S/N = \frac{\sqrt{S}}{1 + \sqrt{S'/S}}$$

where S is the number of detected photons, and S' the number of detected background photons:

$$S = \eta_c \eta \sigma N$$
$$S' = \alpha V N$$
$$\sigma \propto |d \cdot n|^2$$

where d is the absorption dipole moment, n the polarization of the incident beam field, η_c collection efficiency, η fluorescence quantum yield, σ absorption cross section, N the number of incident photons. Parameter α depends on the matrix properties and collection efficiency.

The orientation of molecular dipoles will affect both the absorption cross section and the collection efficiency. In addition, it was pointed out that for large numerical apertures, higher collection efficiency should be expected for studying molecules adjacent to surface or interfaces.

Both absorption and emission spectra can be used to detect single molecules. However, because of low signal-noise ratios, many studies have adopted emission spectra as the main experimental approach. It should be noted that the absorption resolution can be enhanced in the near-field situation [8.17].

In brief, the general conditions for performing single molecule analysis are:

(1) Small excitation volume that contains only one molecule (femtoliters for confocal microscopy, picoliters for flow experiments)

(2) High-efficiency collection optics
(3) High signal-noise ratio of detectors, such as avalanche photodiodes (APD), photomultiplier tubes (PMT), and high-sensitivity CCDs.

8.1.3 Criteria of Single Molecule Identification

Optical characterizations of single molecules by fluorescence, Raman and near-field scanning optical microscopy (NSOM) have been intensively pursued. The endeavor is beneficial to reveal inhomogeneities and dynamics at single molecule level in various environments, such as low temperatures, solutions, and polymer films. The results could serve to develop the conceptual exploration of using single molecules as sensors in nano-environments. This is of particular interest in applications dealing with single biomolecules in solutions.

Fluorescence imaging of single fluorophores have led to detailed information about local environments, such as [8.3–8.5, 8.17–8.19]:

(1) Better knowledge of property distributions in both homogeneous and inhomogeneous systems
(2) Dynamic and statistical information of individual molecules
(3) Possibility to study new effects and intermediate species.

This information can provide important insight into the environment in the immediate vicinity of single molecules, which can not be obtained from ensemble averaged studies. There are several criteria adopted for single molecule detections:

(1) The occurrence of fluorescence signal is proportional to the fluorophore concentration whereas the signal intensity remains unchanged
(2) Photobleaching should occur with characteristic "on" and "off" states
(3) Observation of spectral jumps
(4) Saturation of fluorescence intensity by excitation laser
(5) The overall number of detected photons should coincide with that estimated for a single fluorophore
(6) Antibunching effect from correlated fluorescence spectra.

It has been proposed that single molecule species can be differentiated from their fluorescence lifetime decays in time-correlated single photon counting measurements [8.20]. By combining the time-correlated single photon counting (TCSPC) method and fluorescence correlation spectroscopy (FSC), one is better equipped to identify the fluorophore properties of:

– Absorption spectrum
– Fluorescence spectrum
– Fluorescence quantum yield
– Fluorescence lifetime
– Anisotropy.

Such multiparameter experiments on single molecules are very helpful in differentiating different fluorescence molecules [8.20].

8.2 Single Molecule Imaging in Far-Field Configuration

8.2.1 Imaging by Confocal Fluorescence Microscopy

A confocal microscope uses the same optics for excitation beam focusing and fluorescence detection in reverse pathway [8.21]. A general feature in confocal imaging is the placement of a small pinhole in the illumination as well as the reflected beams. The pinholes are designed to enable imaging of the focal plane of the imaging objective. As shown in Fig. 8.2, the excitation beam is focused onto the sample surface after a dichroic mirror and focusing lens. The probe area at the sample surface is restricted by the diffraction limit, with an approximate diameter of 1 μm and a height of around 2 μm, by spherical aberration of the optical lens. The excitation and detection volumes (typically about 1 fl) are defined at the focal point of the focusing lens. The emitted fluorescence photons pass through the same lens and the dichroic mirror, followed by a pinhole (around 50–100 μm in diameter) and filter, and are eventually collected by a photomultiplier detector (PMD).

Confocal microscopy can be operated to study fluorescence spectroscopy at fixed positions, as well as in large areas in the scanning mode. The application in fluorescence spectroscopy studies is presented in the following sections.

8.2.2 Wide-Field Imaging: Epi-Illumination Microscopy

The imaging capability of a camera is limited by the readout time in the order of kHz in frequency, and optical limitation to a few tens of nanometers in spatial resolution. With a large numerical aperture (NA = $n \sin q$, q is the collection half-angle, and n the refraction index of working medium) of the microscope lens objective, the lateral viewing field is typically around several hundred micrometers and the depth of focus less than a few micrometers.

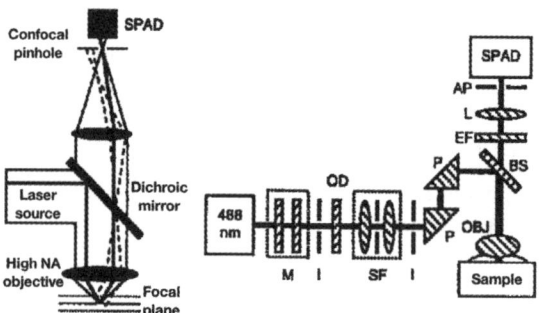

Fig. 8.2. Schematic of confocal fluorescence microscope. Details are provided in the text (extracted from [8.21])

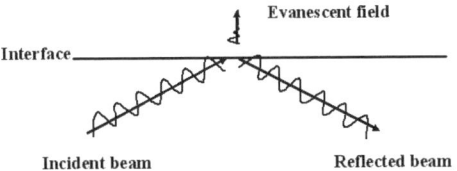

Fig. 8.3. Schematic of total internal reflection configurations

There are three types of excitation configurations in wide-field microscopy:

(1) Epi-illumination
(2) External reflection
(3) Total internal reflection (TIR) illumination

In the total internal reflection scheme (Fig. 8.3), an evanescent field is characterized by exponential decay in intensity [8.22]:

$$I = I_0 \exp\left[-\frac{4\pi n z}{\lambda}\left(\left(\frac{\sin\theta}{\sin\theta_0}\right)^2 - 1\right)^{1/2}\right]$$

where I_0 is the laser beam intensity at the interface, λ the wavelength under vacuum, n the refractive index of the optically dense medium, z the distance from the interface, and θ and θ_0 the beam incident angle and critical angle, respectively. The penetration depth of the evanescent field in TIR configuration is restricted to about a few hundred nanometers. Therefore, the excitation volume will be confined to the vicinity of the liquid–solid interface. This property is highly beneficial to single molecule analysis.

Recently, evanescent excitation by means of a metal-clad configuration of a nanometer-sized metal waveguide was demonstrated [8.23]. The effective detection volume can be as small as zeptoliter (10^{-21} l), and excitation is due to the evanescent light within the nanometer-sized pores (so-called zero-mode waveguide). Such detection assays could enable single molecule studies at relatively high concentrations.

8.3 Low-Temperature Studies of Single Molecules in Solid Matrices

8.3.1 Observation of Single Molecules in Crystalline Matrix

The first evidence of single molecule fluorescence was demonstrated at low temperatures using either emission or absorption spectra. The single molecule of terrylene doped in the para-terphenyl was shown to be a possible venue for realizing single molecule probes. The observation of absorption spectrum of single pentacene molecules doped in the para-terphenyl was carried out in superfluid helium at 1.6 K [8.24, 8.25]. The experiment at low temperatures on

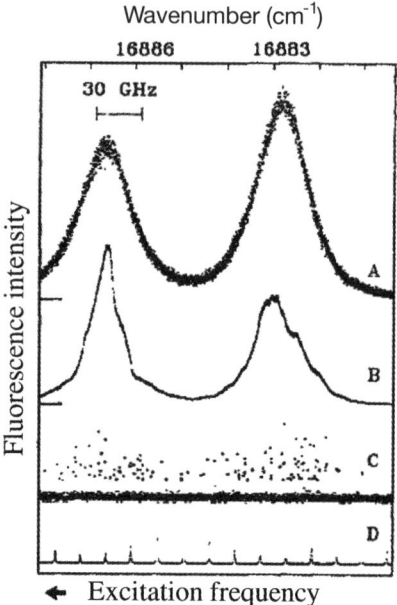

Fig. 8.4. Fluorescence spectra of pentacene in the form of (A) melt-grown crystal, (B) sublimation flake, and (C) single molecules. (D) is the calibration spectrum of an etalon (extracted from [8.27])

terrylene molecules distributed in matrices of p-terphenyl crystals showed the spectral diffusion or random jumps of optical spectrum lines. The molecules at this temperature can be simplified as a two-level system involving only the singlet ground and the excited states. The vibrational relaxation and intercrossing transition are much slower than for the fluorescence transition. The spectrum is typically a single phonon line.

Frequency-modulation Stark double modulation (FMS) and frequency-modulation ultrasound double modulation (FMUS) methods have been used to reduce background signals, yielding similar absorption spectra for single pentacene molecules in the wings of the absorption spectrum [8.24].

Similar behavior was observed for fluorescence excitation spectra of pentacenes doped in the para-terphenyl at 1.8 K (Fig. 8.4) [8.25]. It was found that the spectra peaks can display sudden drops and surges during experiments (Fig. 8.5), which can be attributed to orientation flips of the phenyl ring of the host molecules as a result of optically induced changes in environmental conditions [8.26].

In analyzing fluorescence intensities, a common approach is to obtain the autocorrelation function defined in the following:

$$G(\tau) = \sum_{t=0}^{N-1} g(t)g(t+\tau) \qquad (8.1)$$

8.3 Low-Temperature Studies of Single Molecules in Solid Matrices

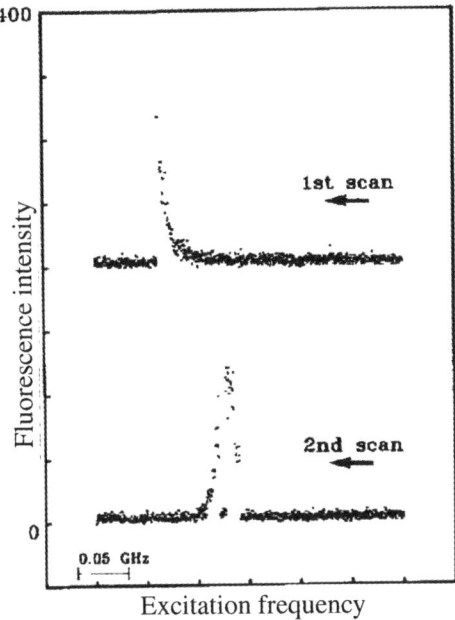

Fig. 8.5. Example of sudden change of excitation spectrum possibly due to hole-burning process (extracted from [8.25])

where N is the total number of time intervals for data collection, and $g(t)$ the detected fluorescence intensity. The results reveal the coherence of the fluorescence emission process, such as the emission lifetime.

It is known that the spectra characteristics of individual terrylene molecules in para-terphenyl matrices are sensitive to the insertion sites. A total of four fluorescent frequencies have been observed and attributed to different immobilization sites in the matrix [8.27]. The transition between the spectral features can be induced by light illumination. Such frequency jumps were attributed to local environmental change in response to the change in geometry of single terrylene molecules in the excited state.

With improved detection sensitivity and high-quality crystals, the emission spectra of single pentacene embedded at O1 and O2 sites in para-terphenyl can be obtained at 1.7 K [8.28]. The spectra characteristics for these sites can be resolved and used for unambiguous identification. It was also suggested that host lattice disorders in the vicinity of the molecular sites could significantly alter the characteristic features of the spectra.

In another type of single molecule study, the fluorescence-based assessment of single perylene molecules in the Shpol'skii matrix of n-nonane led to the identification of a triplet lifetime of 1.1 ± 0.5 ms [8.29], well exceeding the fluorescence lifetime of perylene in ethanol and n-octane of 7 ± 2 ns [8.30].

The so-called Shpol'skii effect was first observed as narrow spectra features of aromatic hydrocarbons dissolved in n-alkane and fast frozen in liquid nitrogen [8.31–8.34]. The multiplet spectra were shown to originate from the different orientation of the guest molecules. The orientation of the fluorophore dipole moment was obtained by adjusting the polarization angle of the excitation beam and recording the fluorescence intensity. The analysis of the multiplet structure of single 2,3,8,9-dibenzanthanthrene (DBATT) molecules in a matrix of n-tetradecane revealed two distinct orientations [8.35] that are related to the two 0–0 bands in the ensemble spectrum. In addition, the spread of orientations within each distribution can be attributed to local fluctuations in matrix orientation (Fig. 8.6).

The 0–0 transition, also called zero-phonon line (ZPL), represents the transition between ground vibrational levels of the electronic ground state and excited state. At far below the Debye temperature, the linewidth of ZPL is narrow due to the weak coupling between electronic excitation and phonons. Under the influence of an external electric field, the position of the ZPL can be shifted as a result of changes in the dipole moment of the excited state. Shifts of ZPL can also be induced by external pressure as a result of pressure-induced compression of the local matrix [8.36].

The observed decay in autocorrelation function provides information on the yield and lifetime of the triplet states, as illustrated in Fig. 8.7 [8.25]. Double exponential decay behavior was reported for pentacene [8.37], terrylene [8.38], terrylene-diimide [8.39], and DBATT [8.40]. The decay constants represent the characteristic dwell time of the triplet states, with possibly two of three sublevels (T_x and T_y levels) having similar dwell times.

In addition, the fluorescence properties of terrylene-diimide can be studied from 1.4 K to room temperature. The vibrational features on the fluorescence spectrum of single terrylene-diimide have also been obtained in a hexadecane matrix [8.39].

Single molecule spectroscopy measurements of the type 2 light harvesting complex (LHC 2) of the purple bacterium *Rhodopseudomonas acidophila* at 1.2 K revealed features associated with the excitation of pigments. The LHC 2 complex consists of two concentric rings with absorption bands at around 800 and 860 nm. The ring of the 800-nm absorption band (B800) is formed with 9 BChl a molecules equally separated along the ring. By contrast, there are 18 BChl a molecules closely packed along the ring with the 860-nm absorption band (B860). By comparison of the fluorescence spectrum of the ensemble and individual LH 2 complex, it was concluded that the excitation is localized on individual BChl a molecules in the B800 ring, and is delocalized among BChl a molecules in the B860 ring [8.41]. The photobleaching of a single BChl a molecule will lead to complete quenching of the ring assembly [8.42].

8.3 Low-Temperature Studies of Single Molecules in Solid Matrices   193

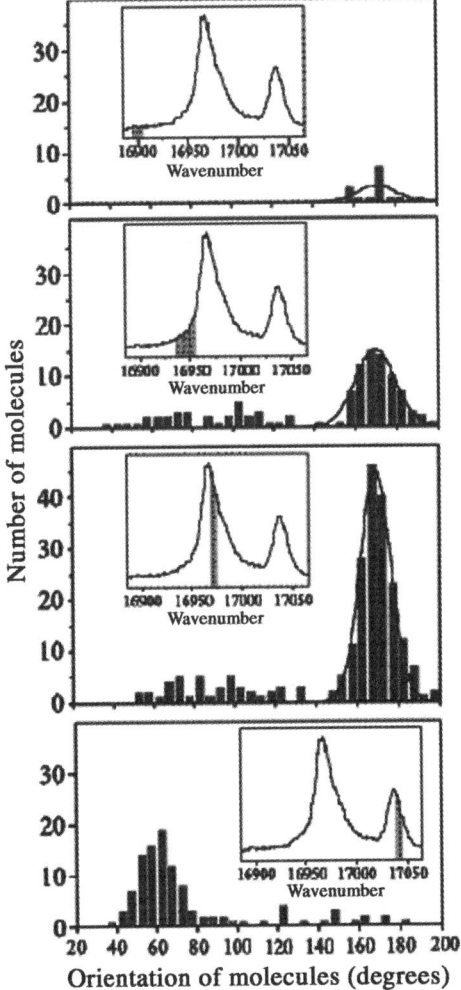

Fig. 8.6. Histograms of the dipole orientation of 456 single DBATT molecules at different frequencies (extracted from [8.35])

8.3.2 Pump–Probe Effects

The pump-probe method has been developed to explore laser–molecule interactions. In this approach, two excitation beams are simultaneously used, one being the pump beam that is near-resonant with the two-level system, and the other beam having weaker intensity than the pump beam. Under the field of the pump beam, the singlet energy levels are shifted as a result of the electric field of the pump beam (also called light-shift effect). The original ground and excited singlet states evolve into a new set of eigenstates (or dressed molecules/atoms), due to molecule(or atom)–field coupling. The experimen-

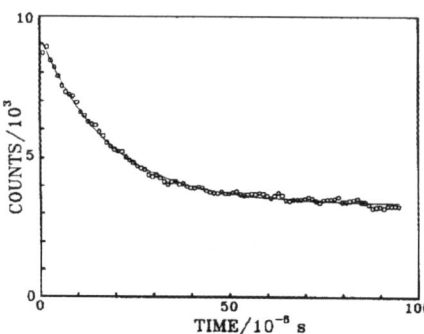

Fig. 8.7. Autocorrelaton function of fluorescence intensity of single pentacence molecules (extracted from [8.25])

tal evidence for such a process is seen through a splitting in the spectrum from the excited state to a third level, the so-called Autler-Townes splitting that can be detected by the absorption spectrum of the probe beam. The frequency difference between the dressed states is called the Rabi frequency. Such investigations are important to the study of molecular non-linear optical properties.

Pump-probe experiments on dibenzanthanthrene (DBATT) embedded in naphthalene crystals also revealed strong light shifts as a function of pump detuning [8.43, 8.44]. A deviation from second-order perturbation was observed when the pump beam was close to the molecular resonance. Furthermore, additional structures were observed that could be attributed to hyper-Raman structures.

When a weak rf power was used, rather than the pump beam, Brunnel et al. [8.45] demonstrated in the system of dibenzanthanthrene (DBATT) in hexadecane (HD) matrix that one could study laser–molecule coupling by measuring the frequency splitting between dressed states, or Rabi transition.

There exist reports of the AC-Stark effect of single terrylene and pentacence molecules in a matrix of para-terphenyl at 1.8 K [8.46, 8.47], as well as terrylene molecules in a polymer host of polyethylene [8.48]. The electric field of the pump beam was shown to have an appreciable effect on the transition frequency, i.e., light shift. The light shift was found to vary linearly with the intensity of near-resonant pump beam. In addition, a dephasing effect of the fluorescence spectrum was observed at single molecule level in condensed matrix environments [8.45, 8.49].

Non-linear optical effects have also been reported for single molecules. For example, two-photon excitation of single diphenyloctatetrane (DPOT) distributed in n-tetradecane (TD) has been demonstrated [8.50]. The zero-phonon line of DPOT at 444 nm was excited by two photons of 888 nm of the excitation laser. In addition, a laser-induced shift of the zero-phonon line was identified and ascribed to the AC Stark effect and C–H band excitation, followed by "fast" energy exchange between the acoustic phonon and the local vibration mode of the DPOT-TD system [8.51]. Two-photon processes of single molecules were also demonstrated in ambient and solvent conditions [8.52, 8.53].

8.3.3 Magnetic Resonance of Single Fluorescence Molecules

Since the unpaired electrons in triplet states lead to a total spin $S = 1$, the spin–spin interaction results in the splitting of the triplet sublevels. The transition between the triplet sublevels can be achieved by magnetic resonance effects induced by the microwave irradiation and observed in the fluorescence spectrum. The results can provide insight into molecular spin dynamics under the influence of external fields.

Optical detection of single molecule magnetic resonance transitions of triplet states of pentacene has been reported [8.54–8.58]. Such studies are complementary to the detection of spin centers by the electron tunneling effect or magnetic force measurements discussed in Chapters 2 and 4. Microwave excitation was introduced during the dark time of single molecule fluorescence cycles to study the magnetic resonance of triplet states (total spin = 1). The triplet sublevels, represented by the T_x, T_y and T_z states (Fig. 8.8), were separated by distances on the order of 0.1 cm^{-1} as a result of dipolar spin–spin interaction.

The sublevels have quite different populations and lifetimes, compared with that of the excited singlet state. T_x and T_y are much more populated than the T_z state, due to the selectivity of the intersystem crossing process [8.59]. By tuning the population of the triplet sublevels with microwaves, one can observe changes in the fluorescence intensity [8.54, 8.55, 8.57].

The pentacene, terrylene molecules in the p-terphenyl matrix showed characteristic asymmetric resonance spectra that can be associated with the second-order hyperfine interaction between the triplet electron spin and the nuclear spin [8.54–8.58]. The hyperfine structures of the electron spin paramagnetic resonance of the triplet state of single pentacene-d_{14} molecules were reported by Köhler et al. [8.60]. The deuterium-substituted pentacene displayed a much reduced linewidth of magnetic transition as a result of the

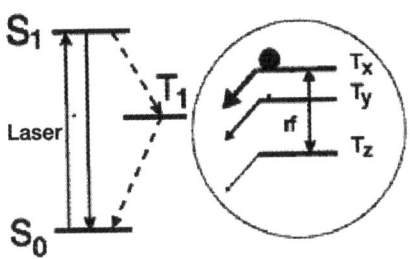

Fig. 8.8. Diagram of energy levels in magnetic resonance measurements of the triple states (extracted from [8.54])

smaller magnetic moment of deuterium, compared with that of hydrogen (about 15%). A similar effect of pentacene-h_{14} molecules interacting with ^{13}C nuclear spin was also observed with different resonance width and positions [8.61].

As an application of the ODMR, the orientation of the single molecule with respect to the external magnetic field (up to 10 mT) can be determined [8.62]. The ODMR effect was also shown to distinguish isotopomers of pentacenece [8.63].

8.4 Single Fluorescence Molecules in Liquid Conditions

8.4.1 Experimental Considerations

Upon studying single molecules in solution conditions, experimental designs different from that of low temperatures have been developed. The detection of single molecules in solution environments has been achieved in several approaches with minimized detection volumes [8.64, 8.65]:

(1) By using the hydrodynamic focusing effect, the sample stream is narrowed to diameters of 1–20 μm by parallel high-speed sheath flow. The laser is typically focused to a diameter of about 10 μm, which will result in a detection volume of 1–10 pl. This will allow reliable detection of the analyte molecules present in the excitation volume.
(2) Using free-falling microdroplets with diameters in the micrometer range can also lead to detection volumes around 100 fl or less. The analyte molecules are dissolved in the droplet. An electrodynamic trap is used to control the microdroplet in the laser field.
(3) Minimized detection volumes can also be obtained in microcapillaries and microstructures with diameters of a few micrometers. The probe volume can be on the order of a few picoliters.
(4) Confocal excitation can achieve a volume of about 1 fl, or 1 μm^3.
(5) Total internal excitations and high-resolution microscopy for studying single fluorophores at surfaces.

Additional improvement of signal-noise ratio can be obtained by photobleaching the solvent prior to introducing analyte species.

Data analysis can be carried out using autocorrelation, as presented in Eq. (7.1) or by weighted-quadratic-sum filter for analyzing individual fluorescence bursts [8.65]:

$$S(t) = \sum_{t=0}^{k-1} \omega(\tau) g(t+\tau)^2$$

where $\omega(\tau) = (\tau+1)/k$ for $\tau = 0$ to $k-1$, otherwise $\omega(\tau) = 0$.

Different from the criteria for single molecule identification at low temperatures, the identification of single molecules can be made with the following signature:

(1) Fluorescence spectra
(2) Fluorescence lifetime
(3) Burst size
(4) Time-resolved fluorescence anisotropy.

Detection of single dye molecules (rhodamine derivatives) with fluorescence above 600 nm was achieved with a CW diode laser or helium neon laser in conjunction with a confocal microscope [8.66]. The fluorescence background from impurities was significantly reduced in this frequency regime. The use of CW diode and helium neon lasers can help reduce equipment costs.

8.4.2 Examples of Fluorescence of Single Molecules in Solutions

Identification of Single Fluorophores

Identification of single fluorescence molecule events has been reported in several studies. For example, individual chemifluorescence reaction events between radicals of 9,10-diphenylanthracene (DPA) have been observed by analyzing the intensity trajectories of the photon emissions [8.67]. The histogram of the time intervals between photon emissions was shown to follow a Poisson distribution, which suggests the nature of individual events. Fang and Tan [8.68] used the evanescent wave produced by total internal reflection of a laser inside an optical fiber (both cylindrical and squared-shaped) to observe individual rhodamine 6G (R6G) molecules, and demonstrated the linear dependence of the number of observed dye molecules on the bulk concentrations.

With high detection sensitivity of a confocal fluorescence microscope, the trajectories of single fluorescence molecule can be followed in solutions, including R6G in ethanol and water, and fluorescein in mercaptoethanol solution. The observed averaged dark time between emissions revealed the triplet state lifetime (4–6 μs) of single molecules and the time scale for diffusional re-crossing of the detection volume [8.69].

Conformation Effect of Single Fluorophores

The effect of internal structures on molecular fluorescence has been demonstrated on single molecule level. Detailed analysis of single molecule fluorescence of tetrahedrally coordinated oligophenylenevinylene (TOPV) revealed significant fluctuation in polarization axis in association with the motion of the central carbon atom, suggesting that luminescence is dependent on the conformation of the molecule [8.70]. By analyzing the fluorescence distribution of single fluorophores of tetramethylrhodamine linked to DNA oligonucleotides, a wide range of fractions of long lifetime decay was revealed. This was explained by the conformational transitions of individual oligonucleotides, and was not observable in ensemble average studies [8.71, 8.72]. Conformational dynamics of single tRNAPhe molecules was also studied in terms of their fluorescence lifetime distributions [8.73]. A burst-integrated fluorescence lifetime scheme (BIFL), which combines the macroscopic detection time and arrival time of the photons, was proposed to study the single molecule conformational dynamics [8.74].

Adsorption and Diffusion of Fluorophores

Many interesting results have been reported on the adsorption and diffusion of single molecules. The single green fluorescence proteins (GFPs) showed greater intensity fluctuations than that of the organic fluorophores [8.75]. Studies of autofluorescent proteins (eCFP, eGFP, eYFP) adsorbed on phospholipid membranes showed a typical emission rate of about 3,000 photon ms^{-1}, saturation intensities in the range 6–50 kW cm^{-2}, and photobleaching yields of 10^{-4} to 10^{-5}. Diffusional properties of single eYFP molecules in free and anchored states have also been studied [8.76].

Epi-illumination microscopy was also applied to determine the binding of oligonucleotides or ligands with probe and target sequences labeled with different fluorophores. Dual-wavelength fluorescence labeling was shown to distinguish the specifically bound molecules from the physisorbed species [8.77, 8.78].

The intensity trajectory of single diIC$_{18}$ molecules was seen to be affected by the microenvironment of lipid membrane DPPC [8.79]. For lipid monolayers of DPPC, the characteristic fluctuation time changes from 440 ms to over 1 s with increasing surface pressure, whereas for the lipid bilayer the fluctuation time is on the order of 2 s. The cause of the intensity fluctuation was ascribed to the twisting motion of the diIC$_{18}$ molecule that is affected by the lateral fluctuations of lipid tailgroups.

Chemical Reactivity of Single Fluorophores

The reactivity of single molecules is one of the exciting topics in single molecule studies. This information could have impacts on the fundamental understanding of the reaction pathways. Epifluorescence microscopy with TIR excitation was used to directly visualize individual ATP dissociation and association events by single-headed myosin enzyme molecules (S-1) in solutions [8.80]. The attachment of single fluorophores to the myosin enzyme molecules was identified by the discrete distribution of fluorescence intensity. The dissociation rate was determined from the lifetime measurement of the fluorescent spots.

The fluorescence behavior of cholesterol oxidase (CO$_x$) enzyme molecules is characterized by the state of the flavin adenine dinucleotide (FAD). FAD is fluorescent in the oxidized state and fluorescence is "off" for the reduced state FADH$_2$. The enzymatic activity can be studied by assessing the intensity trajectories with stochastic on and off emissions [8.81]. The analysis of the fluorescence intensity trajectories of single CO$_x$ molecules on 2 mM 5-pregene-3β-20α-diol substrate revealed a disordered distribution of the reaction rate. The cause for such static disorder could be different conformers,

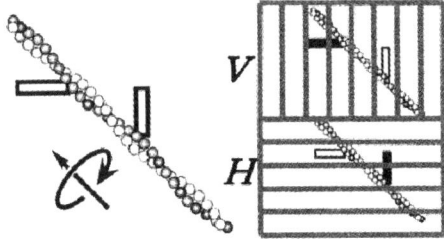

Fig. 8.9. Schematic of the rotational movement of an actin filament in the presence of myosin molecules. The excitation was circularly polarized. The detection of the rotation was achieved by monitoring the linear polarized components in vertical and horizontal directions (extracted from [8.84]).

post-translation modification, or proteolytic damages to the key residues of the enzyme. A memory effect was observed for the adjacent on-times, and this effect was diminished for events separated by 10 turnovers. The dynamic disorder of the reaction rate can be analyzed from the autocorrelation of on-times of the single enzyme molecules.

Single molecule reactivity of the enzyme lactate dehydrogenase (LDH-1) in electrophoretic flow can be resolved by fluorescence spectroscopy [8.82]. A fourfold difference of activity was observed among the single enzyme molecules, possibly related to molecular conformations. Large variations in single molecule enzymatic activity were also observed for alkaline phosphatase [8.83].

Movement of Molecular Motors

In another case of single molecule studies, the one-dimensional sliding of myosin molecules along an actin filament has been directly visualized with an inverted epifluorescence microscope [8.84]. The actin filament has a right-handed helical structure with a pitch of about 72 nm (Fig. 8.9). Under the given experimental conditions, the fluorescence (5-iodoacetamidotetramethylrhodamine)-decorated actin filaments were adsorbed on top of the surface of a heavy meromyosin (HMM) layer in buffer solutions. The polarization of the emitted fluorescence was observed to display periodic variations as a result of orientation alternation of single fluorescence molecules (Fig. 8.10). The observed coherent change in fluorescence polarization was attributed to the axial rotation of actin filaments during sliding motion. The periodicity of the polarization variation was about 1 μm, suggesting that the myosin enzyme protein molecules may skip many pitches along the actin axis during sliding, and the rotational torque is not significant in actin–myosin interactions.

Both orientational ordered and disordered movement patterns of single kinesin head protein along microtubules were identified depending on the stages of ATP hydrolysis observed by epifluorescence polarization microscopy [8.85].

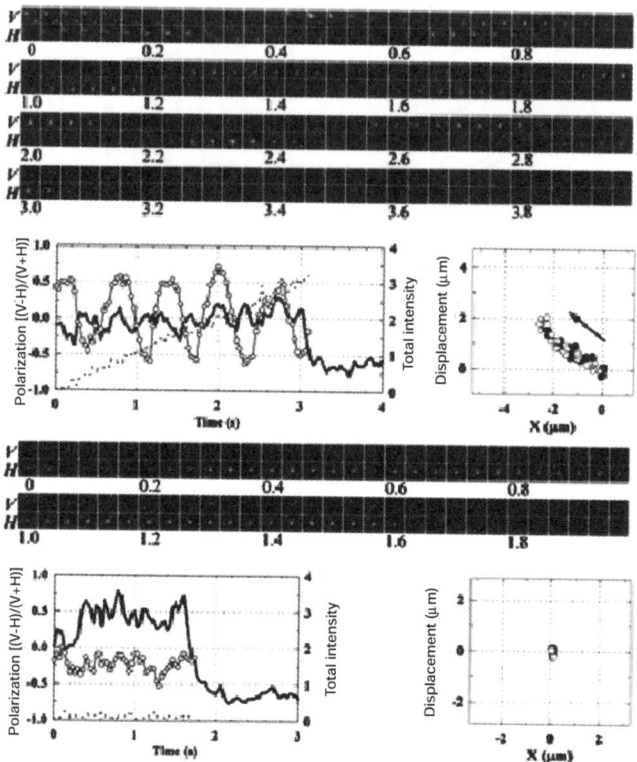

Fig. 8.10. Time series of periodic variations of observed intensity for vertical and horizontal polarized components (extracted from [8.84])

The rotation of a molecular motor, F1-ATPase, was also directly visualized by a far-field microscope [8.86]. High-precision experiments on enzyme movements have been carried out based on various optical techniques, such as video-enhanced difference interference contrast microscopy [8.87].

8.4.3 Single Molecule Diffusions in Living Cells

Attempts to study single fluorescent molecules inside living cells have progressively advanced in the past few years. By tightly focusing the laser beam into a living cell, one can observe photon-burst of fluorophores inside the cells. The main obstacles for studying single molecules inside living cells are cellular autofluorescence, such as for flavinoids in the visible regime. Several approaches can be used to reduce the cellular fluorescence background:

(1) Decrease of the excitation volume [8.88, 8.89]
(2) Utilization of time gating with a fluorophore of much longer fluorescence lifetime than those of flavins [8.90, 8.91]

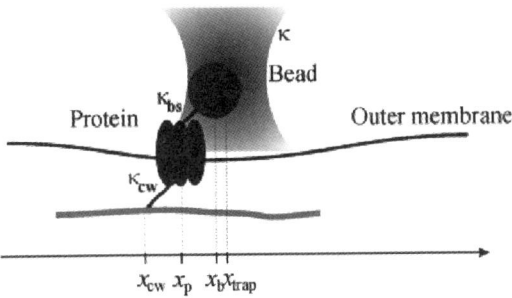

Fig. 8.11. Schematic of an optical tweezers-assisted method for studying diffusion of single protein molecules anchored in the membrane of the cell wall (extracted from [8.98])

(3) Short photobleaching treatment with intense light pulse before actual experiments.

The intracellular diffusion rates of organic dyes (R6G and synthetic oligonucleotides) and fluorescently labeled biomolecules (transferrin, an iron transport protein) were found comparable to those in aqueous solution [8.92]. This is consistent with the results of earlier studies [8.93, 8.94]. By contrast, macromolecules were found to have lower mobility inside cells than was the case in water.

Single protein molecules of R-phycoerythrin (RPE), an autofluorescent protein with high quantum yield and extinction coefficient, were imaged by epifluorescence microscopy in the cytoplasm and nucleoplasm of TC7 cells. The protein molecules move slowly enough to allow one to collect multiple frames of images to identify their trajectories inside the cell. A broad distribution of diffusion constants of single RPE molecules was obtained, the values being 2.7 mm^2 s^{-1} in cytoplasm and 1.1 mm^2 s^{-1} in nucleoplasm [8.95]. The binding and diffusion behavior of single receptor molecules to the living cell membrane was demonstrated with an objective-type total internal reflection fluorescence microscope [8.96, 8.97]. The diffusion of single protein molecules of l-receptor in the outer membrane of living *Escherichia coli* bacteria was studied with the aid of optical tweezers (Fig. 8.11) [8.98]. The protein molecule was attached to a bead that can be manipulated by the optical tweezers. The measured diffusion constant of the protein molecules was $1.5 \pm 1.0 \times 10^{-9}$ cm^2 s^{-1}.

In another study, green fluorescent proteins (GFP) attached to the cell membrane protein E-cadherin were observed by the evanescent field excitation method. The E-cad-GFP were found to be in the form of oligomers, based on the observed higher fluorescence intensities, compared to that expected for monomers. The oligomers diffuse either in Brownian motion or confined/stationary mode [8.99]. The progress on single autofluorescence protein molecules has been reviewed recently [8.100].

The potential of single molecule detection techniques in molecular biology was demonstrated in the study of infection pathways of individual adeno-associated virus adsorbing and penetrating living cell membranes [8.101]. The results could help reveal the details of virus–cell interactions.

8.4.4 Single-Pair FRET

Fluorescence resonance energy transfer (FRET) is a unique method to measure the separation variation between donors and acceptors. The concept is based on the energy transfer in donor–acceptor pairs due to dipole-induced dipole interaction. In this method, both the donor and acceptor are attached to a single molecule, and only the donor is activated by the excitation beam. The change of the donor–acceptor separation is subtly reflected in the emission intensities of the donor and acceptors. Such experiments are powerful in studying intramolecular structural variations, and revealing the reaction kinetics of single biomolecules under physiological conditions.

The technical aspects and applications of single molecule FRET have been described in several review papers [8.102–8.104]. The labeling methods to attach fluorophore to target molecules are summarized in the following:

(1) Single fluorophore as a point light source. This is applicable to study the diffusion of single molecules.
(2) Co-localization of non-interacting fluorophores on different macromolecules. Such labeling enable studies of the kinetics of association, binding and turnover events.
(3) Co-localization of interacting fluorophores on one and the same molecule or different molecules. This would allow one to study fluorescence resonance energy transfer (FRET) mechanisms.
(4) With the capability of analyzing emission polarizability, the angular motion of molecules can be studied.

A green fluorescent donor dye and a red fluorescent acceptor dye were attached to cysteine residues introduced at the amino and carboxy termini of a cold-shock protein from the hyperthermophilic bacterium *Thermotoga maritima* (CspTm), as illustrated in Fig. 8.12. The donor was excited by laser irradiation and the energy transfer to the acceptor, which could be observed as fluorescence emission of both acceptor and donors (Fig. 8.13) [8.105]. The energy transfer process is sensitive to the separation between donor and acceptor, as shown in Förster's equation [8.107]:

$$E = \frac{1}{1 + (\frac{R}{R_0})^6} = [1 + \gamma I_d/I_a]^{-1}$$

where R_0 is the distance of 50% energy transfer, dependent mainly on fluorophore properties and relative dipole orientation, and γ is a correction factor accounting for the wavelength dependence of the detection efficiency and the quantum yields. I_d and I_a are corrected emission intensities of donor and

8.4 Single Fluorescence Molecules in Liquid Conditions

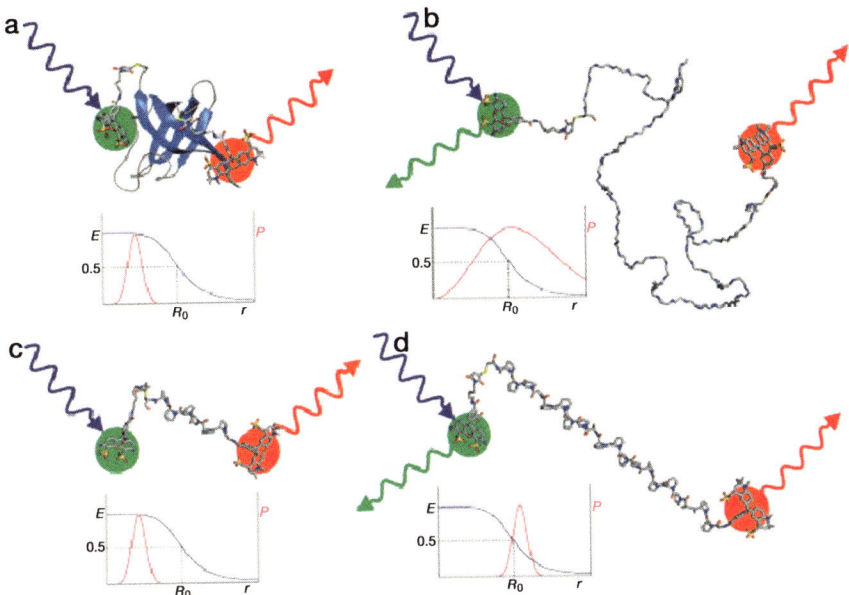

Fig. 8.12. Schematics of the protein structures and the dye labels of donor and acceptor groups. **a** Folded protein structure. **b–d** Unfolded proteins (extracted from [8.105])

acceptor, respectively [8.102]. The apparent FRET efficiency for each burst can be estimated as the ratio of acceptor counts to the sum of acceptor and donor counts. A wide range (16–85%) of single-pair energy transfer efficiency has been reported [8.5].

Intramolecular single-pair FRET analyses on the surface-bound enzyme staphylococcal nuclease (SNase, which catalyzes the hydrolysis of DNA and RNA) revealed the anticorrelated emission of donor and acceptor spectra (Fig. 8.14) [8.106]. The acceptor emission prevails at early stages, followed by a switching to donor emission-dominated spectra (Fig. 8.14b). The transition is an indicator of photobleaching of acceptors:

$$R_0^6 = 8.79 \times 10^{23} (n \cdot \varphi_D \cdot J \cdot \kappa)^2$$

where n is the index of refraction of the medium, φ_D the donor quantum yield, κ^2 ($=2/3$) the orientational factor, and J the spectral overlap of donor emission and acceptor absorption spectra.

It has been established that the presence of SNase active-site inhibitor deoxythymidine diphosphate (pTp) significantly increases the characteristic time scale for the fluctuation of energy transfer efficiency, as a result of hindered rotational dynamics and temporal fluctuations [8.106].

The measured distribution of energy transfer efficiencies of single CspTm molecules suggests an estimated upper limit of reconfiguration time of about

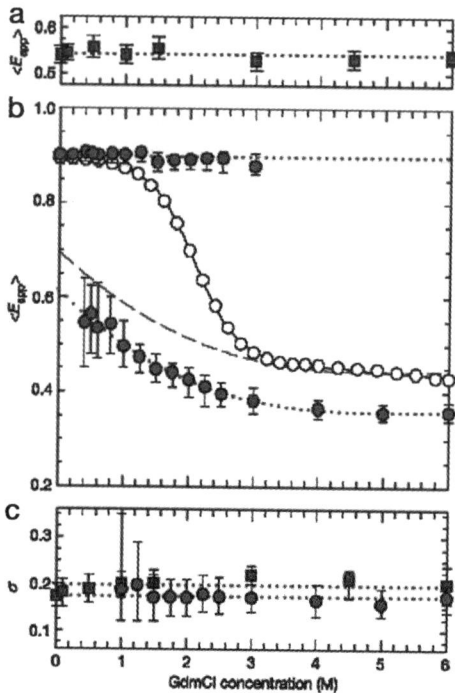

Fig. 8.13. Solvent (guanidinium chloride, GdmCl) effect on the FRET efficiency of CspTm proteins. *Open circles* are ensemble data, and *filled circles* are single molecule data. The change in FRET efficiency represents variations in end-to-end distance (extracted from [8.105])

25 ms. This provides an example of using single molecule fluorescence spectra to study the dynamics of folding processes of proteins [8.105].

spFRET studies of single protein chymotrypsin inhibitor 2 (CI2) molecules freely diffusing in solutions have been carried out [8.108]. The CI2 molecules were site-specifically labeled with dye pairs of Cy5 (acceptor) and tetramethylrhodamine (TMR, donor). The effect of the denaturant concentration, guanidinium chloride, on the folding–denaturization process can be directly observed through the histograms of FRET efficiencies. The time scale for interconversion between folded and denatured states of the single protein molecules was found to be less than the diffusion-limited observation time (about 1 ms). It was also demonstrated that the distributions of FRET efficiency can be converted into potential energy functions, providing helpful insights into protein folding pathways.

Fluorescence studies of single enzyme molecules have greatly enhanced the capability of examining the kinetic properties of enzymatic processes, such as in cholestrol oxidase experiments. Transient intermediates can be

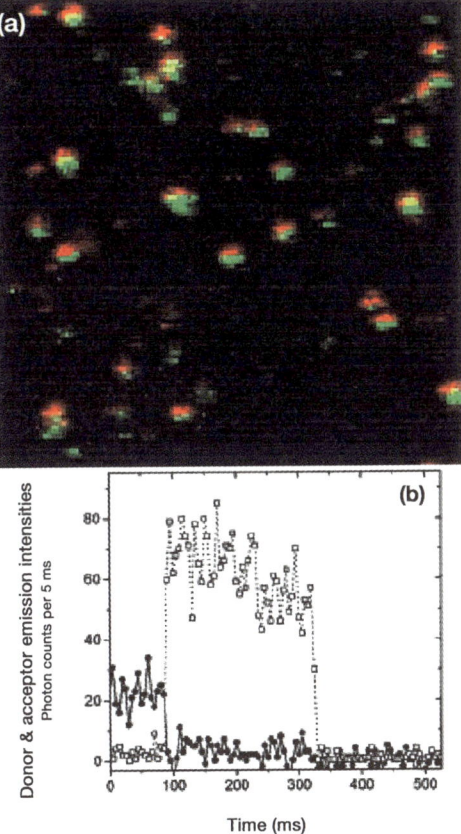

Fig. 8.14. a Dual-color image of doubly labeled SNase enzyme molecules. Donor emission is represented by *green* and acceptor emission is represented by *red*. Excitation laser is 514 nm (Ar$^+$ laser, 15 µW). b Emission spectra of donor (*squares*) and acceptor (*circles*). The inverse correlation of the emission intensities is evident (extracted from [8.106])

revealed during real-time observations, and time-dependent trajectories can yield valuable dynamic information [8.109].

The single-pair fluorescence resonance energy transfer (spFRET) method was applied to study the stability of hairpin DNA structures immobilized on functionalized glass surfaces [8.110]. It was found, for a hairpin structure with a loop size of 40 adenosines and a stem of seven or nine bases, that the open-state lifetime is much longer than the closed-state lifetime. In addition, it has been confirmed that the closed-state lifetime is dependent on the stem length and independent of the loop structures.

By modifying a single-stranded DNA with a fluorophore at one end and a quencher at the opposite end, a target-specific method was developed. Such probe molecules, also called molecular beacon DNA probes, can be used to effectively detect the DNA hybridization process at single molecule level [8.111].

As an example, individual three-helix junction RNA molecules were labeled with fluorescein dye (donor) and Cy3 dye (acceptor) at two separate ends. The three-helix molecules were immobilized on the surface through biotin–streptavidin interaction. There is a binding site in one of the helix arms that can bind ribosomal protein S15. The binding of the S15 protein can induce the folding of the junction molecule. The measured fluorescence distributions clearly revealed the difference between folded and open conformations as a result of S15 protein binding [8.112]. In addition, the folding effect due to the presence of divalent ions Mg^{2+} was studied. The observed conformational change of the junction molecule under the influence of Mg^{2+} concentration was used to estimate the dissociation constant. It was shown that the dynamics of the conformation change under protein or Mg^{2+} can be followed in real time.

The dependence of the FRET efficiency on the donor–acceptor separation was directly studied by using a series of oligonucleotides end-labeled with TMR (donor) and Cy5 (acceptor) [8.113]. The single oligonucleotide molecules were in free diffusion state in buffer solutions. The fluorescence bursts passing through the excitation volume were recorded to obtain the FRET efficiencies. It was demonstrated that the observed FRET efficiency histograms can also be used for ratiometric studies of the subpopulation species.

The FRET efficiency measurement of the protein construct cameleon YC2.1 was found to be highly dependent on the Ca^{2+} concentration in the environment [8.114]. This protein construct has been used as a ratiometric indicator of Ca^{2+} in solutions. The detailed analysis of the histograms of FRET efficiency at different Ca^{2+} concentrations revealed fluctuations due to local calcium concentration, which affects the binding kinetics [8.114, 8.115]. The emission ratio of Detran-SNARF-1 pairs immobilized in agarose gel was shown to be sensitive to pH values [8.116]. The binding of ligand–receptor pairs of biotin–streptavidin was studied by the FRET method. The rhodamine (donor) labeled biotin and cyanine (acceptor) labeled biotin molecules were coadsorbed on the streptavidin pre-covered substrate. Dual-wavelength imaging by epifluorescence microscopy helped identifying the colocalization of biotin ligands. The observation of FRET effects by illuminating at donor excitation wavelength confirmed the co-localization of the ligand pairs [8.117]. It was further demonstrated that by measuring polarizations of the ligand–receptor pairs formed by biotin and streptavidin, the dynamics of energy transfer can be studied. The observation has implications in assessing orientational behavior within the ligand–acceptor pair.

8.5 Single Molecules in Other Support Media

8.5.1 Single Molecules in Polymer Hosts

Polymers, such as poly(methyl methacrylate) (PMMA) or polystyrene (PS), have been used as matrix material for a number of single molecule studies. The free volume of the polymer matrix makes it possible to immobilize individual molecules at room temperature. The study of the carbocyanine probe molecule, 1,1'-dioctadecyl-3,3,3',3'-tetramethylindocarbocyanine perchlorate (DiIC$_{18}$), revealed little translational and rotational diffusion [8.118]. The results could also be considered as a reflection of local heterogeneity and dynamic properties.

The single carbocyanine molecule, 1,1'-didodecyl-3,3,3',3'-tetramethylindocarbocyanine (DiIC$_{12}$), embedded in poly(methyl methacrylate) (PMMA) film was resolved by epi-illumination microscopy [8.119]. An appreciable shift of emission spectrum peak position of up to 30 nm was observed and attributed to environmental perturbation of the radiative emission. A Gaussian-like lifetime distribution was obtained for the single molecules located at the polymer–air interface, with a correlation to the fluorescence peak wavelength. By contrast, the control experiment at hydrocarbon oil–polymer interfaces produced only a scattered distribution for fluorescence lifetime. The observed variation of fluorescence lifetime was suggested to be an effect of the presence of the dielectric interface, which modifies the radiative component of the decay process through the expression [8.120]:

$$\frac{1}{\tau} = \frac{1}{\tau_{non-rad}} + \frac{1}{\tau_{rad}} \left[\frac{L_{//}(z)}{L_\infty} \sin^2 \theta_e + \frac{L_\perp(z)}{L_\infty} \cos^2 \theta_e \right]$$

The dipole molecule is located at distance z from the interface with azimuthal emission angle θ_e from the interface normal. $\tau_{non-rad}$ and τ_{rad} are non-radiative and radiative lifetimes, respectively. The quantities $L_{//}/L_\infty$ and $L_\perp/L_\infty$ represent normalized radiated power, and become unity when the molecule is far away from the interface. It can be seen that the emission lifetime depends on the molecular orientation, and this could lead to new experimental approaches to estimate emission orientation. The radiative rate is proportional to the third power of the emission frequency [8.5].

It was found both theoretically and experimentally that the emission behavior of molecules can be appreciably affected by the presence of a proximal metal probe. The enhancement of fluorescence may be associated with the local electric field [8.121, 8.122] or localized plasmon, as in the case of gold-coated glass tips [8.123]. Not surprisingly, the orientation of the fluorescence molecule is found critical to the simulation of the observed apparent fluorescence distribution patterns.

Single molecules of crystal violet (CV) dispersed in a matrix of poly(methyl methacrylate) (PMMA) displayed distinctive site-dependent fluorescence characteristics shown in Fig. 8.15, namely, a strong fluorescence intensity/relatively long-lifetime spot and low intensity/short-lifetime spot [8.124].

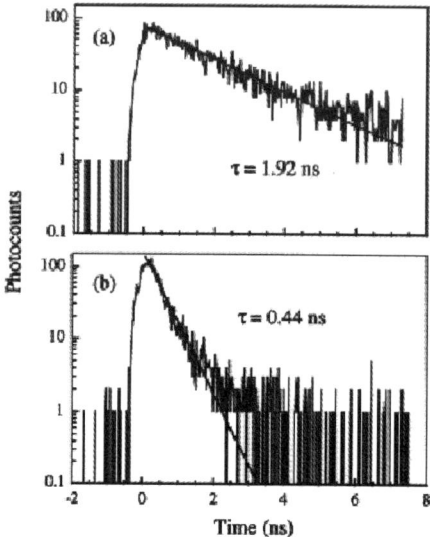

Fig. 8.15. Site dependence of fluorescence decay lifetime of CV fluorophores embedded in PMMA matrix (extracted from [8.124])

The difference was attributed to the local free volume variation of the polymer matrix.

The rotation behavior of single rhodemine 6G (R6G) molecules in poly (methylacrylate) (PMA) with glass transition temperature of $Tg = 8\,°C$ shows characteristics of multiple time scales in autocorrelation function at temperatures close to Tg [8.125]. The single R6G molecules have a rather low lateral diffusion constant in the polymer melt, and the rotation of the molecules was measured in terms of linear dichroism. The measured fluorescence intensity displayed sudden changes, reflected as distinct characteristics in the autocorrelation function. This is a direct indication that single molecule dynamics are sensitive to local environmental changes.

The measured spectral diffusion of single tetra-tert-butylterrylene (TBT) embedded in poly(isobutylene) (PIB) thin films at 1.4 K revealed a wide range of spectral linewidths and distinct dependence on the excitation intensity (Fig. 8.16) [8.126]. The autocorrelation function changes from monoexponential step at low excitation intensity to logarithmic decay at high excitation intensity. This was is attributed to the coupling of dopant molecules with multiple two-level systems (TLS) of the matrix molecules at high excitation intensities.

By measuring the autocorrelation function of fluorescence of single terrylene distributed in a matrix of polyethylene, one can investigate the stability of the TLS, and their coupling strength with phonons [8.127]. (The stability of some TLSs was found to be on the order of a few hours.)

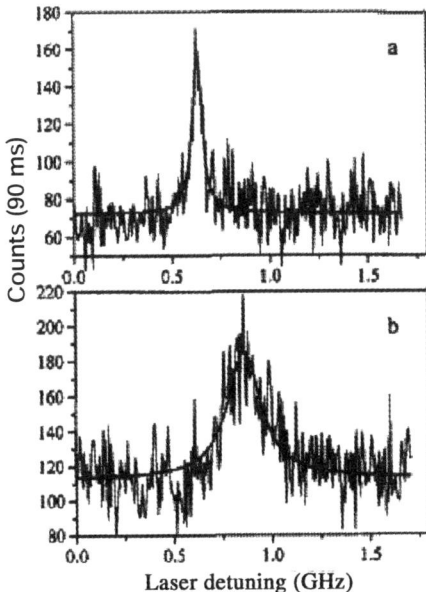

Fig. 8.16. Example of spectral diffusion of TBT showing different linewidths (extracted from [8.126])

Single molecule studies have also been performed on co-polymer poly(p-phenylene vinylene) (PPV) and poly(p-pyridylene vinylene) (PPyV). Both segments of the co-polymer are conjugated species and known for light-emitting properties. Statistical analysis of fluorescence behavior of the single molecules revealed three characteristic, discrete levels of fluorescence intensity [8.128]. It was concluded that the discrete fluorescence intensities could reflect the quenching effect of photochemical defects during the migration of excitons within the polymer chain. This observation could provide clues for the intramolecular electronic energy transfer between segments of polymers.

Single molecule spectroscopy was applied to study the guest–host interaction between cyanine dye (pinacyanol) and second-generation polyphenylene dendrimer [8.129]. Different association structures were observed depending on the guest to host ratio. The molecules were embedded in a matrix of polyvinylalcohol (PVA). By modulating the incident beam polarization, a coherent modulation of fluorescence intensity corresponding to single dye molecules was identified. In addition, a reduced triplet lifetime was observed for the dye molecule encapsulated in the dendritic host, compared with dye molecules in the PVA host. The difference was attributed to the larger free volume for the dendritic host.

The orientation of fluorophore components in bichromophoric molecules embedded in polymer hosts was be analyzed using the modulated fluorescence intensity traces (MFITs) method [8.130]. In this approach, the polarization direction of the excitation laser was rotated at various frequencies, and the fluorescence polarization of single molecules was recorded accordingly. The measured azimuthal angle between the chromophores in the polarization plane was approximated by Monte Carlo simulations.

Intramolecular vibrational features of single fluorophores in a matrix of crystallite and polymer were obtained by the fluorescence excitation technique. This approach resembles the resonant Raman method in that the molecules are electronically excited. The downward shift of long-axis ring expansion peaks for the amorphous sites was ascribed to the greater free volume of the polymer matrix [8.131, 8.132].

Significant on/off blinking of individual green fluorescent protein (GFP) molecules, S65/S72A/T203F (T203F) and S65G/S72A/T203Y (T203Y), differing in terms of hydroxy group near the fluorophore, in polyacrylamide gels was found at room temperature during illumination by 488-nm light [8.133]. The autocorrelation times of both T203Y and T203F were shown to depend on the illumination intensity. The difference in the magnitude of this dependence is indicative of the effect of hydroxyl groups in the process of chromophore activation/inactivation. In addition, the non-fluorescent dark state could be switched into emission state by illumination of a 405-nm wavelength laser. A three-state reaction mechanism was proposed to account for the switching behavior of GFPs.

8.5.2 Lateral Diffusion Behavior of Single Molecules

The diffusion trajectories of fluorophores in both two- and three-dimensional matrices can be recorded by observing the fluorosphore positions. The study of the diffusion of individual molecules could lead to useful insight into the matrix environment in the direct vicinity of single molecules. The measurements can also be adapted to study single molecule diffusion inside living cells. Quantitative measurements can be performed in terms of: :

- Diffusion of ensembles of single molecules,
- Single molecule diffusion trajectory in two dimensions, and
- Three-dimensional diffusion trajectory deduced from intensity variation.

Real-time diffusion of rhodamin-6G (R6G) and single-stranded DNA molecules in water was measured by using an optical microscope equiped with an intensified charge-coupled device (ICCD) camera with TIR excitation configuration. A generally smaller diffusion coefficient and longer lifetime were found for the R6G–DNA complex, compared with those of the dye molecules [8.134]. The three-dimensional diffusional trajectories of single Nile red molecules embedded in a matrix of poly(acrylamide) (PAA) gel have been observed in the view field [8.22].

Protein molecules of concanavaline A conjugated with 5- (and 6-) carboxytetramethylrhodamine succinimidyl ester (ConA-TAMRA) in solutions on fused-silica surfaces were studied with single molecule accuracy with TIR excitation configuration [8.135]. The observation of the residence time distribution confirmed that the protein molecules were attracted to the surface by long-range electrostatic interaction, rather than direct immobilization on the substrate surface.

The diffusion of single fluorescence-labeled lipid molecules in phospholipid membranes was shown to be characterized by high positional precision [8.136]. As an example, two distinct mobility components for the diffusional motion of single lipid molecules were shown to have values of 4.4 and 0.07 μm^2 s^{-1}, by using the probability distribution function method [8.137]. The same molecule, when embedded in a polymer matrix, was constrained in a corral of radius about 140 nm at a time scale below 100 ms.

In an effort to distinguish single fluorophores, a formalism was described based on conditional probability density for identifying co-localized fluorophores [8.138]. The probability of N fluorophores in a signal of intensity $I \pm \Delta I$ can be expressed as:

$$p(N \mid I \pm \Delta I) = \frac{p(I \pm \Delta I \mid N)}{\sum_n p(I \pm \Delta I \mid n)}$$

$$p(I \pm \Delta I \mid N) = \sqrt{\frac{1}{2\pi \Delta I^2}} \int dI' \, \rho_N(I') \exp\left[-\frac{(I'-I)^2}{2\Delta I^2}\right]$$

$\rho_N(I)$ can be calculated recursively, given the known probability density function for single fluorophores:

$$\rho_N(I) = \int dI' \, \rho_1(I') \rho_{N-1}(I-I')$$

The reliability function can be expressed as:

$$r(N \mid I \pm \Delta I) = \frac{p(N \mid I \pm \Delta I) - p(N \pm 1 \mid I \pm \Delta I)}{p(N \mid I \pm \Delta I) + p(N \pm 1 \mid I \pm \Delta I)}$$

The criterion for fluorophore assignments is $r > 0.25$. The approach was applied to the stoichiometry analysis of the aggregation of fluorescence-labeled biotin–streptavidin complexes in a phospholipid membrane [8.139].

The lateral diffusion of single fluorescent–labeled lipid molecules (TMR/POPE, 1-palmitoyl-2-oleolyl-sn-glycero-3-phosphoethanolamine (POPE) derivatized with 5-(and 6-)carboxy-tetramethylrhodamine succinimidyl ester (TMR)) at a length scale of 100 nm was observed to be in agreement with the random walk model, and differed from the ensemble averaged behavior. The measurement was based on the mean-square-distance versus measurement time delays. The measurement of individual TMR/POPE molecules revealed an exponential decay behavior typical for the random-walk model. The exponential decay factor showed a corresponding diffusion coefficient ($1.4 \pm 0.3 \times 10^{-8}$ cm^2 s^{-1}) similar to the value obtained from the slope of the ensemble averaged statistical plot ($1.42 \pm 0.23 \times 10^{-8}$ cm^2 s^{-1}). The diffusion coefficient also revealed inhomogeneities in the fluidic lipid membrane.

Similar studies have been performed on phospholipid bilayers and polymer-supported phospholipid monolayers by measuring the probability density functions [8.19].

The observation of fluorescence intensities of sparsely distributed rhodamine B molecules on silicon oxide surfaces resulted in quantized histograms, and such features were also used in estimating fluorophore numbers within each observation spot [8.140].

Similar studies on the interface effect have been performed with adsorbed dye molecules on surfaces of glass or silica, both with and without covalent attachments. Generally, the data demonstrate reduced fluorescence lifetime due to charge transfer to and from the localized states on the surface [8.141–8.144]. The heterogeneous structure of surfaces can result in non-exponential decay behavior. In addition, the fluorescence quenching effect of energy quencher molecules (azulene) was studied using surface-adsorbed single molecules of tetramethyl rodamine (TMR) [8.143].

The measurement of the emission spectra of single cresyl violet molecules adsorbed on indium tin oxide (ITO) surfaces showed a strong variation in fluorescence lifetime, which was ascribed to heterogeneous interactions between the adsorbate and substrate [8.145].

8.5.3 Fluorescence from Single Atomic Clusters and Defects

Fluorescence of single atomic clusters of Ag_n ($n = 2$–8) was observed from photoillumination [8.146] and electrical activation (by applying either DC or AC voltages) [8.147]. Similar electroluminescence was observable with Cu_n clusters at higher dc voltages. The emission characteristics of single semiconducting nanocrystals have also been reported [8.148, 8.149].

Individual nitrogen-vacancy defects in diamonds have been identified as stable luminescent centers. The optically detected magnetic resonance of individual N-V centers indicated different splitting parameters, which could be attributed to the strain-induced differences among the centers [8.150].

Single photon emission from single molecules of terrylene in para-terphenyl matrix [8.151], dibenzanthanthrene (DBATT) in hexadecane (HD) matrix [8.152], and nitrogen-vacancy (NV) centers in diamonds [8.153] were demonstrated.

The study of NV centers in nanocrystalline diamonds showed photostability similar to that of bulk diamonds [8.154].

8.6 Tip-Induced Single Molecule Fluorescence

Tunneling electrons from STM tips have been shown as a source for exciting photon emission of C_{60} molecules adsorbed on Au(110) 1×2 surfaces [8.155]. The mapping of the photon emission from the C_{60}-covered surface can be correlated with the STM topography image recorded quasi-simultaneously, with a spatial resolution of about $4\,\text{Å}$. The contrast of emission mapping does not depend on the tip polarity at ± 2.8 V.

Fluorescence excitation was also achieved by injection of energetic tunneling electrons. This excitation was localized at submolecular level with the aid of an STM tip. The fluorescence of a C_{60} monolayer on a Au(111) surface was measured at 78 K at tip bias voltage 2.3 V. A spectral peak centered at around 1.65 eV was ascribed to the fluorescence from C_{60} molecules. This value is close to the energy difference between HOMO and the Fermi level (about 1.7 eV) [8.156].

Tip-induced fluorescence was shown in Zn(II)-etioporphyrin I adsorbed on NiAl(110) surfaces [8.157]. It was found that the fluorescence spectra are sensitive to molecular conformation. It is probable that the combination of spatial resolution of the STM technique and single molecule spectroscopy will yield further information useful in understanding the nature of molecular fluorescence. In a related study, fluorescence of Cu-tetra-[3,5-di-t-butylphenyl] porphyrin (Cu-TBPP) was observed with a high bias voltage around 6.0 V and tunneling current of 5.0 nA [8.158].

8.7 Dynamics of Single Polymeric Molecules Studied by Fluorescence Microscopy and Related Techniques

As an extension of fluorescence microscopy, it would be worthwhile to switch attention to studies focusing not on the behavior of single fluorescence molecules, but rather on single molecules labeled with multiple fluorescence molecules. The experimental techniques are also expanded to include those results that are of high relevance to the fluorescence observations. These investigation mostly involve polymeric molecules and biomolecules, and have been extensively pursued using fluorescence microscopy. The advantage of fluorescence microscopy in terms of large viewing field, contrasting with atomic force microscopy, makes it a unique technique for studying the dynamics of long molecules. The importance of such studies can be found in fields such as polymer physics and biophysics, for topics such as the reptation process, entropic force, and conformational variations of single molecules. This has been presented in many review articles and books. Here, we document several case studies using mainly DNA molecules, to exemplify progress of more general interest.

8.7.1 Dynamics of Single Macromolecules in Solutions

It has long been realized that an entangled chain macromolecule should move along a path defined by its own contour under topological confinements, as described in the reptation model [8.159–8.162]. The observation of single macromolecules under various confinement conditions has enabled direct verification of theoretical predictions.

The direct observation of self-diffusion of DNA molecules using fluorescence microscopy in solutions confirmed the agreement with the reptation model. However, a deviation from the standard scaling exponent was observed for short DNA segments over a range of concentrations, indicative of microscopic effects at single molecule level [8.163]. By using fluorescence correlation spectroscopy, the dynamics of DNA molecules was shown to be consistent with the model for semiflexible chains under hydrodynamic interaction. The autocorrelation function was shown to be sensitive to the fluorescence label densities [8.164].

The effort to detect and separate molecular species at single molecule level has been seen for many years. Representative studies can be found in the development of electrophoresis techniques. The pursuit of single molecule resolution in the electrophoretic separation process has stimulated exciting advances in this field.

The migration of polymers (DNA, polyelectrolyte molecules) under the influence of an electric field has been studied both theoretically and experimentally over decades. The understanding of electrodiffusion mechanisms is crucial in developing techniques for separating molecules of different mass (electrophoresis). The important parameters of the migration process are the gyration radius of the molecules, and the characteristic pore size of the gels. A number of studies have introduced model structures such as two-dimensional arrays of posts prepared by microlithographical methods [8.165–8.167]. The results confirm that the DNA molecules are moving in a free-draining state when not attached to the posts, and the time duration for molecules remaining attached to the posts was shown to be a function of molecular length. It was also pointed out that simple post arrays would not lead to a significant range of electrophoretic mobility [8.165]. In a different design of post array that includes narrow entrance and branching points, the mobility was shown to have strong dependence on the molecular length [8.165].

The combination of confocal microscopy with capillary gel electrophoresis was shown to enhance the on-column sensitivity of DNA molecules [8.168]. Several improvements have been implemented to improve signal-noise ratios, including optimizing the intercalating dyes and digital photon counting interface. A thiazole orange derivative (TO6) was found to be a sensitive intercalator. The capability of counting single DNA molecules in the flow measurements was supported by the autocorrelation function of the dilution series recording the fluorescence bursts of the pass molecules. An on-line sensitivity of a few hundred M13 DNA (7,250 bp) molecules was demonstrated.

Single molecule fluorescence detection was shown applicable in the quantitation of β-actin DNA (838 bp) labeled by YOYO-1. By consecutive recording of the molecular distributions, the molecular migration mobility was obtained and used as a parameter for screening DNA molecules from impurities [8.169]. It was also reported that the presence of PEO could significantly reduce the impurity-associated background fluorescence. With appropriate criteria, it was shown that the observed β-actin DNA population contained in the detection volume is proportional to the bulk concentration.

The hydrodynamic screening effect in the electrophoretic movement of single DNA molecules was studied using micrometer-sized arrays and the DNA molecules symmetrically hooked to the posts. A generalized depen-

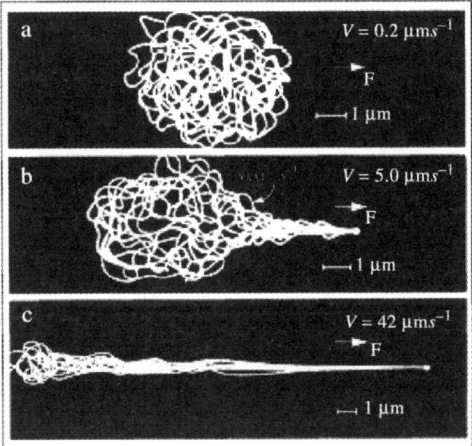

Fig. 8.17. Fluorescence image of DNA molecule under different pulling speeds of a magnetic bead. **a** 0.2 μm s^{-1}; **b** 5.0 μm s^{-1}; **c** 42 μm s^{-1} (extracted from [8.175])

dence of the collision time and travel distance on the impact parameter was described [8.170]. It was found that the thinner array led to higher extensions and slower relaxation rate under flow conditions [8.171]. The observation confirms that the total drag force for polyelectrolyte molecules is independent of the molecular conformation, and the hydrodynamic interaction can be adjusted by the level of confinement.

Dynamic behavior of single DNA molecules in shear flow has been studied with fluorescence microscopy. Fluorescence-labeled lambda bacteriophage DNA (λ-DNA) has been studied in various flow conditions by means of video fluorescence microscopy. A steady shear flow can induce temporal fluctuations in the conformation of single λ-DNA molecules [8.172].

In elongational flows, or flow with a velocity gradient along the flow direction, the stretching behavior of λ-DNA molecules has been shown to be related to various conformations, such as dumb-bell, half dumb-bell, coiled, folded and kinked molecules, in both steady and sudden flows [8.173, 8.174].

The friction coefficient of single DNA molecules in solution was measured by the magnetic bead method [8.175]. The DNA molecule was attached to the magnetic bead by means of a linker of streptavidin, and pulled by an external field (Fig. 8.17). The concurrent fluorescence microscopy observation showed strong conformation change at different pulling velocities. The measured friction coefficient was constant at low velocity, and became nonlinearly dependent on high pulling velocities when the contours of the DNA molecule was highly deformed.

8.7.2 Single Molecules Moving Through Channels

The transportation of single molecules across membranes can be monitored as an ionic current signal [8.176–8.178]. Although this is not a fluorescence detec-

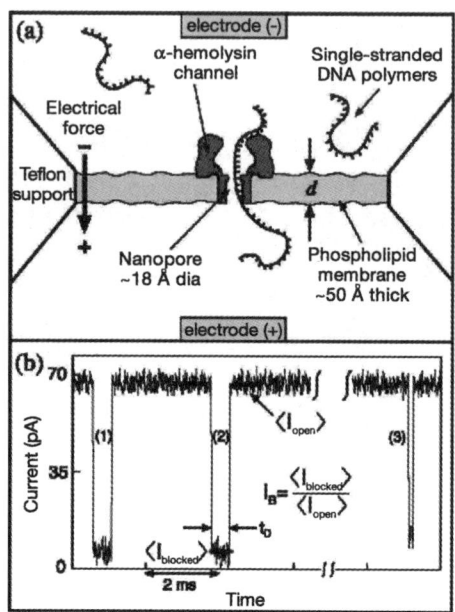

Fig. 8.18. a Schematic of DNA molecules passing through a lipid bilayer. **b** The translocation of single DNA molecules is reflected in the sudden blockade of ionic current (extracted from [8.180])

tion approach, we consider it as of great importance to the study of macromolecule dynamics, and therefore this information is presented as a complementary section. The transportation of single-stranded oligonuncleotides through nanopores embedded in phospholipid bilayers can be monitored by sudden changes in ionic current across the bilayer (Fig. 8.18) [8.179, 8.180]. The nanopores were formed by self-assembling α-hemolysin (α-HL) in the bilayers. The probability of the molecule entering via the nanopore opening can be derived from the distribution of the time intervals between successive entering events, and was shown to depend strongly on the electric potential and temperature, but only weakly on the sequence of the DNA molecules. The interaction between the oligonucleotides and the channel was evidenced by the sharp transition of translocation time as the molecular length approaches that of the channel stem. A dependence of translocation duration time on the DNA sequence was also observed. These data could provide useful insight into transmembrane channeling processes. The driving force for single-stranded DNA through α-HL nanopores can be estimated by [8.180]:

$$F \simeq zeV/d$$

where ze is the effective charge per base, V the voltage drop across the nanopore, and d the distance between adjacent bases (about 4 Å).

Strong dependence of the measured ionic current level and apparent translocation velocity on the oligomer length (or base number N) was identified for $N < 12$. This is equivalent to the α-HL pore length. In addition, the translocation velocity was found to depend non-linearly on the applied voltage, considered to be associated a with a possible confinement-induced drag force.

It was shown that when cyclodextrins were introduced to the inside of the nanopores of α-HL, selectivity in ligand bindings can be achieved [8.181, 8.182]. The kinetics of binding events of the analyte to cyclodextrins can be characterized by the stochastic behavior of ionic currents.

By covalently attaching single-stranded DNA to the nanopore formed by α-HL, the kinetics of duplex formation can be studied as the second oligonucleotide molecule enters the nanopore and is translocated through the membrane under the electric field [8.183]. Enthalpy barriers for unzipping events of double-stranded DNA oligomers through α-HL nanopores were obtained from the distribution of blockade time durations [8.184]. The effects of temperature, voltage, and ionic concentrations on the unzipping rate were also demonstrated.

In another study, microlithographically fabricated channel structures were used to investigate the channeling process of single DNA molecules [8.185, 8.186]. The real-time observation of the migration process of long-chain DNA molecules through an entropic trap formed by a 90-nm channel revealed that the escape energy barrier is independent on the chain length. This is because the escape of the polymer from the entropic trap is initiated by stretching a small part of the polymer, rather than the whole molecule. The measured mobility increases with the polymer chain length. This is because bigger polymers have more monomers facing the channel to form beachhead for escape.

Separation effects based on entropic recoiling of long DNA molecules passing through nanofluidic channels have been observed [8.186, 8.187]. The recoiling of long-strand DNAs was controlled by a pulsed electric field. Whereas the short-stranded DNA may be fully entrapped inside the channel, a portion of long DNAs may remain outside the channel. Once the electric field is removed, the long DNAs undergo a recoiling process that enables the separation of DNA molecules of different contour lengths. A multistage setup involving similar interfaces was proposed to further enhance the separation power.

8.7.3 Migration of DNA Molecules on Flat Surfaces

The conformation relaxation and self-diffusion dynamics of long DNA molecules adsorbed on cationic lipid bilayers was directly visualized using fluorescence microscopy [8.188]. The DNA molecules (λ-phage) can be modeled as strongly adsorbed to the cationic surface, and the diffusion is restricted in the two-dimensional regime. The diffusion behavior can be characterized by

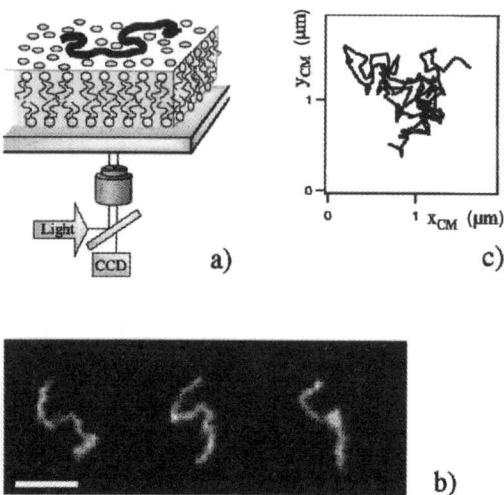

Fig. 8.19. a Schematic of the setup for studying DNA movement on a lipid bilayer surface **b** Fluorescence images of 10,090-bp λ-DNA segments (bar equals 10 μm) at 2-seconds intervals. **c** Center of mass movements (extracted from [8.189])

the movements of the center of mass, and was found to be in good agreement with Rouse dynamics (Fig. 8.19) [8.188].

Fractioning effects have also been found for DNA molecules on silicon oxide and silane-modified Si surfaces [8.189]. These two surfaces represent different levels of interactions between DNA and support surfaces. The polar silicon oxide surface interacts with DNA more strongly than does the silane-modified Si surface. The ability to separate molecules was attributed to the friction between the surface and adsorbed DNA segments. A qualitative illustration can be proposed by dividing the surface-bound DNA structure into two portions, a part that adsorbs directly to the surface and a part that extends into solution. Under the influence of an applied electric field, when considering independently the electrophoretic motion of the adsorbed portion or the portion in solution, information on the free draining process is gained. The molecular dynamics simulation results showed that the corresponding mobilities are independent of the chain length, and the mobility is higher for DNAs in solution than is the case for strong interacting surfaces. This situation corresponds to a strongly interacting surface (where DNAs are fully adsorbed to the surface) or a weak interacting surface (where DNAs mostly extend into the solution). It is important to note that in the intermediate interacting regime, the surface-bound DNA will have both the adsorbed part and the part that extends into solution. The segment of adsorbed DNA is qualitatively proportional to the total chain length, leading to different mobilities. This study indicates that flat surfaces can be utilized to separate molecular motions of surface-bound polymeric species.

8.7.4 Single Molecule Condensation of DNA

The condensation of DNAs into compact structures is a common process for native genomes. The controlled bending and kinking of DNAs are possible with enzymes [8.190], ions [8.191, 8.192] or polyamines [8.193, 8.194]. For example, it was found that the Cro protein can induce DNA bending, and the bend angles of DNA were different for specific and non-specific sites [8.190]. It is known from extensive studies that many factors influence the packing processes. There is evidence suggesting, that naturally occurring polyamines in living cells, such as spermine, spermidine and putrescine, can exert a substantial effect on the packaging process of DNAs in bacteria [8.195] and viruses [8.196]. However, the role of polyamines in the higher-order condensation of chromatin as well as their effect on compacted DNA expression are yet to be fully understood.

Many studies [8.193, 8.194] have demonstrated that in vitro polycations (including polyamines) can condense DNAs into three main types of structures, depending upon the concentration and length of the DNA molecules: (1) in extremely dilute solutions, long DNA molecules undergo a monomolecular collapse; (2) in very dilute solutions, microaggregates form with short or long molecules, and remain in suspension. Experimental data show that certain condensed structures such as toroids and rods [8.197–8.200] can be induced by polyamines in DNA solutions.

The condensation of DNA molecules has been studied by a number of techniques. With fluorescence microscopy, real-time observations of DNA condensation under the influence of solvents have provided information on the energy landscape of the coiled state–globule state transition of single DNA molecules (Fig. 8.20) [8.201]. The interaction of fluorescent (YOYO-1) labeled λ-DNA with a silica surface, as well as an alkane thiol-modified Au surface by evanescent field excitation revealed that the molecular conformation and adsorption behavior depends on pH and solvents, suggesting a combination of hydrophobic (non-polar) and electrostatic interactions in the adsorption process [8.202]. Whereas a random coil structure of λ-DNA is prevalent at neutral and basic pH, a condensed state results at pH < 3.0 on fused-silica surfaces.

In the case of toroidal structures, and the important role of polyamines in vivo, understanding the organization of DNA within toroidal condensates and the dynamics of their formation became the focus of further studies. Toroidal structures have been extensively investigated by using electron microscopy [8.203], light scattering [8.204], circular dichroism (CD) and hydrodynamic measurements [8.197, 8.205]. Experimental data reveal that although the toroid sizes from independent preparations vary considerably [206, 8.208], average toroid size is surprisingly unaffected by the length of DNAs being condensed. DNA molecules ranging from 400 bp in size to genomic length can experience monomolecular or multimolecular aggregation to produce toroids similar in size [8.193]. Quantitative analysis has suggested that free energies, including electrostatic energy, bending energy, hydration energy and mixing energy, may be the main driving forces underlying DNA condensation.

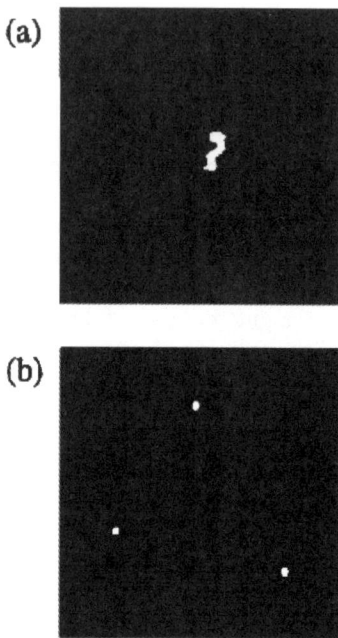

Fig. 8.20. Fluorescence image of DNA molecules in **a** coiled state in solutions without alcohol, and **b** globule state in 40 vol% 2-propanol solutions (extracted from [8.201])

Among these factors, electrostatic repulsion is dominant, and condensation cannot occur unless ∼90% or more of the charge along the DNA backbone is neutralized [8.204, 8.209]. In-depth and detailed discussions can be found in many studies [8.210, 8.211].

Based on the biological implication of toroidal condensation, it was postulated that the DNAs in the toroid should be arranged with certain characteristics, as explained by a spool-like model [8.212, 8.213], which postulates that DNAs are wound in an orderly way to form a toroid, and by the constant radius of curvature model [8.214], which postulates that the DNAs are circumferentially wound with a constant diameter to form a toroid. Noguchi et al. [8.215] suggested that the formation of toroids occurs as a coil–globule transition of polymers, with the folding process of toroid accompanied by the partly coiled DNA chain. Recently, Dunlap et al. [8.216] used atomic force microscopy (AFM) to examine DNA in incomplete condensates induced by polyethylenimine, and identified, for the first time, individual DNA strands in the condensed state. In this experiment, DNAs were observed as being clearly arranged in parallel.

Moreover, some experimental data indicate that local ordering exists in DNA–spermidine condensates. Very dilute solutions of high-molecular weight DNA in the presence of the tetravalent cation spermine show CD spectra characteristic of a helical cholesteric supramolecular ordering of DNA [8.217]. X-ray diffraction studies of the structure of the aggregates formed by spermidine or by trivalent cation spermidine with high-molecular weight DNA molecules [8.218] revealed a strong equatorial reflection, which corresponds to a 25.5-Å periodicity. These results suggest the formation of a hexagonal lattice (crystalline or liquid crystalline). It was also discovered that over a large range of spermidine and DNA concentrations, short DNA molecules can form liquid crystalline phases [8.194,8.219]. This result raised the questions of whether both the toroids and rods have similar local ordering, how the ordering structures are arranged, and whether these structures are microdomains of a liquid crystalline or crystalline phase. In another study, by using atomic force microscopy, both monomolecular (type I) and multimolecular (type II) toroids were observed in different spermidine–DNA concentrations [8.220]. These data reveal highly localized inhomogeneity of the toroidal structures.

9 Single Molecule Fluorescence Imaging and Spectroscopy: Near-Field Studies

9.1 Near-Field Scanning Optical Microscopy

9.1.1 Introduction of Near-Field Effect

Conventional optical microscope has numerous advantages that have made it the most popular imaging system. Even with the proliferation of different types of microscopes today, there is no other microscope that can match all the advantages of optical microscopy. However, a primary disadvantage is a fundamental limit to the resolving ability of conventional optical microscopy. The spatial resolution is about $\lambda/2$, the well-known diffraction limit [9.1], on the order of half a wavelength. This has greatly limited the application of optical microscopy.

Optical microscopy and spectroscopy have long been key techniques in medicine, biology, chemistry and materials science. There are a few advantages of optical microscopy and spectroscopy:

Universality: All materials and samples attenuate light and have spectroscopic states. Optical microscopy can be used for observing a wide variety of biological and chemical samples.

Non-destructiveness: Optical microscopy can operate with any transparent fluid medium between the objective and the sample (e.g., air, water or oil), so the sample can be viewed in its native environment. When non-ionizing, visible light is used, radiation damage is negligible. Most chemical reactions are not perturbed by long-wavelength light.

Convenience: In most cases, no sample preparation is needed to view specimens with optical microscopy. However, labeling, staining and sectioning are sometimes required. Optical microscopy is inexpensive and simple to operate. Also, optical microscopy is usually safe, and precautions are limited mostly to protective eyeglasses.

Real-Time Observation: The speed of optical microscopy is limited only by signal-to-noise ratio considerations, and dynamic processes can be studied with optical microscopy. By using ultrafast light sources, the speed can be extended even into the femtosecond time domain. Thus, biological phenomena, chemical reactions, crystallization, etc., can be observed under the microscope as they occur in situ or in vivo.

In the past two decades, the idea of active subwavelength-sized light sources [9.2–9.4] led to single molecule detection and imaging [9.5–9.7]. The

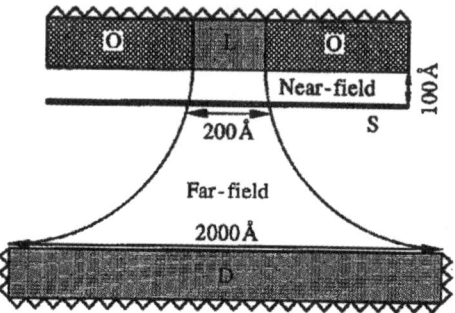

Fig. 9.1. Schematic of near-field optics configuration

realization of better spatial resolution by smaller light sources has led to the concept of near-field optics (NFO). The principle underlying this concept is schematically shown in Fig. 9.1. The near-field apparatus consists of a near-field light source, sample, and far-field (or near-field) detector. To form a subwavelength optical probe, light is directed to an opaque screen containing a small aperture. The radiation emanating through the aperture and into the region beyond the screen is at first highly collimated, of dimension equal to the aperture size, which is independent of the wavelength of the light employed. This occurs only in the near-field regime. To generate a high-resolution image, a sample has to be placed within the near-field region of the illuminated aperture. The aperture then acts as a subwavelength-sized light probe that can be used as a scanning tip to generate an image. Therefore, this is called near-field scanning optical microscopy (NSOM) [9.2–9.10].

There are several contrast mechanisms proposed for near-field optical microscopy that permit clear imaging of features in a broad range of samples. These mechanisms, in addition to absorption and scattering, include fluorescence microscopy [9.8–9.10], polarized-light microscopy, phase contrast microscopy, and differential-interference microscopy [9.11–9.14]. These are not easily accomplished by other techniques such as electron microscopy or X-ray crystallography.

Unlike scanning tunneling microscopy (STM) or atomic force microscopy (AFM), imaging in NSOM is via the interaction of light with the surface by either a simple refraction/reflection contrast, or by absorption and fluorescence mechanisms. The advantages of NSOM are its non-invasive nature, ability to view non-conducting and soft surfaces, and the addition of an optical spectral dimension. The latter does not exist in either STM or AFM. The potential for extracting spectroscopic information from a nanometer-sized area makes it particularly attractive for biomedical research and materials science.

There are two signal detection modes for NSOM operations. One is using the aperture as an illumination source [9.15]. The evanescent field at the aperture interacts with the sample, and the light transmitted through sample is collected through optical lenses as far-field signals (illumination mode).

The other approach is to use far-field light to illuminate the sample, and the NSOM probe is used to collect the generated near-field signal (collection mode) [9.16].

The two prevalent operation modes of traditional optical microscopy are transmission and luminescence. For NFO chemical and biological applications, these will probably continue to be the most important modes, especially if one includes spectroscopy measurements. Also, various reflection and collection mode NSOM techniques have been devised [9.8–9.10]. Figure 9.2 gives several arrangements used in NSOM. There are several contrast methods: absorption, refractive index, reflection and fluorescence (luminescence), and not all of these are well understood. One can also count polarization [9.11,9.12] and spectroscopy as separate modes of contrast. Furthermore, there are a large number of quantum effects, such as energy transfer, energy down-conversion and energy quenching [9.10]. The simplest optical contrast mechanisms in the near-field regime, e.g., refractive index, are not yet fully understood [9.17], and thus they are still under intensive study. Actually, the microscopic quantum effects are better understood than the mesoscopic (near-field) optical interactions. The most important consideration for sample preparation is surface roughness. It is limited by the probe shape in a way similar to all scanning probe techniques. The sample thickness is an important factor for all transmission (forward-scattering) modes of operation, but not for the reflectance (back-scattering) and some "collection" modes. The near-field approach couples the optical resolution with the distance from the probe; the higher the desired resolution, the thinner the required sample. On the other hand, the contrast mode (absorption, refractive index) may limit the thickness of the sample. Even the fluorescence mode may be limited by the thinness of the sample i.e., the absorption cross section. However, this can be overcome by intensity, by auxiliary fluorophores, or by quantum mechanisms (energy transfer). Thus, various luminescence modes appear to be the most promising for forward-scattering near-field microscopy.

In brief, NSOM can be applied in studies such as [9.9]:

- Absorption imaging
- Polarization imaging
- Refractive imaging
- Reflectivity imaging
- Fluorescence imaging
- Excitation spectra
- Emission spectra
- Fluorescence lifetime

The best resolution to date has been claimed [9.9] to be about 12 nm (with 514-nm light). Presumably this was achieved with a 20-nm-diameter aperture. A signal of 50 nW has been claimed for an 80-nm aperture [9.9]. Also, NSOM has successfully been applied in single molecule detection [9.5–9.7]. The integration of NSOM with fluorescence spectroscopy has enabled spatial resolution typically around 100 nm or less, in addition to spectroscopic

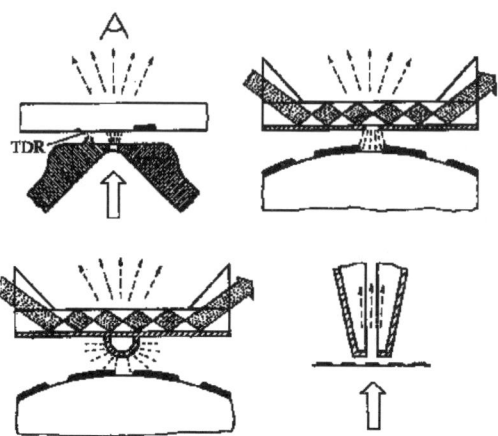

Fig. 9.2. Illustration of experimental setups for NSOM operation

analyses. Such improvements have far-reaching implications in fluorescence studies of molecular aggregates, photochemical and photophysical processes at single molecule level [9.18].

9.1.2 NSOM Probe Designs

By scanning an optical probe on a sample surface, near-field scanning optical microscopy (NSOM) can provide images with a resolution higher than the Abbe diffraction limit [9.3]. The resolution of the image is generally considered to depend only on the probe size and the probe–sample distance [9.9]. Therefore, controlling the probe in close proximity to the sample surface is an essential issue for successful NSOM work. The probe aperture is typically between 30 and 100+ nm in diameter. The obtained optical information is unique among SPM techniques. Extensive applications in biomolecules research have been actively pursued world wide.

The light source, which is the "heart" of the NFO technique, has to be (1) small, (2) intense, (3) durable, and (4) spatially controlled. As is well known, the size of the tip determines the resolution, provided that the tip can be scanned close enough to the sample. There are two major types of probe materials used in NFO: metal-coated glass micropipettes, and nanofabricated optical fiber tips [9.8–9.10]. Optical fiber tips and micropipettes are easily fabricated to sizes of approximately 50 nm, and the smallest nanofabricated optical fiber tip reported to date is about 20 nm [9.9]. The fabrication of miniaturized optical probes has assisted the development and application of NFO in a wide variety of fields.

Near-field optical nanoprobes can be classified into three categories: passive optical probes, such as coated micropipettes or small holes on screens [9.8, 9.19, 9.20], semi-active light sources, such as optical fiber tips [9.3, 9.9, 9.10], and active light sources, such as nanometer crystal light sources

[9.2, 9.4, 9.21]. Compared to a hollow micropipette tip, a nanofabricated optical fiber tip is a "semi-active" photon tip that is orders of magnitude brighter, easily coupled to an optical source, and at least as mechanically sturdy as a micropipette. It is interesting to note that the top of a fiber tip is very resistant to breakage. The photochemical stability of optical fiber tips is excellent under very intense illumination. It was found that heat could damage the aluminum coating at the tip. Both types of probes have been made around 500 Å in diameter without difficulties in applications as light sources.

Optical fiber tips have been used in many fields, and can be fabricated either by heating and stretching or by chemical etching. The apparatus for fiber-tip pulling usually consists of a micropipette puller and a heating unit (CO_2 infrared laser or electric filament). The CO_2 laser beam or the electric filament in the puller is used to heat the optical fiber for the pulling process. The laser beam is reflected by a mirror, and directed to heat the optical fiber fixed on the puller. The details of the pulling setup and procedures can be found in a number of publications [9.3, 9.4, 9.9]. By using appropriate program parameters and laser power, optical fibers can be tapered to subwavelength diameters. After pulling, the optical fiber tip is coated with aluminum by vapor deposition to form a small aperture. The procedure of vacuum deposition of metals is well known but far from trivial. A specially built high-vacuum chamber is employed for coating these pulled fiber tips: only the fiber-tip sides are coated with aluminum, leaving the end face as a transmissive aperture. To transform it into a light source, a visible or UV laser beam is coupled to the opposite end of the pulled tip. This probe delivers light very efficiently, since most of the radiation is bound to the core up to a few micrometers away from the tip. A randomly chosen 0.1-µm optical fiber probe gives 10^{12} photons per second [9.21]. Using the same puller, glass micropipettes have been pulled with different subwavelength diameters. Both optical fiber tips and micropipettes have been used as optical nanoprobes.

The metal-coated fiber tip can be further polished by focused ion beam (FIB) treatment to obtain a highly uniform, flat end of the probe. Such treatment can reduce the uncertainty caused by the metallic grains in the coating [9.22–9.25]. The aperture diameter of probes polished by FIB treatment is in the range 35–100 nm. The imaging of single fluorescence molecules of R6G and perylene orange with such probes has been demonstrated [9.26]. In the case of single perylene orange molecules, it was shown that the electric field at the boundary between aperture and coating may cause additional features in the apparent molecular topography.

With the help of electron beam lithography [9.27, 9.28] and reactive ion etching [9.29], a nanometer-sized aperture can be fabricated in a standard AFM cantilever, which allows sensing the normal force while introducing illuminating light through the hole. The induced luminescence can also be collected through the probe. Such probe configuration benefits from typical AFM capabilities in approaching and topography imaging. Metallic tips (chromium) with hollowed apertures were also fabricated with microlithography techniques (Figs. 9.3 and 9.4), and can potentially be used for simulta-

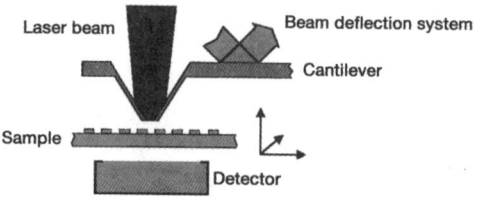

Fig. 9.3. Schematic of detection configuration using a microlithography fabricated NSOM cantilever probe (extracted from [9.30])

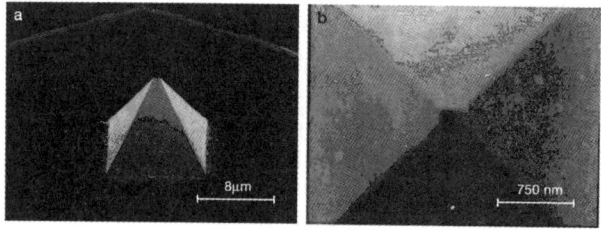

Fig. 9.4. a and **b** are SEM images of the probe apex. The aperture is about 130 nm. (extracted from [9.30])

neous measurements of optical mechanical, magnetic and thermal properties of the samples [9.30].

By bending the aperture end of optical fibers, a cantilever-shaped probe can be formed. Efficient control of probe–sample separation can be achieved with normal force as control parameter (Fig. 9.5) [9.31–9.33]. This approach introduces the AFM contact or tapping mode feedback for probe approaching and probe–sample separation control. Cantilever-shaped optical fibers can also be applied to study Raman spectroscopy [9.32]. However, one should note that the cantilever probes often suffer from bending losses that can reduce the light throughput efficiency. In order to compensate such losses, higher excitation power should be used. It is interesting to note that the combination of optical and force detection could lead to new detection functions for studying effects such as light-induced protein conformation dynamics [9.34, 9.35].

In another study, a rectangular-shaped cantilever mounted with an aluminum-coated quartz tip was explored as NSOM probe, and the probe–sample separation was controlled as constant height in AFM operations [9.35]. Fluorescence-labeled protein goat anti-rat antigen can be clearly resolved individually using such probes.

The apertureless approach uses the refractive medium to focus the incoming beam as well as signal collection [9.36]. Since this method does not require probe fabrication, the uncertainties of detailed tip geometry are of less significance.

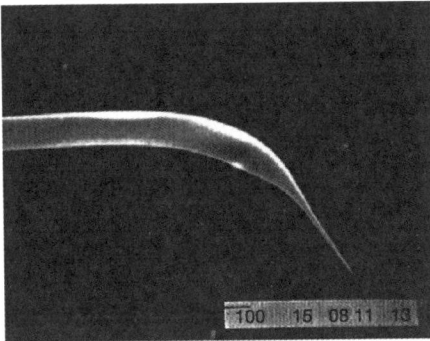

Fig. 9.5. A cantilever-shaped NSOM probe (extracted from [9.31])

9.1.3 Approaching Modes

As presented in a number of reports, a widely employed technique to control the probe–sample distance is that based on a shear force mechanism [9.37, 9.38], by which oscillation of a tapered fiber optic probe excited at its resonance frequency is damped as the probe approaches the sample surface. The decrease in amplitude of the vibration probe can be detected and used as a distance control signal.

The other cause for uncertainties in probe approaching by the shear force mechanism is the detection of change in vibration amplitude, which has been tackled in two different ways. The first is by means of an optical method. The more widely used system is based on focusing a laser beam on a modulated fiber optic probe, and measuring the laser spot diffracted by the probe with a split detector. The signal obtained from the detector is proportional to the vibration amplitude, and can be used in a feedback loop to perform distance control [9.39].

The second approach is based on various non-optical methods using a standard tuning fork [9.40], a piezo tube [9.41], piezo plate [9.42], and bimorph cantilever [9.43]. The most obvious advantages of non-optical methods are: they do not require an additional laser source, preventing stray light from disturbing the measurement of the NSOM signal; they are not necessary for accurate alignment of the external optics with respect to the probe; and they are easily applied to an NSOM to operate at low temperatures, under vacuum or in liquids. As mentioned in the preceding section, the cantilever-shaped probes can be adapted to atomic force microscopy by using the normal force as approaching control parameter. Other non-optical methods for approaching include capacitance detection [9.44], and the impedance method [9.45].

The probe–sample approach can also be achieved by combining the optical detection function with STM and AFM. For a cantilever-shaped probe, the AFM approaching modes can be readily adapted. For a gold-coated fiber probe, one can use the tunneling current as the control parameter to adjust probe–sample separations [9.15, 9.16].

9.2 Near-Field Scanning Optical Microscopy and Spectroscopy

Advances in nanometer-resolved microscopy, spectroscopy and chemical sensor probes promise to bring us much closer to an ultimate goal in chemical analyses, i.e., the non-invasive detection of a single molecule, radical or ion, the determination of its precise coordinations, and the characterization of its structural conformations, as well as its internal dynamics and energetics, as a function of time and environmental perturbations. Near-field optical microscopy and spectroscopy is a new tool providing hope for highly improved imaging of thin organic films at microscopic scale [9.46] and at single molecule level [9.5–9.7, 9.18, 9.33]. Near-field spectroscopy has also been applied within living cells [9.47].

9.2.1 Near-Field Optical Microscopy

Techniques such as STM and AFM come close to the ideal of single molecule electrical and mechanical detection but there are still some challenges that are particularly acute for soft organic/biological molecules. The detection of optical properties of single molecules below diffraction limitation could lead to more detailed information on the internal energetic structures of the molecules.

Individual carbocyanine dye molecules in a submonolayer spread have been imaged with NSOM [9.6]. About two dozen isolated dye molecules can be imaged within seconds. The imaging resolution is about 50 nm, and the molecular location is resolved within about 25 nm in the horizontal plane, and 5 nm in the vertical direction. Furthermore, the much smaller molecular transition dipole is a point detector mapping out the electric-field distribution of the near-field light source. In addition to imaging individual dye molecules, one can also obtain information on the orientation of these molecules (via polarization and transition dipole fitting).

The microscopic diffusion behavior of rhodamine-6G (R6G) molecules embedded in a thin film matrix of polyvinylbutyral (PVB) was studied by NSOM. The dye molecules were excited by the evanescent field of the NSOM probe (17 nW at $\lambda = 514.5$ nm) [9.48]. Two distinctively different types of behavior were revealed. At submicrometer scale, a non-uniform diffusion behavior was identified that can be attributed to the reptation-like motion of macromolecules (Fig. 9.6). Observations at a larger scale showed the diffusion is dominated by random-walk behavior. The orientation-dependent fluorescence intensity was ascribed as a source of inhomogeneous emission.

Using cantilevered probes, the fluorescence molecules 1,1'-dioctadecyl-3,3,3',3'-tetramethylindocarbocyanine perchlorate (diIC$_{18}$) within a matrix of L-α-dipalmitoylphosphatidyl-choline (DPPC) monolayer can be resolved in ambient and water conditions [9.49]. Since in this operation mode the probe–sample distance is controlled by the tapping mechanism, the drive amplitude may have some effect on the NSOM image resolution. It was found

9.2 Near-Field Scanning Optical Microscopy and Spectroscopy

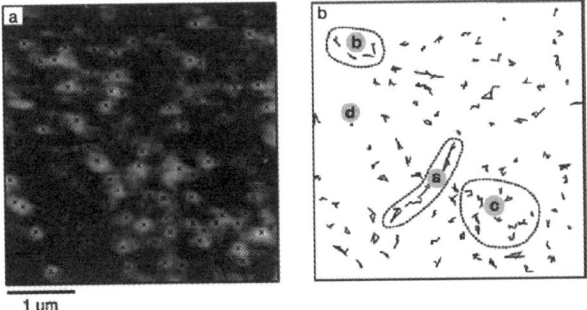

Fig. 9.6. a NSOM fluorescence image of dispersed R6G molecules, and **b** the trajectories of the molecules from seven consecutive images (extracted from [9.48])

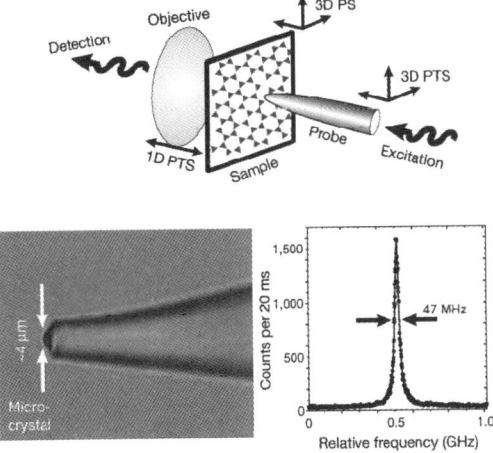

Fig. 9.7. Schematic of combined NSOM and confocal microscope using single fluorescence molecules as the light source (extracted from [9.50])

that by reducing the drive amplitude, the aperture of the cantilevered probe can remain within the collimation zone, generating high-resolution images of soft sample surfaces.

The micrometer-sized crystal p-terphenyl with doping concentration of terrylene was attached to an optical fiber probe suitable for standard NSOM operation in combination with a confocal microscope, as shown in Fig. 9.7 [9.50]. Both topography and optical (using the fluorescence of single terrylene molecules) images of a two-dimensional array of aluminum islands can be obtained with such probes. The contrast of the sample features displays a variety of patterns for different tip–sample separations. Theoretical simulations of photonic local density of states based on Green's dyadic formalism suggest that the observed patterns result from the variation of molecular orientations [9.51].

9.2.2 Near-Field Optical Spectroscopy

Both topographic and spectroscopic information can be obtained with near-field optics configurations. Near-field scanning optical spectroscopy (NSOS) [9.10, 9.43] is based on NSOM. NSOS inherits all the advantages of NSOM, adds a spectral dimension to the near-field optics technique, and can be used to obtain spectra of various nanostructures, such as nanocrystals and quantum wells. The ability to obtain spectroscopic information at a nanometer-sized resolution makes NSOS very promising for a wide variety of scientific research topics. Examples include the detection of fluorescent labels on biological samples, and isolating local nanometer-sized heterogeneities in microscopic samples. The luminescence spectrum of single quantum wells obtained by near-field excitation revealed additional fine features in the spatial distribution of internal sites [9.52]. Researchers have also studied microscopic crystals in order to demonstrate that nanoscopic inhomogeneities can be detected in what might at first appear to be a homogeneous sample [9.43].

The ultimate goal of the NSOS method is to obtain spectroscopic information with single molecules. The NSOS apparatus is quite similar to that of NSOM [9.10]. In NSOS, an optical probe with an emissive aperture that is submicrometer in size is positioned such that the sample is within the near-field region. With piezoelectric control of the fiber tip, the tip can be accurately positioned over a fluorescing region of the sample, and a spectrum recorded. Excitation of the sample can either be external, with detection through the fiber tip, or by dithering fiber tip itself, with subsequent detection of the emitting photons. This means that it is not necessary for the sample to be of any particular thickness or opacity, but it should nevertheless be a relatively smooth surface. Optical probes used in NSOS are the same nanometer-sized optical fiber light sources for sensor preparation [9.9, 9.10]. The experimental apparatus for measuring fluorescence spectra with high spatial resolution is sketched in Fig. 9.8. Here, a 422-nm line from a HeCd laser is coupled into an optical fiber with a high-precision coupler. The fiber tip is mounted in a hollow tube of piezoelectric material, which is positioned by the usual STM control electronics.

In a typical experimental setup for NSOS, a sample (deposited on a glass slide) is mounted on the near-field microscope such that it is perpendicular to the exciting tip, and the entire apparatus rests on the base of an inverted frame microscope for external collection. Excitation of the sample via the fiber tip generates fluorescence that is collected by an objective, filtered (to remove laser light), and collimated before exiting the microscope. The fluorescence is then focused onto an optical multichannel analyzer (OMA), and thus the data are collected and analyzed in a computer.

A variety of samples have been studied to date. For example, films of a 1.0 wt.% mixture of tetracene in polymethylmethacrylate were prepared by spin-coating a dichloromethan solution on a glass slide [9.43]. Tetracene/PMMA films examined under the fluorescence microscope show microaggregates of tetracene with an average size of about 10 μm embedded in the polymer. The background fluorescence from the film appears greenish-

9.2 Near-Field Scanning Optical Microscopy and Spectroscopy 233

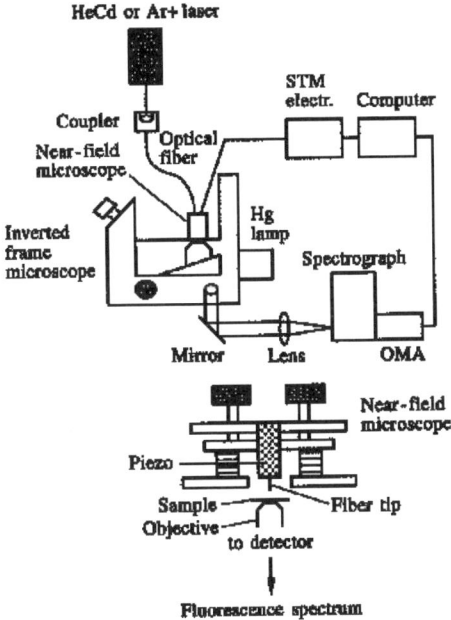

Fig. 9.8. An example experimental setup for near-field spectroscopic measurements (extracted from [9.43])

yellow, and is presumed to be from either isolated molecules, or crystals smaller than the minimum size that can be resolved with a conventional microscope. What is surprising about the aggregates is that this fluorescence ranges in color from green to yellow to red. Thus, the macroscopic fluorescence spectrum obtained with conventional (far-field) light source excitation is very broad, containing contributions from the background and all colors of aggregates. With NSOS, it is then easy to excite a specific aggregate and record the fluorescence spectrum of individual clusters.

By obtaining the characteristic spectra of a single-molecule or molecular aggregate, the mechanism of light–matter interaction may be different in the far- and near-field regimes, leading to different spectral selection rules and, in particular, to an enhanced cross section of light absorption (and thus fluorescence) [9.9]. These phenomena are an extra bonus for near-field detection. Similar SMD work has been done on rhodamine-6G molecules. The photophysics and photochemistry of this molecule have been investigated on the single molecule basis [9.7]. In short, the NSOM approach to SMD permits the determination and localization of single-dye molecules.

Near-field spectroscopy on single molecules of 1,1′-dioctadecyl-3,3,3′,3′-tetramethyl indocarbocyanine (diI) dispersed in PMMA film in ambient conditions revealed inhomogeneous fluorescence spectral properties, such as shifts of peak position, spectra shape and width, as well as time dependence

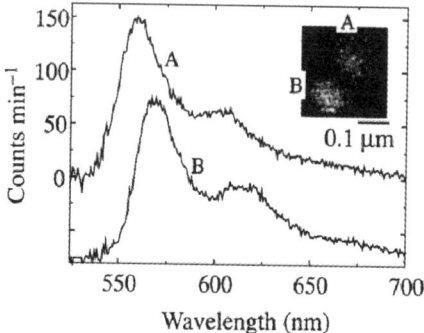

Fig. 9.9. Fluorescence spectra of two adjacent diI molecules (extracted from [9.53])

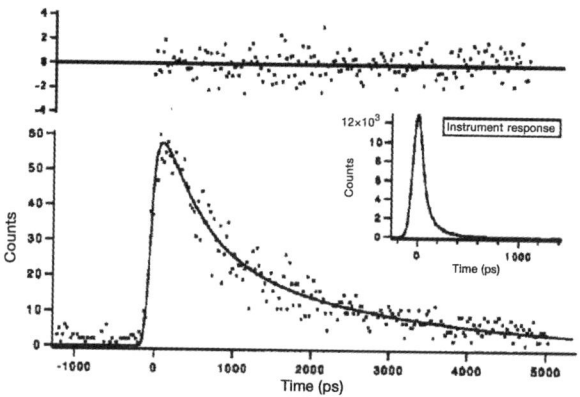

Fig. 9.10. NSOM measured fluorescence decay of single LHC II molecules dispersed in membrane bilayer. *Upper spectrum* is the residue signal and the insert is the instrument response function (extracted from [9.56])

(Fig. 9.9) [9.53]. The cause of the variation could be associated with the local polarization environment. For example, the narrow width spectrum peak could indicate that the fluorophore is located in a more rigid matrix environment, in contrast to that of the broader spectrum peak.

Fluorescence intensity jumps of single sulforhodamine 101 molecules horizontally adsorbed on a glass surface were shown to change coherently with the modulation of parallel polarized excitation by using near-field spectroscopy [9.54]. The perpendicular polarization excitation had no effect on the fluorescence spectrum. The intensity jump was thus attributed to the spectral fluctuation, and any effect of molecular reorientation could be excluded.

In a related study, the spectral fluctuation of single sulforhodamine 101 molecules immobilized in PMMA matrix was analyzed with an epi-illumination microscope [9.55]. A characteristic double exponential behavior was observed in the autocorrelation function of the emission trajectories. It was further identified that the fast component is independent of the excita-

9.2 Near-Field Scanning Optical Microscopy and Spectroscopy

tion rate and wavelength, whereas the slow component displays discernible dependence on the excitation parameters. The authors concluded that the fast component reflects spontaneous fluctuation at room temperature, which could be caused by intramolecular conformational change or changes in intermolecular nuclear coordinates. The slow component is a photo-induced fluctuation possibly related to the photoexcitation of the molecule, or radiationless relaxation from the triplet state. The individually dispersed lightharvesting complex (LHC II) in a thylakoid membrane bilayer was examined with NSOM operated in shear force mode [9.56]. The fluorescence decay spectrum obtained with time-correlated single photon counting at the positions of the LHC II revealed a double exponential with lifetime of 450 and 2.7 ps (Fig. 9.10).

The formation of fluorescent fibers as a result of fibrous assembly of pseudo-isocyanine dye (PIC) 1,1'-diethyl-2,2'-cyanine and poly(vinyl sulfate) (PVS) was studied with NSOM. The polarized fluorescence spectrum revealed that the fluorophores are aligned along the main axis of the fiber. A crosslinking mechanism was proposed to describe the assembly and formation of larger fibers from cationic dye and anionic polymer molecules [9.57].

The metallic coating of the NSOM tip was found to have appreciable effect on the spectral characteristics of fluorophores [9.18, 9.54, 9.58]. The example can be best seen in the measurement of the decay behavior of individual sulforhodamine 101 molecules [9.54, 9.58] and rhodamin 6G molecules [9.59], which can be approximated as flat-lying dipoles. The lifetime of the molecule was seen to be longest when the NSOM tip was over the center of the molecule, and to decrease as the tip moved away from the molecule while keeping a constant height of 5 nm. By contrast, an increase of lifetime was observed as the NSOM tip, positioned 30 nm above, moved away from the molecule. Such dependence of fluorescence lifetime on the geometric position of the NSOM tip was interpreted on the basis of theoretical simulation (finite-difference-time-domain, FDTD method) of a horizontal oscillating dipole in the vicinity of the metallic medium [9.58]. It was found that the radiative energy transfer is dominant at larger tip–sample separation, whereas nonradiative energy dissipation is more pronounced at small tip–sample separation.

The static Stark effect was observed as the electric field induced a shift of fluorescence line for pentacene in p-terphenyl matrix [9.60]. The static electric field was introduced by the metal coating of the NSOM tip. Similar effects were demonstrated for pentacene and terrylene using a dielectric microsphere under optical excitation [9.61].

9.2.3 Fluorescence Resonance Energy Transfer (FRET) Studied by NSOM

Fluorescence resonance energy transfer (FRET) between a donor molecule and an acceptor molecule can be characterized by the transfer efficiency introduced in the preceding chapter.

NSOM has been used to observe the fluorescence of single donor–acceptor pairs consisting of oligonucleotide TMR-10-TR (TMR: tetramethylrhodamine, donor molecule; TR: Texas Red, acceptor molecule; 10: number of DNA base pairs) [9.62]. The target molecules were immobilized on aminopropyl-silane (APS) treated glass surfaces. The fluorescence spectra of single donors and acceptor molecules were obtained, as well as their photobleaching behavior.

The energy transfer efficiency can be estimated based on the intensity change in the emission spectrum.

1. Donor photobleaches first:

$$E = \frac{I_D - I_{D_A}}{I_D - \alpha I_{D_A}}$$

2. Acceptor photobleaches first:

$$E = \left[1 + \left(\frac{I_{D_A}}{I_{A_D}}\right)\left(\frac{\Phi_A}{\Phi_D}\right)\right]^{-1}$$

where I_{D_A} is the integrated donor emission before acceptor photobleaching, I_D is the integrated donor emission after acceptor photobleaching, α (<1) the fraction of acceptor absorption that remains after photobleaching, I_{A_D} the integrated area of the sensitized emission of the acceptor, and Φ the quantum yield for the donor-only or acceptor-only complexes.

The images are recorded in an array of 128×128 pixels, with an integration time per pixel of 10 ms. Both images were obtained **simultaneously** by separating the emission with a dichroic mirror at 600 nm, and collecting the signals on two APD detectors (Fig. 9.11). The two color images are overlaid, and their red-green-blue values are added together to form the composite image of Fig. 9.11c. There is a considerable amount of cross-talk between the channels, due to dichroic polarization sensitivity and to spectral overlap of the two fluorophores. To minimize this cross-talk, narrow bandpass filters are placed in front of the detectors. The number of red, green, and yellow spots can be used as a crude estimate of the degree of DNA hybridization on the dry surface [9.62].

In another approach, the donor and acceptor molecules are decorated to the NSOM probe and sample (glass slide) surfaces separately [9.63, 9.64]. By pressing the probe to the sample surface using the shear-force method, the separation between donor and acceptor molecules can be controlled and the accompanying fluorescence intensity can be monitored, reflecting the efficiency of energy transfer between donor and acceptor species. In such studies, the acceptors (OM57: 1-butyl-3,3-dimethyl-2-[5-(1-butyl-3,3-dymethyl-3H-benz[e]indolin-2-yliden)-1,3-pentadienyl]-3H-benz[e]indolium perchlorate) are attached to the probe apex, so that the FRET process involves only the donor molecules (DCM: 4-dicyanmethylene-2-methyl-6-(p-dimethylaminostyryl)-4H-pyran) directly underneath the probe apex.

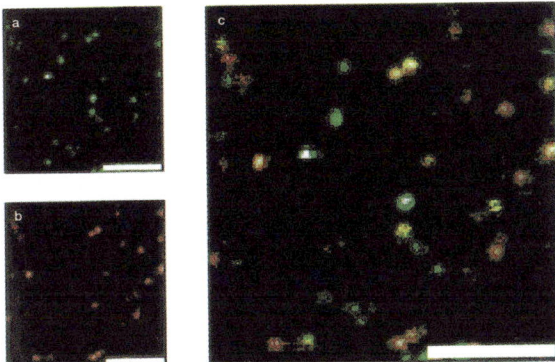

Fig. 9.11. Two-color near-field scanning fluorescence images of doubly labeled DNA molecules, obtained with linearly polarized excitation light (along the white 1-mm scale bar). The donor channel image (**a**) is colored in green. The acceptor channel image (**b**) is colored in red (extracted from [9.62])

9.3 Other Near-Field Optical Microscopy

9.3.1 Near-Field Optical Chemical Sensors

Microspectroscopy has often been utilized for chemical analysis and biological intracellular analysis. For instance, a reagent is introduced into a cell (e.g., with a micropipette under the microscope) and the color or spectrum of the cell changes and provides information about the pH or the calcium content of the cell. Fiber optical chemical sensors have been developed for such measurements [9.65]. However, their spatial resolution is limited by the physical size of the optical fiber, typically 100 µm. A spin-off from NSOM technology is the development of submicrometer, subwavelength near-field optical chemical sensors (NOCS) [9.4, 9.21]. Their chemical preparation by photopolymerization is based on near-field optical excitation, which limits the size of the produced probe [9.66]. In addition, the sensing occurs in the near-field regime of the optical excitation, thereby highly increasing the sensitivity per photon and per sensor molecule. This near-field operation has decreased the volume needed for non-destructive analysis to well below a femtoliter [9.21]. Such a subwavelength pH sensor is schematically shown in Fig. 9.12.

The first biological application of submicrometer NOCS [9.4, 9.66] was demonstrated for 10- and 12-day-old rat conceptuses. The NOCS consist of an aluminized fiber tip with a co-polymer supertip containing the pH-sensitive dye [9.21]. The analysis is based on ratios of fluorescence intensities at different wavelengths of the same spectrum, or on ratios of fluorescence intensities at different wavelengths of two different spectra obtained by two different excitations (ratios of ratios), providing for internal calibration [9.4]. Intra-embryo pH values were 7.55 for 10-day rat conceptuses, and 7.27 for 12-day rat conceptuses. These values are in good agreement with the reported

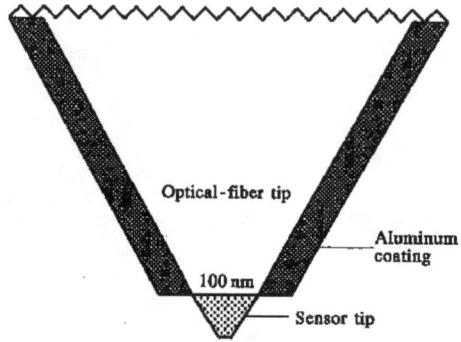

Fig. 9.12. Schematic of a chemical sensor derived from an NSOM probe

results for "homogenized" rat conceptus samples, where more than 1,000 embryos had to be crushed. By contrast, only a single embryo was needed for the pH measurements via miniaturized NOCS. In addition, chemical dynamic alterations in pH of intact rat conceptuses, in response to strong variation under environmental conditions, have been recorded. This is the first time that such an experiment has been carried out on a single, live rat embryo. The ability of the sensors to measure pH changes, in real time, in the intact rat conceptus demonstrates their potential application for dynamic analysis in small multicellular organisms and single cells.

Compared to conventional devices [9.65], a thousand-fold miniaturization of immobilized optical fiber sensors, a million-fold or more sample-size reduction and at least a hundred-fold shorter response time have been achieved by combining nanofabricated optical fiber tips with near-field photopolymerization. Also, the submicrometer sensors have improved the detection limits by a factor of a billion [9.21].

9.3.2 Scanning Exciton Microscopy

The concept of active light sources enables a totally new mode of NSOM, based not on the blocking or absorption of photons but rather on quenching directly the energy quanta that otherwise would have produced photons. For instance, a thin, localized gold film (or cluster) can quench an excitation (or exciton) that would have been the precursor of photons. Furthermore, a single atom or molecule on the sample could quench (i.e., by energy transfer) the excitations located at the tip of the light source. For simplicity, we assume that the active part of the light source is a single atom, molecule or crystalline site, serving as the "tip of the tip". This quenching energy transfer from the excitation source's active part (donor) to the sample's active part (acceptor) may or may not qualify technically as an NSOM technique. However, it is currently our best option for single-atom or molecule resolution and sensitivity. This technique is basically a quantum optics microscopy. It has

9.3 Other Near-Field Optical Microscopy

been called scanning exciton microscopy (SExM) [9.67], and also molecular exciton microscopy (MEM) [9.68].

MEM is conceptually quite similar to STM. The excitons "tunnel" from the tip to the sample. However, there is no driving voltage or field. Rather, it is the energy-transfer matrix element that controls the transfer efficiency. Its unusual matrix elements allow us to reach one of the highest sensitivity to distance, higher than that of STM and comparable to that of AFM. In addition, the most striking result of this direct energy transfer is its ultrahigh sensitivity to isolated or single molecular chromophores. The quantum-optics energy transfer is highly efficient within the range of the Forster radius.

Thus, a single excitation can be "absorbed" by the sample acceptor. By contrast, based on the Beer-Lambert law, about a billion photons are needed to excite a single acceptor in the absence of other acceptors. Furthermore, as the distance range is limited to about 10 nm for the direct energy transfer, MEM is as much a near-field technique as STM or AFM, i.e., very sensitive in the single digit nanometer range, and much less sensitive beyond 10 nm. However, in combination with conventional NSOM, the range can be extended to about 200 nm. Thus, MEM is a technique able to "zoom in" from macroscopic to nanoscopic distances. Obviously, such a "zooming in" enhances the speed of operation. It also allows use to use a much more universal range of samples, from metal spheres and clusters to soft, in-viva biological units. In addition, MEM can use fluorophores, metallic clusters, etc., to enhance contrast, sensitivity and resolution with the help of NSOM. It can also be used in conjunction with lateral force feedback, similarly to NSOM.

10 Surface-Enhanced Raman Scattering (SERS) of Single Molecules

10.1 Introduction of SERS Effect

The principle of Raman spectroscopy is based on the inelastic scattering of photons due to molecular vibrational energy states [10.1]. The molecule under excitation radiation can be excited to a virtual state, and followed by relaxation to a vibrational level by emitting a photon. In the elastic process, i.e., the final state is the same as the initial state prior to excitation, the frequency of the emitted photon is the same as the excitation, and the process is so-called Rayleigh scattering. In the case where the final state differs from the initial state, the difference between the emitted photon and the excitation photon is called the Raman shift. If the final vibration level (ground level) is lower than the initial one (first excited level), then there will be a blue shift in emitted photon frequency. This is the so-called anti-Stokes line. By contrast, if the final state (first excited level) is higher than the initial level (ground level), a red shift will occur that is called the Stokes line. An illustration of the Raman scattering process is given in Fig. 10.1. The relative intensities of Stokes and anti-Stokes lines reflect the population ratio of the ground and first excited vibration levels. The application of Raman spectroscopy has been established as an important venue for studying intramolecular vibrational characteristics.

The Raman scattering cross section of single molecules is typically on the order of 10^{-29} cm^2, compared to 10^{-16}–10^{-19} cm^2 for fluorescence cross sections [10.1]. Therefore, a large enhancement of the Raman signal is critical to the practical detection of molecules. Large enhancement effects from roughened metal surfaces were first reported in 1974 by Fleischmann et al. [10.2]. It was soon suggested that either the local electric field of the excitation photon or the resonance Raman scattering contributed to the observed enhancement of the Raman signal [10.3,10.4]. Extensive investigations since then have been focused on revealing the mechanisms of such enhancement effects. The findings have been grouped in two major enhancement mechanisms, i.e., electric field effect or chemical effect [10.5–10.7]. Recent studies on single molecules adsorbed on nanometer-size metal particles have stimulated renewed interest on this topic.

Regarding the electric field effect, note that the Raman signal intensities at Stokes frequency and anti-Stokes frequency depend on the excitation intensity linearly and quadratically, as a result of the population of first excited

10 Surface-Enhanced Raman Scattering (SERS) of Single Molecules

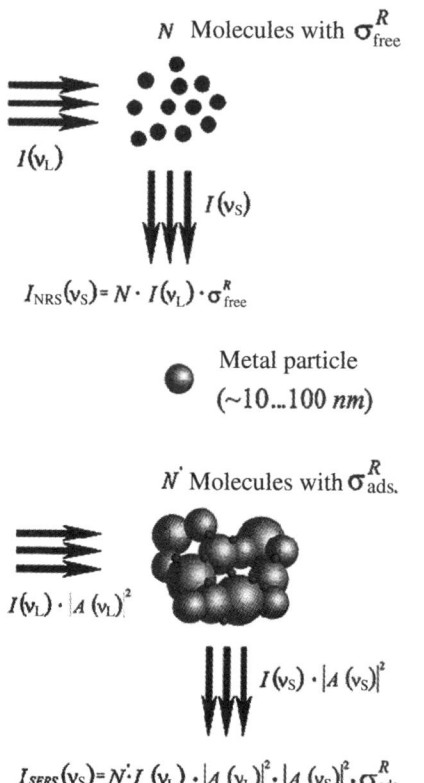

Fig. 10.1. Schematic illustration of normal Raman scattering (*upper panel*) and surface-enhanced Raman scattering (*lower panel*) (extracted from [10.7])

vibrational state [10.8]. The electric field inside the sphere and at the surface of a metallic sphere (dielectric constant $\varepsilon = \varepsilon' + i\varepsilon''$) surrounded by medium with dielectric constant ε_m ($\varepsilon_\mathrm{m} = 1$ in vacuum) can be written as:

$$E_\mathrm{in} = \frac{\varepsilon - \varepsilon_\mathrm{m}}{\varepsilon + 2\varepsilon_\mathrm{m}} E_0 \tag{10.1}$$

$$E_\mathrm{sur} = \frac{3\varepsilon}{\varepsilon + 2\varepsilon_\mathrm{m}} E_0 \tag{10.2}$$

where E_0 is the incident field. For plasmon resonance, one would expect ε' close to -2, leading to a large electric field at the surface. The imaginary part ε'' of the dielectric constant for noble metals of Ag, Au, and Cu is relatively small ($\varepsilon = -2 + 0.2i$ for Ag at 380 nm [10.9]). These factors add up to result in a greatly enhanced electric field at the surface of noble metals.

The enhancement factor can be expressed in a general form as:

$$^\mathrm{em}G_\mathrm{SERS} \approx \left| \frac{E(r_\mathrm{m}, \omega)}{E_0(\omega)} \right|^4$$

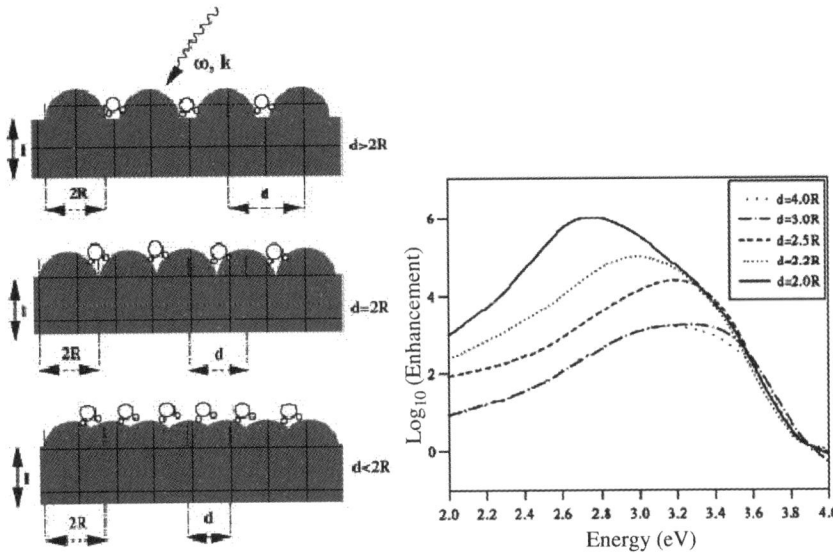

Fig. 10.2. Illustration of EM effect on SERS. The enhancement is shown to be dependent on the excitation energy and the separation between local protrusions. **a** Illustration of different intersphere separations. **b** Calculated enhancement of electric field at the cylinder surfaces (extracted from [10.13])

It should be noted that the shift of Raman frequency and the molecular position relative to the metal surface could also affect the actual enhancement factor [10.10, 10.11]. For a simplified model of a sphere surface, the considerations lead to the expression [10.11]:

$$^{em}G_{SERS} = \left[\frac{\varepsilon(\omega_L) - \varepsilon_m}{\varepsilon(\omega_L) + 2\varepsilon_m}\right]^2 \left[\frac{\varepsilon(\omega_S) - \varepsilon_m}{\varepsilon(\omega_S) + 2\varepsilon_m}\right]^2 \left(\frac{r}{r+d}\right)^{12} \quad (10.3)$$

where ω_L and ω_S are the laser frequency and Stokes frequency, respectively, r the sphere radius, and d the molecule–sphere distance. Enhancement of single particles was shown to be around 10^{11} or less [10.12]. The magnification of the electric field in the contact regions of metallic objects was shown to lead to an enhancement factor of around 10^6–10^7 (Fig. 10.2) [10.13].

On the other hand, it was also observed that for molecules having similar Raman scattering cross sections, very different enhancement factors could be obtained under comparable experimental conditions. The well-know example is that adsorbed CO molecules have much stronger (about 2 orders of magnitude) Raman scattering than that of N_2, whereas they have similar Raman scattering cross sections in free state. This has been attributed to the possible charge exchange between the adsorbed molecules and the metal substrate, either from the HOMO to the Fermi level of the metal, or from the Fermi level of the metal to the LUMO of the molecule, as illustrated in Fig. 10.3. Another relevant observation is the potential dependence of Raman scattering

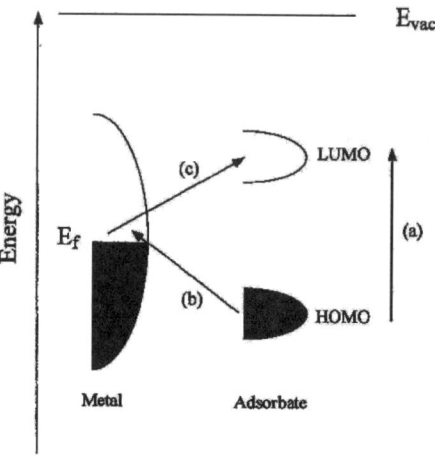

Fig. 10.3. Schematic of charge exchange between the substrate and the orbitals of adsorbed molecules (extracted from [10.14])

under electrochemical conditions. This evidence led to the consideration of effects due to the chemical nature of the molecular adsorbate, referred to as "chemical enhancement effect" [10.14]. The theoretical analysis of the chemical enhancement effect requires considering the integrated molecule–metal system, rather than the individual components [10.15].

Improved signal-to-noise ratios can be obtained for studies at low temperatures in liquid nitrogen [10.16]. The dielectric constant depends on temperature, as a result of electron collision frequency [10.17, 10.18].

10.2 SERS of Single Molecules

10.2.1 Single Particle SERS Effect

The SERS of individual rhodamine 6G (R6G) molecules on single Ag nanoparticles at a dye–particle ratio of 1:10 revealed a surprisingly high enhancement factor on the order of 10^{14} to 10^{15} [10.19]. The number of observed "hot" spots is roughly one in 10^3 particles. In addition, the polarization of the incident beam was related to the particle orientations. AFM observations showed a range of 110–120 nm in diameter for most hot particles. Other forms of single particles or small aggregates were also observed. The spectral fluctuation behavior, with peak position changes as large as 10 cm^{-1} and typically on the order of seconds, was observed to vary slightly among hot particles.

Several types of Ag nanoparticles show very high enhancement factors for SERS studies of single R6G molecules, including rod-like and faceted particles, as well as particle aggregates [10.19]. The concentration of R6G was below 10^{-10} M to ensure less than one molecule on average being adsorbed

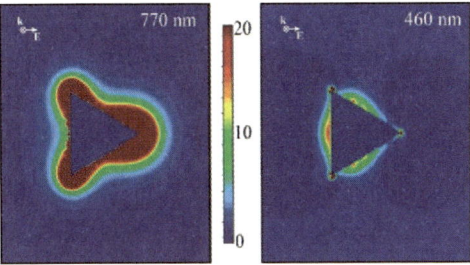

Fig. 10.4. Calculated electrical field enhancement contours of a triangular Ag prism. The incident beam passes midway across the prism (extracted from [10.20])

onto a single nanoparticle. The observed SERS features from in-plane C–C stretching modes at 1,657, 1,578, 1,514, 1,365, 1,310, and 1,184 cm^{-1} are similar to the SERS spectra in the bulk state and solution resonant Raman of R6G. The observed SERS of single R6G molecules also depend on the excitation polarization, and have spectral jumps (in both frequency and intensity) characteristic of single molecule measurements. The reported high enhancement factor was on the order of 10^{14} to 10^{15}, and is attributed to the absence of population-average effects.

Simulations of the enhancement effect of single nanoparticles with spheres, spheroids, and triangular shapes were performed using modified long wavelength approximation (MLWA) and discrete dipole approximation (DDA) methods [10.20]. The results clearly indicate that the enhancement of the electromagnetic field is high near particle surfaces. For non-spherical nanoparticles, such as spheroids and triangular-shaped particles, the maximum enhancement occurs at the sites of a curved surface, such as tips or protrusions (Fig. 10.4). It was also shown that the dielectric effect of the surrounding solvent, as well as the supporting substrate also contribute to the shift of local surface plasmon resonance frequency.

The enhancement effect of Ag nanoparticles fabricated by the nanosphere lithography method was studied using $Fe(bpy)_3^{2+}$ as model system [10.21]. The enhancement factor for trangular-shaped Ag particles of 117-nm width and 27-nm height was estimated as around 1×10^8, and the enhancement factor for SERRS is greater than 7×10^9.

A blue shift of as much as 40 nm was observed as the spacing was reduced between the periodic arrays of Ag nanoparticles (both circular and triangular-shaped) [10.22]. This is in contrast to the generally observed red shift for planar arrays as a result of dipole interactions. The long-range coupling mechanism of radiative dipole interaction and retardation effect was proposed to explain the observed blue shift in the periodic arrays.

10.2.2 SERS of Nanoparticle Aggregates

Evidence of single molecule Raman scattering spectra was demonstrated in experiments with an average of 0.6 crystal violet dye molecule in a detec-

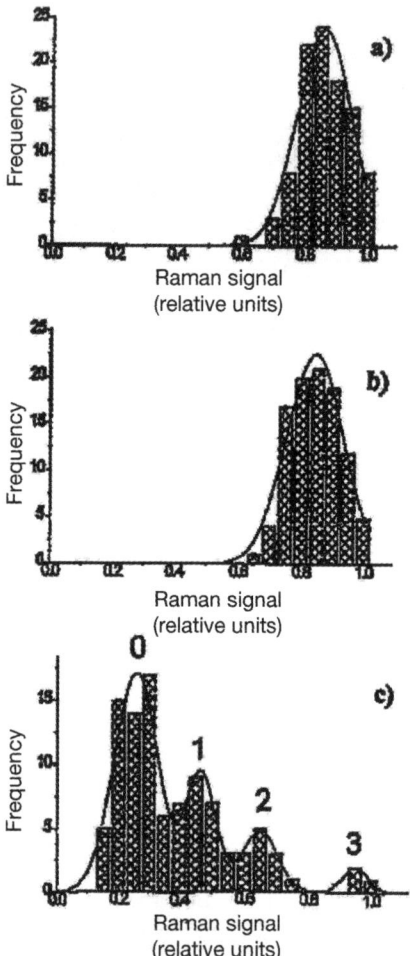

Fig. 10.5. a Distribution of Raman signal intensity at 1,030 cm^{-1} for 100 crystal violet (CV) molecules in the detection volume. **b** SERS intensity at 1,174 cm^{-1} for six CV molecules. **c** 0.6 CV molecules in the detection volume at 1,174 cm^{-1}. The discrete distribution is evident in **c** (extracted from [10.23])

tion volume of about 30 pl [10.23], with excitation wavelength of 830 nm. This study was performed in the solution state with the Cl$^-$ concentration an order of magnitude higher than that in [10.19] and the dye–particle ratio at 1:100. The effective cross section of the crystal violet dye molecules is around 10^{-17} to 10^{-16} cm^2 per molecule. A distinctive discrete distribution of SERS measurements at 1,174 cm^{-1} characteristic line shows the effect of 0–3 molecules in the detection volume (Fig. 10.5), which was not observable when the concentration corresponded to six molecules in the detection volume.

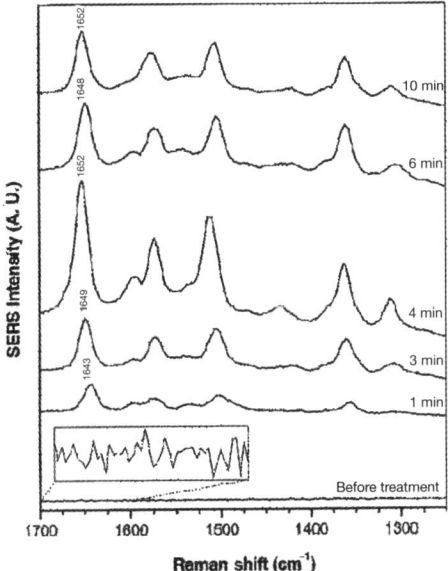

Fig. 10.6. Time-dependent R6G SERS spectrum on a single Ag particle at 10 mM bromide concentration (extracted from [10.24])

Another important source of SERS is associated with the activation effect of ions at particle surfaces [10.24]. Systematic comparison of single molecule SERS of rhodamine 6G on Ag nanoparticles found that halide ions (Cl^-, Br^-, I^-) could contribute as much as 10^2 to 10^3 to the reported enhancement factor. The typical blinking behavior for single molecule detection was also observed in the study (Fig. 10.6). By contrast, thiosulfate ions can destroy the SERS activity. Other ions, such as citrate, sulfate and fluoride, have little effect on SERS activity.

The Rayleigh spectra of Ag nanoparticles were studied by using dark-field optical microscope, and the correlation with Raman activity was explored. No consistent correlation could be established between SERS active particles and the corresponding Rayleigh scattering spectra [10.25]. The proposed mechanism points to the charge exchange interaction between chemisorbed organic molecules and the Ag particles.

It was identified by using atomic force microscope (AFM) that the junction sites between Ag nanoparticles are predominantly SERS active of rhodamin 6G (R6G) molecules (Fig. 10.7) [10.26].

The depolarization of the R6G molecules at the "hot" spots was studied by measuring the dependence of s and p polarization intensities of the emitted fluorescence on the sample orientation.

$$\rho = \frac{I_{ss} - I_{sp}}{I_{ss} + I_{sp}}$$

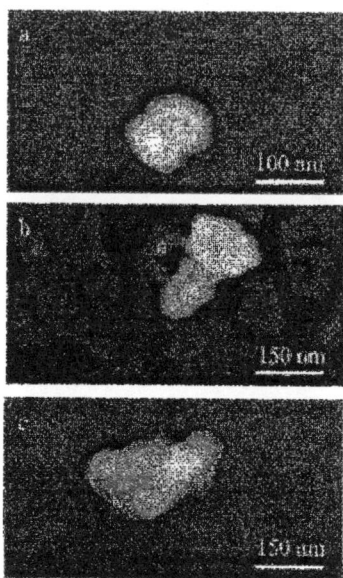

Fig. 10.7. Examples of SERS-active Ag aggregates observed by AFM (extracted from [10.26])

where I_{ss} is the s-component of the emission intensity, and I_{sp} is the p-component of the emission component. Excitation intensity is s-polarized.

It was found that both Raman peak intensity and the continuum portion of the signal varied coherently with sample orientation (Fig. 10.8), suggesting that the molecule was attached to the "hot" spot with fixed orientation [10.27, 10.28]. The continuum portion was ascribed to the electronic Raman scattering from the defect sites of the Ag clusters.

Kneipp et al. [10.29] studied non-linear Raman effects of crystal violet on Ag colloids at single molecular level. The surface-enhanced hyper-Raman spectrum (SEHRS) was shifted relative to the second harmonic of excitation frequency, and can be recorded as the second-order diffraction of monochromator. The intensity of SEHRS depends quadratically on the excitation intensity [10.29, 10.30]. A strong enhancement effect was observed, with an enhancement factor on the order of 10^{20} using near-infrared excitation, compared to the reported enhancement factor of about 10^{14} for crystal violet on Ag nanoparticles. Enhanced Raman signals at anti-Stokes frequency were also found to depend on laser intensity quadratically as expected.

Using near-infrared (830 nm) excitation, SERS of 1 or 2 1,1'-diethyl-2,2'cyanine or pseudoisocyanine (PIC) molecules (absorption band at 530 nm) adsorbed on colloidal Ag particles (diameter 15–40 nm) in methanol solution were obtained. The enhancement factor was estimated at about 10^{13} [10.31].

Detailed study on the intensity fluctuation of SERS spectra of single R6G on Ag nanoparticles revealed a clear dependence on excitation laser power

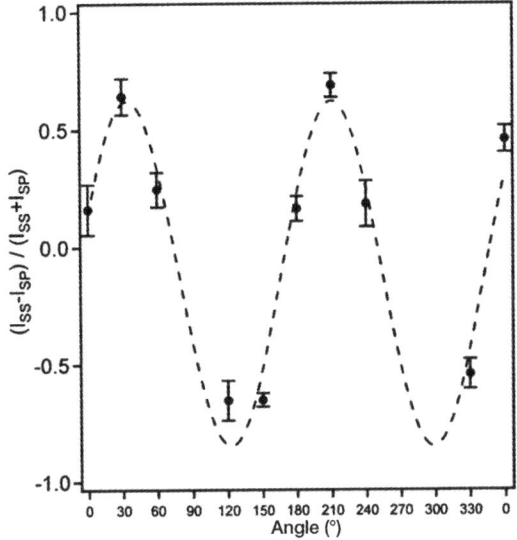

Fig. 10.8. Angular variation of the polarization parameter from R6G adsorbed on Ag nanoparticles (extracted from [10.27])

and anion concentration [10.32]. The fluctuation of the C–H bend vibration bands was found to be more apparent than the in-plane stretch vibration bands. Seeing that thermal-stimulated processes could be excluded based on the estimate of temperature rise by incident laser, the lateral diffusion of adsorbed molecules was suggested as the main cause of the observed spectral fluctuation. Furthermore, the driving force of the diffusive motion can be associated with the possible effect of charge transfer between the molecule and metal substrate.

Single dye molecules of bis(benzimidazo)-perylene (AzoPTCD) were dispersed in a matrix of Langmuir-Blodgett film of eicosanoic or arachidic acid ($C_{19}H_{39}COOH$). SERRS spectra were obtained on a film of about one AzoPTCD per square micrometer supported on Ag thin films [10.33]. The comparison of SERRS and resonant Raman spectrum (RRS) revealed characteristic differences in the relative intensity (Fig. 10.9), fwhm, and wavenumbers, possibly due to the excimer excitation present in the molecular aggregates.

Resonant excitation can be achieved by either plasmon resonance of the colloidal particles or the adsorbed molecules. Furthermore, both excitations could be achieved in specific systems. Photobleaching behavior of single R6G molecules adsorbed on Au nanoparticles was identified under double resonance condition, i.e., both R6G molecule and Au nanoparticles are simultaneously resonant [10.34]. This was achieved by selecting the specifically sized colloidal particles displaying overlapping absorption spectra with the adsor-

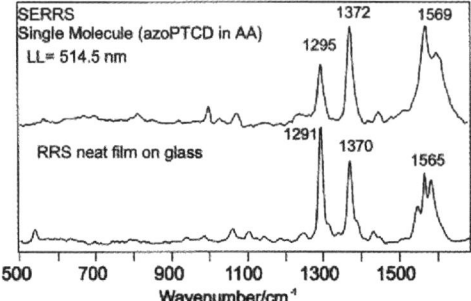

Fig. 10.9. SERRS of single AzoPTCD molecules and RRS of neat AzoPTCD film on glass (extracted from [10.33])

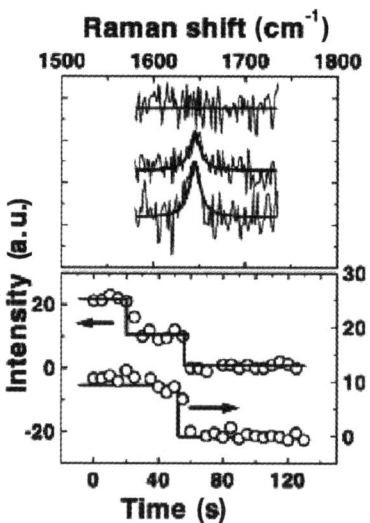

Fig. 10.10. Step-like behavior of SERRS signal duration for single R6G molecules (extracted from [10.34])

bate molecule. The resulting SERRS intensity displayed discontinuous distribution of duration, which is characteristic of single molecules (Fig. 10.10).

The spectral fluctuation of SERS under low coverage of R6G or cytochrome c on Ag nanoclusters has studied with a setup consisting of an AFM and a scanning confocal Raman microscope [10.35]. The setup allows in situ analysis of the characteristics of individual Ag clusters with an overlapping accuracy of 100 nm. The results revealed both photo-induced and spontaneous SERS fluctuations for R6G molecules, whereas mostly photo-induced fluctuations were observed for cytochrome c molecules. There are also large differences in the fluctuation behavior among clusters.

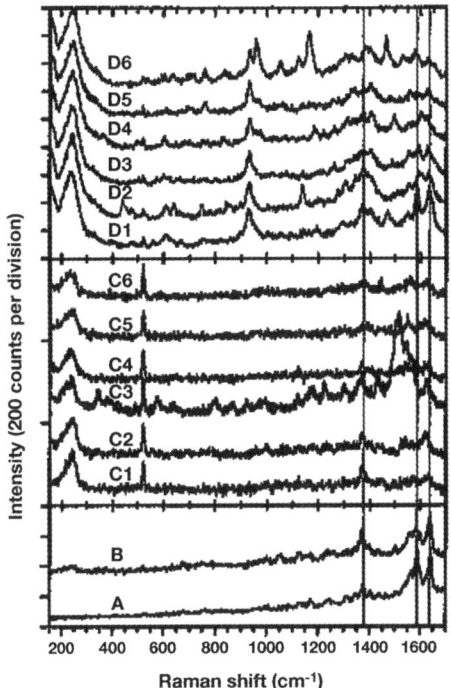

Fig. 10.11. Raman spectra of (A) Hb crystal and (B) dense Hb film, and SERS signal for single Hb molecules at "hot" sites. C1–C6 and D1–D6 are two time series of SERS. The excitation intensity was 1 mW in A and 1 µW in B–D (extracted from [10.37])

The possible causes of spectral fluctuations include:

(1) Charge transfer between adsorbate and substrate;
(2) Inhomogeneity in surface work functions;
(3) Rotational and lateral motions of the adsorbates.

Such fluctuations could be more pronounced in single molecule studies. A number of SERS studies have been seen focused on individual biomolecules. SERS of adenine molecules adsorbed on Ag nanoparticles was observed by Ishikawa et al [10.36]. The blinking behavior of the SERS signal suggested the observation was from individual adenine molecules, even though the concentration was relatively high (about 10^{-6} M).

Single protein molecules of hemoglobin (Hb) showed significant SERS signals when immobilized between Ag particles approximately 100 nm in diameter [10.37]. The Hb molecule contains four Fe-protoporphyrin prothetic hemegroups embedded in polypeptide chains. There are α- and β- type subunits coexisting in the protein. The concentration of Hb was 1×10^{-11} M, and the molecule to particle ratio was about 1:3 to achieve single molecule distribution within a probe area 1 mm in diameter. The hot spots of enhanced

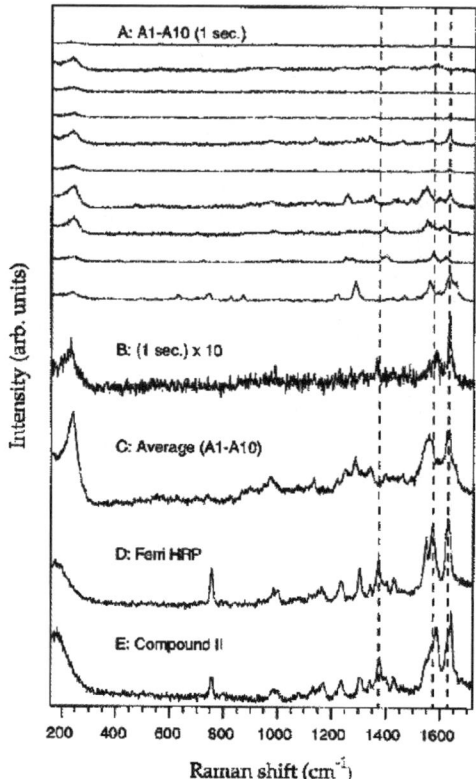

Fig. 10.12. SERS of single HRP molecules shows temporal fluctuations (A1–A10). The spectra D and E are ordinary resonance Raman spectra provided for comparison (extracted from [10.38])

Raman signal came predominantly from the sites connecting two or three Ag particles. In addition, the Raman active dimers with their dimer axis parallel to the incident beam electric field produced highest SERS signal intensity. The effective enhancement factor was estimated at about 10^{10}. In addition, appreciable spectral fluctuations of peak intensity and positions were observed for single Hb molecules (Fig. 10.11). There could be multiple factors associated with the fluctuations, such as variations of protein conformation, adsorption state, and particle motion.

The observed enhancement effect was accounted for by considering the amplification of the electric field due to the coupling between incident light and surface plasmon of the particle surface. By reducing the separation between particles, further enhancement of the local electric field between the particles can be obtained. Theoretical simulations using incident beams of ω_I = 514.5 nm and $\omega_\nu = 1,500$ cm^{-1} led to an optimized particle size of around 90-nm diameter.

SERS studies on horseradish peroxidase (HRP) adsorbed on Ag nanoparticles revealed similar spectral fluctuation behavior. Appreciable shifts of marker mode frequencies at 1,375, 1,575 and 1,630 cm^{-1} were found at the excitation wavelength of 514.5 nm (Fig. 10.12) [10.38]. Spectral fluctuation of tyrosine (Tyr) molecules on immobilized Ag colloidal particles also revealed abrupt spectral changes [10.39]. The addition of chloride anion was found to have minimal effect on the Tyr SERS features.

The scattering cross section σ can be estimated from the following expression:

$$\sigma = 4\pi I A/(\Omega \varepsilon N \phi)$$

where I is the number of recorded counts, A the sample area, Ω the measurement solid angle of the spectrometer, ε the ratio of the recorded counts to the collected photons, N the number of sampled molecules, and Φ the incident photon flux.

A value of 8×10^{-30} cm^2 was estimated for the cross section of Tyr molecules in dense state, and of about 10^{-17} cm^2 from SERS of single Tyr molecules. The ratio between the two cross sections results in an enhancement factor of about 10^{12}.

10.3 Tip-Induced SERS

The combination of scanning probe microscopy and Raman scattering has the unique advantage of studying local regions at nanometer scale, and ultimately single molecules. The tip apex is unambiguously the location for enhancement effects due to localized electric fields. The high spatial resolution of the SPM technique could provide important structural information about the specific sites with enhancement effect. Another attractive merit of the tip method is that it may help differentiate the electromagnetic effect from the chemical effect, since the tip–surface separation can be readily adjusted to tune the local electric field strength.

Using the apex of a metal tip or metal-coated AFM cantilever as an enhancement source, several groups have tried to combine STM and AFM with SERS detection or tip-enhanced-Raman-scattering (TERS). So far, the enhancement is reported at around 10,000 for molecules with large Raman cross sections [10.40]. A recent study of CN- ions on Au(111) surfaces and malachite green isothiocyanate (MGITC) on Au(111) and Pt(110) surfaces using Au or Ir tips achieved an enhancement factor of 4×10^5 and 10^6, respectively [10.41].

The enhancement effect from metal-coated AFM cantilevers was demonstrated recently. This approach is based on the general model for SERS that the plasmon excitation of microscopic metal particles is a contributing factor for the SERS effect. Anderson [10.42] used a gold-coated AFM cantilever in combination with a Raman spectral instrument. The gold-coated probe was shown to generate detectable SERS signals on sulfur thin film, and also as a SERS substrate by attachment of small amounts of C_{60} molecules.

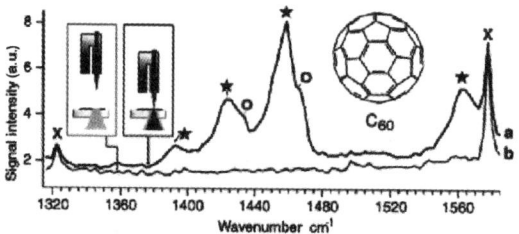

Fig. 10.13. Tip-induced enhancement of Raman signal of C_{60} film when probing the area in the presence of an AFM tip approached with the tuning fork method (spectrum a). The control spectrums is recorded without AFM tip (spectrum b). *Stars* mark the C60 modes due to charge transfer with the gold tip (extracted from [10.43])

An enhancement factor of about 10^4 was estimated for the sulfur thin film. The enhancement effect can be optimized with silver, gold, and copper with grain diameters in the range of 10–200 nm. An enhancement factor of 2,000 was estimated using a Ag-coated AFM tip of diameter about 50 nm over a molecular film of brilliant cresyl blue (BCB) adsorbed on glass slides [10.43]. In a similar experiment, a Au tip with diameter below 20 nm produced an enhancement factor of about 40,000 over C_{60} films (Fig. 10.13).

The excitation laser can also pass through thin metal films to generate back-scattered signals. Cantilever-shaped optical fiber tips have been developed to incorporate both AFM and Raman capabilities, as demonstrated in commercial systems of Nanonics Imaging Ltd. (www.nanonicsimaging.com). The AFM mode provides local topography information whereas Raman (far-field and near-field) can provide spectroscopic characteristics of the same surface position.

10.4 Near-Field SERS

In a parallel approach of studying SERS effects with the scanning probe method, NSOM has been applied to obtain localized topography, fluorescence as well as Raman information. The improved spatial resolution of NSOM could help in detailed analysis of surface properties at nanometer scale. In one study, near-field SERS of cresyl fast violet (CFV) and p-aminobenzoic acid on a silver film support was demonstrated using an excitation wavelength of 488 nm. The Raman scattering signal (or fluorescence signal) was collected through an optical fiber, showing a number of distinct features (Fig. 10.14a) [10.44]. Common to the spectra in Figs. 10.15b and 10.15c are the broad feature around 450 cm^{-1} attributed to fiber glass, and the 800 cm^{-1} peak corresponding to SiO_4. The aromatic stretch modes are represented by the sharp peaks at 1,510 and 1,640 cm^{-1} for CFV, together with the ring breathing mode at 1,195 cm^{-1} (Fig. 10.14b). For CFV molecules, the

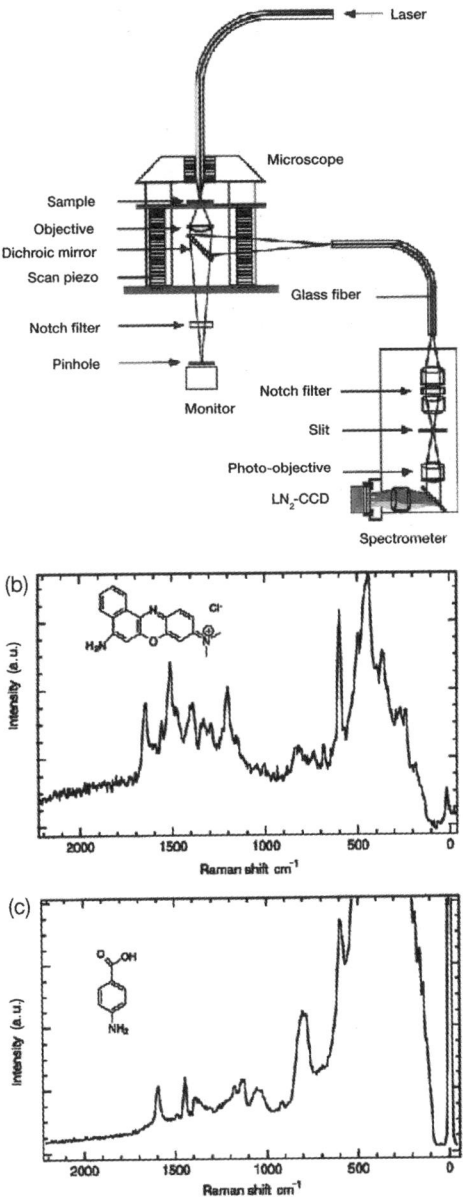

Fig. 10.14. a The combined setup of NSOM and Raman spectroscopy. The near-field excitation is introduced through the NSOM fiber Near-field SERS of **b** cresyl fast violet and **c** p-aminobenzoic acid obtained at Ag thin film surface (extracted from [10.44])

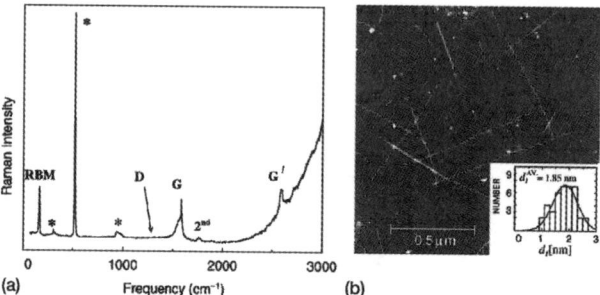

Fig. 10.15. a Raman spectroscopy of single nanotubes at the excitation wavelength of 785 nm. The features marked by "*" are ascribed to the Si/SiO$_2$ substrate. **b** AFM image of the isolated CNTs (extracted from [10.49])

excitation resonance enhancement also contributes to the observed SERS effect. p-Aminobenzoic acid is characterized by values of 1,450 and 1,600 cm^{-1} (deformation mode of amino group). The near-field SERS was also demonstrated for brilliant cresyl blue (BCB) labeled DNA [10.45], cresyl fast violet and rhodamine 6G [10.46]. The substrate was silver film and the excitation was the near-field at the apex of an NSOM tip. The observed near-field spectrum is very similar to the far-field one. The spatial resolution was shown to be about 100 nm with this setup. A high local Raman enhancement factor of 10^{13} was recorded at some surface sites.

Near-field-induced Raman scattering of Rb-doped KTiOPO$_4$ (KTP or potassium titanyl phosphate) was demonstrated using on a commercial NSOM [10.47]. Similar studies were performed on polydiacetylene nanocrystals, diamond films and polyphenylenevinylene (PPV) films using integrated NSOM and Raman spectroscopy [10.48]. So far, most of the reported SERS studies using probe methods are at nanometer scale, and it remains a challenging task to observe SERS effects of single molecules by this approach. Improvements of probe apertures and sample preparations will be needed to advance the capability of resolving single molecule SERS effects.

10.5 Raman Spectroscopy of Carbon Nanotubes

Raman spectroscopy on isolated carbon nanotubes and bundles has provided useful information on the vibration and electronic properties of the material (Fig. 10.15) [10.49–10.51]. The transition between valence and conduction bands is known to strongly enhance the Raman signal. The G-band line shape can be a characteristic feature for the electronic properties (Fig. 10.16).

The radial breathing mode (RBM) (ω_{RBM}) is known to be proportional to the inverse of the diameter of a single-walled nanotube (D):

$$\omega_{\text{RBM}} = \frac{223.5}{D} + 12.5 \tag{10.4}$$

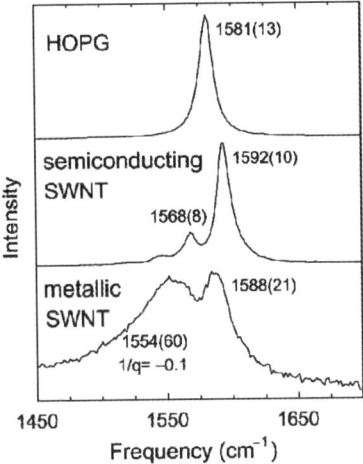

Fig. 10.16. G-band characteristics of graphite and carbon nanotubes. The linewidths are given in parenthesis (extracted from [10.49])

Therefore, ω_{RBM} can be used as a unique and direct measure of structural parameters of single-walled carbon nanotubes. In a recent study, both fluorescence and Raman spectra of single carbon nanotubes have been reported [10.52]. The detection of single molecules was supported by the observation of identical polarization dependence for the fluorescence and Raman spectra. No spectral jumps of the fluorescence spectra were observed. The Raman spectra revealed the radial breathing mode (RBM) (around 300 cm^{-1}) of the nanotube, and can be correlated with the diameter of the CNT.

The results of SERS effects at single molecule level presented in this chapter may be only a partial reflection of the effort invested in this direction. Since Raman spectroscopy is an important aspect for molecular recognition, there is bound to be continued interest in pursuing the investigation on vibrational properties of single molecules. This may prove to be particularly rewarding, considering the fast development of detection techniques such as probe microscopies on single molecules.

References

Chapter 2

2.1   J.G. Simmons, J. Appl. Phys. 34, 1793(1963)
2.2   J.G. Simmons, J. Appl. Phys. 34, 2581(1963)
2.3   C.K. Chow, J. Appl. Phys. 36, 559(1965)
2.4   J. Bardeen, Phys. Rev. Lett. 6, 57(1961)
2.5   W.A. Harrison, Phys. Rev. 123, 85(1961)
2.6   A. Zawadowski, Phys. Rev. 163, 341(1967)
2.7   H. Albrecht, G. Keller, F. Thieme, Surf. Sci. 69, 677(1977)
2.8   T.E. Feuchtwang, Phys. Rev. B10, 4121(1974); Phys. Rev. B10, 4135(1974); Phys. Rev. B12, 3979(1975)
2.9   R.C. Jaklevic, J. Lambe, Phys. Rev. Lett. 17, 1139(1966)
2.10 D.J. Scalapino, S.M. Marcus, Phys. Rev. Lett. 18, 459(1967)
2.11 J. Lambe, R.C. Jaklevic, Phys. Rev. 165, 821(1968)
2.12 J.A. Appelbaum, W.F. Brinkman, Phys. Rev. 186, 464(1969)
2.13 A.D. Brailsford, L.C. Davis, Phys. Rev. B2, 1708(1970)
2.14 L.C. Davis, Phys. Rev. B2, 1714(1970)
2.15 J. Klein, A. Léger, M. Belin, D. DéeFourneau, M.J.L. Sangster, Phys. Rev. B7, 2336(1973)
2.16 J.R. Kirtley, P.K. Hansma, Phys. Rev. B12, 531(1975); Phys. Rev. B13, 2910(1976)
2.17 E.L. Wolf, Principles of Electron Tunneling Spectroscopy, Oxford University Press, New York, 1985
2.18 J.W. Gadzuk, Phys. Rev. B1, 2110(1970)
2.19 E.W. Plummer, R.D. Young, Phys. Rev. B1, 2088(1970)
2.20 M.M. Dignam, R.C. Ashoori, H.L. Stormer, L.N. Pferffer, K.W. Baldwin, K.W. West, Phys. Rev. B49, 2269(1994)
2.21 L. Wang, J.K. Zhang, A.R. Bishop, Phys. Rev. Lett. 73, 585(1994)
2.22 G. Binnig, H. Rohrer, C. Gerber, E. Weidel, Phys. Rev. Lett. 49, 57(1982); G. Binnig, H. Rohrer, Physica 127B, 37(1984)
2.23 C.L. Bai, Scanning Tunneling Microscopy and its Application, Springer, Berlin Heidelberg New York, 1995
2.24 J.A. Stroscio, W.J. Kaiser, eds., Scanning Tunneling Microscopy, Academic Press, San Diego, 1993
2.25 H.J. Güntherodt, R. Wiesendanger, eds., 1994, Scanning Tunneling Microscopy I, 2nd edn., Springer, Berlin Heidelberg New York, 1994

2.26   J. Tersoff, D.R. Hamann, Phys. Rev. B31, 805(1985)
2.27   C.J. Chen, Phys. Rev. B42, 8841(1990); Phys. Rev. Lett. 65, 448(1990)
2.28   N.D. Lang, Phys. Rev. Lett. 56, 1164(1986)
2.29   G. Binnig, H. Rohrer, Angew. Chem. Eng. 26, 606(1987)
2.30   P.K. Hansma, J. Tersoff, J. Appl. Phys. 61, R1(1987)
2.31   N.D. Lang, Phys. Rev. B34, 5947(1986)
2.32   R.M. Feenstra, J.A. Stroscio, A.P. Fein, Surf. Sci. 181, 295(1987)
2.33   N.D. Lang, Phys. Rev. B37, 10395(1988)
2.34   C. Wang, B. Giambattista, C.G. Slough, R.V. Coleman, M.A. Subramanian, Phys. Rev. B42, 8890(1990)
2.35   L. Olesen, M. Brandbyge, M.R. Sørensen, K.W. Jacobsen, E. Læsgaard, I. Stensgaard, F. Besenbacher, Phys. Rev. Lett. 76, 1485(1996)
2.36   B. Marchon, P. Bernhardt, P.M. Bussell, G.A. Somorjai, M. Salmeron, W. Siekhaus, Phys. Rev. Lett. 60, 1166(1988)

Chapter 3

3.1   J.K. Gimzewski, E. Syoll, R.R. Schlittler, Surf. Sci. 181, 267(1987)
3.2   P.H. Lippel, R.J. Wilson, M.D. Miller, Ch. Wöll, S. Chiang, Phys. Rev. Lett. 62, 171(1989)
3.3   S. Chiang, Chem. Rev. 97, 1083(1997)
3.4   R. Strohmaier, J. Petersen, B. Gompf, W. Eisenmerger, Surf. Sci. 418, 91(1998)
3.5   J.S. Foster, J.E. Frommer, Nature 333, 542(1988)
3.6   D.R.E. Smith, H. Horber, Ch. Gerber, G. Binnig, Science 245, 43(1989)
3.7   D.P.E. Smith, J.K.H. Horber, G. Binnig, H. Nejoh, Nature 344, 641(1990)
3.8   J.P. Rabe, S. Buchholz, Science 253, 424(1991)
3.9   D.M. Cyr, B. Venkataraman, G.W. Flynn, A. Black, G.M. Whitesides, J. Phys. Chem. 100, 13747(1996); L. Giancarlo, D. Cyr, K. Muyskens, G.W. Flynn, Langmuir 14, 1465(1998)
3.10   F.R.F. Fan, J. Kwak, A.J. Bard, J. Am. Chem. Soc. 118, 9669(1996)
3.11   A.A. Gewirth, B.K. Niece, Chem. Rev. 97, 1129(1997)
3.12   O.M. Magnussen, Chem. Rev. 102, 679(2002)
3.13   R. Wiesendanger, H.-J. Güntherodt, eds., Scanning Tunneling Microscopy III, Springer, Berlin Heidelberg New York, 1993
3.14   C.J. Chen, Introduction to Scanning Tunneling Microscopy, Oxford University Press, New York, 1993
3.15   P. Sautet, Chem. Rev. 97, 1097(1997)
3.16   A. Yazdani, D.M. Eigler, N.D. Lang, Science 272, 1921(1996)
3.17   C.L. Claypool, F. Faglioni, W.A. Goddard III, H.B. Gray, N.S. Lewis, R.A. Marcus, J. Phys. Chem. B101, 5978(1997)
3.18   F. Faglioni, C.L. Claypool, N.S. Lewis, W.A. Goddard III, J. Phys. Chem. B101, 5996(1997)

References

3.19 W. Liang, M.H. Whangbo, A. Wawkuschewski, H.J. Cantow, S.N. Magonov, Adv. Mater. 5, 817(1993)
3.20 J.K. Spong, H.A. Mizes, L.J. LaComb, M.M. Dovek Jr, J.E. Frommer, J.S. Foster, Nature 338, 137(1989)
3.21 S.B. Lei, C. Wang, X.L. Fan, L.J. Wan, C.L. Bai, Langmuir 19, 9759(2003)
3.22 R.E. Weber, W.T. Peria, Surf. Sci. 14, 13(1969)
3.23 L. Giancarlo, D. Cyr, K. Muyskens, G.W. Flynn, Langmuir 14, 1465(1998)
3.24 Y. Jugnet, F.J.C.S. Aires, C. Deranlot, L. Piccolo, J.C. Bertolini, Surf. Sci. 521, L639(2002)
3.25 P. Thostrup, E. Christoffersen, H.T. Lorensen, K.W. Jacobsen, F. Besenbacher, J.K. Nørskov, Phys. Rev. Lett. 87, 126102(2001)
3.26 J.A. Stroscio, D.M. Eigler, Science 254, 1319(1991)
3.27 M.-L. Bocquet, P. Sautet, Surf. Sci. 360, 128(1996)
3.28 G. Meyer, B. Neu, K.H. Rieder, Chem. Phys. Lett. 240, 379(1995)
3.29 P. Sprunger, F. Besenbacher, I. Stensgaard, Surf. Sci. 324, L321(1995)
3.30 L. Bartels, G. Meyer, K.H. Rieder, Appl. Phys. Lett. 71, 213(1997)
3.31 J.R. Hahn, W. Ho, Phys. Rev. Lett. 87, 196102(2001)
3.32 J.R. Hahn, H.J. Lee, W. Ho, Phys. Rev. Lett. 85, 1914(2000)
3.33 B.N.J. Persson, A. Baratoff, Phys. Rev. Lett 59, 339(1987)
3.34 B.N.J. Persson, Phys. Scr. 38, 282(1988)
3.35 N. Lorente, M. Persson, Phys. Rev. Lett. 85, 2997(2000)
3.36 F.E. Olsson, N. Lorente, M. Persson, Surf. Sci. 522, L27(2003)
3.37 H. Ohtani, R.J. Wilson, S. Chiang, C.M. Mate, Phys. Rev. Lett. 60, 2398(1988)
3.38 S. Chiang, R.J. Wilson, C.M. Mate, H. Ohtani, J. Microscopy 152, 567(1988)
3.39 P.S. Weiss, D.M. Eigler, Phys. Rev. Lett. 71, 3139(1993)
3.40 P. Sautet, M.L. Bocquet, Phys. Rev. B53, 4910(1996)
3.41 P. Sautet, M.L. Bocquet, Surf. Sci. 304, L445(1994)
3.42 J.I. Pascual, J.J. Jackiw, Z. Song, P.S. Weiss, H. Conrad, H.-P. Rust, Phys. Rev. Lett. 86, 1050(2001)
3.43 J.I. Pascual, J.J. Jackiw, Z. Song, P.S. Weiss, H. Conrad, H.-P. Rust, Surf. Sci. 502, 1(2002)
3.44 R. Smoluchowski, Phys. Rev. 60, 661(1941)
3.45 J.I. Pascual, J.J. Jackiw, K.F. Kelly, H. Conrad, H.-P. Rust, P.S. Weiss, Phys. Rev. B62, 12632(2000)
3.46 A.J. Fisher, P.E. Blöchl, Phys. Rev. Lett. 70, 3263(1993)
3.47 I. Stensgaard, L. Ruan, E. Lægsgaard, F. Besenbacher, Surf. Sci. 337, 190(1995)
3.48 S.J. Stanick, M.M. Kamna, P.S. Weiss, Science 266, 99(1994)
3.49 S.J. Stanick, M.M. Kamna, P.S. Weiss, Surf. Sci. 41(1995)
3.50 J. Yoshinobu, H. Tanaka, T. Kawai, M. Kawai, Phys. Rev. B53, 7492(1996)
3.51 K. Morgenstern, S.W. Hla, K.H. Rieder, Surf. Sci. 523, 141(2003)

3.52 C. Ludwig, R. Strohmaier, J. Petersen, B. Gompf, W. Eisenmenger, J. Vac. Sci. Technol. B12, 1963(1994)
3.53 X. Lu, K.W. Hipps, X.D. Wang, U. Mazur, J. Am. Chem. Soc. 118, 7197(1996)
3.54 X. Lu, K.W. Hipps, J. Phys. Chem. B101, 5391(1997)
3.55 K.W. Hipps, X. Lu, X.D. Wang, U. Mazur, J. Phys. Chem. 100, 11207(1996)
3.56 M. Nakamura, Y. Morita, H. Tokumoto, Appl. Surf. Sci. 113/114, 316(1997)
3.57 M. Kanai, T. Kawai, K. Motai, X.D. Wang, T. Hashizumi, T. Sakura, Surf. Sci. 329, L619(1995)
3.58 R. Möller, R. Coenen, A. Esslinger, B. Koslowski, J. Vac. Sci. Technol. A8, 659(1990)
3.59 S. Yim, T.S. Jones, Surf. Sci. 521, 151(2002)
3.60 J.J. Cox, S.M. Bayliss, T.S. Jones, Surf. Sci. 425, 326(1999)
3.61 M. Böhringer, R. Berndt, W.-D. Schneider, Phys. Rev. B55, 1384(1997)
3.62 J.-Y. Grand, T. Kunstmann, D. Hoffmann, A. Haas, M. Dietsche, J. Seifritz, R. Möller, Surf. Sci. 366, 403(1996)
3.63 X.H. Qiu, C. Wang, Q.D. Zeng, B. Xu, S.X. Yin, H.N. Wang, S.D. Xu, C.L. Bai, J. Am. Chem. Soc. 122, 5550(2000)
3.64 M. Lackinger, M. Hietschold, Surf. Sci. 520, L619(2002)
3.65 K. Walzer, M. Hietschold, Surf. Sci. 471, 1(2001)
3.66 R. Strohmaier, C. Ludwig, J. Petersen, B. Grompf, W. Eisenmenger, J. Vac. Sci. Technol. B14, 1079(1996)
3.67 K.W. Hipps, D.E. Barlow, U. Mazur, J. Phys. Chem. B104, 2444(2000)
3.68 D.E. Barlow, K.W. Hipps, J. Phys. Chem. B104, 5993(2000)
3.69 M. Pomerantz, A. Aviram, R.A. McCorkle, L. Li, A.G. Schrott, Science 255, 1115(1992)
3.70 T.A. Jung, R.R. Schlittler, J.K. Gimzewski, Nature 386, 696(1997)
3.71 T.A. Jung, R.R. Schlittler, J.K. Gimzewski, H. Tang, C. Joachim, Science 271, 181(1996)
3.72 T. Yokoyama, S. Yokoyama, T. Kamikado, Y. Okuno, S. Mashiko, Nature 413, 619(2001)
3.73 Y. Okuno, T. Yokoyama, S. Yokoyama, T. Kamikado, S. Mashiko, J. Am. Chem. Soc. 124, 7218(2002)
3.74 X.H. Qiu, G.V. Nazin, A. Hotzel, W. Ho, J. Am. Chem. Soc. 124, 14804(2002)
3.75 J. Gaudioso, W. Ho, J. Am. Chem. Soc. 123, 10095(2001)
3.76 J. Gaudioso, L.J. Lauhon, W. Ho, Phys. Rev. Lett. 85, 1918(2000)
3.77 J.T. Kim, T. Kawai, J. Yoshinobu, M. Kawai, Surf. Sci. 360, 50(1996)
3.78 Y.Z. Li, J.C. Patrin, M. Chander, J.H. Weaver, L.P.F. Chibante, R.E. Smalley, Science 252, 547(1991); Smalley, Science 253, 429(1991)
3.79 H. Xu, D.M. Chen, W.N. Creager, Phys. Rev. Lett. 70, 1850(1993)

3.80 J.G. Hou, J.L. Yang, H.Q. Wang, Q.X. Li, C.G. Zeng, L.F. Yuan, B. Wang, D.M. Chen, Q.S. Zhu, Nature 409, 304(2001)
3.81 J. Weckesser, C. Cepek, R. Fasel, J.V. Barth, F. Baumberger, T. Greber, K. Kern, J. Chem. Phys. 115, 9001(2001)
3.82 P.W. Murray, I.M. Brookes, S.A. Haycock, G. Thornton, Phys. Rev. Lett. 80, 988(1998)
3.83 C.G. Zeng, B. Wang, B. Li, H.Q. Wang, J.G. Hou, Appl. Phys. Lett. 79, 1685(2001)
3.84 X.-D. Wang, T. Hashizume, H. Shinohara, Y. Saito, Y. Nishina, T. Sakurai, Phys. Rev. B47, 15923(1993)
3.85 X.Yao, T.G. Ruskell, R.K. Workman, D. Sarid, D. Chen, Surf. Sci. L85(1996)
3.86 P.W. Murray, J.K. Gimzewski, R.R. Schlittler, G. Thornton, Surf. Sci. 367, L79(1996)
3.87 J.K. Gimzewski, S. Modesti, R.R. Schlittler, C. Chavy, Phys. Rev. Lett. 72, 1036(1994)
3.88 C. Joachim, J.K. Gimzewski, R.R. Schlittler, C. Chavy, Phys. Rev. Lett. 74, 2102(1995)
3.89 L.F. Yuan, J.L. Yang, H.Q. Wang, C.G. Zeng, Q.X. Li, B. Wang, J.G. Hou, Q.S. Zhu, D.M. Chen, J. Am. Chem. Soc. 125, 169(2003)
3.90 D. Wang, J. Zhao, S.F. Yang, L. Chen, Q.X. Li, B. Wang, S.H. Yang, J.L. Yang, J.G. Hou, Q.S. Zhu, Phys. Rev. Lett. 91, 185504(2003)
3.91 K.F. Kelly, D. Sarkar, G.D. Hale, S.J. Oldenburg, N.J. Halas, Science 273, 1371(1996)
3.92 J.G. Kushmerick, K.F. Kelly, H.-P. Rust, N.J. Halas, P.S. Weiss, J. Phys. Chem. B103, 1619(1999)
3.93 V.M. Hallmark, S. Chiang, J.K. Brown, Ch. Wöll, Phys. Rev. Lett. 66, 48(1991)
3.94 T. Sleator, R. Tycko, Phys. Rev. Lett. 60, 1418(1988)
3.95 C. Ludwig, B. Gompf, J. Petersen, R. Strohmaier, W. Eisenmenger, Z. Phys B Condens. Matter 93, 365(1994)
3.96 K. Glöckler, C. Seidel, A. Soukopp, M. Sokolowski, E. Umbach, M. Böhringer, R. Berndt, W.-D. Schneider, Surf. Sci. 405, 1(1998)
3.97 E. Umbach, C. Seidel, J. Taborski, R. Li, A. Soukopp, Phys. Status Solidi B192, 389(1995)
3.98 R. Strohmaier, C. Ludwig, J. Petersen, B. Gompf, W. Eisenmenger, Surf. Sci. 351, 292(1996)
3.99 F. Rosei, M. Schunack, P. Jiang, A. Gourdon, E. Lægsgaard, I. Stensgaard, C. Joachim, F. Besenbacher, Science 296, 328(2002)
3.100 M. Schunack, L. Petersen, A. Kühnle, E. Lægsgaard, I. Stensgaard, I. Johannsen, F. Besenbacher, Phys. Rev. Lett. 86, 456(2001)
3.101 M. Schunack, E. Lægsgaard, I. Stensgaard, I. Johannsen, F. Besenbacher, Angew. Chem. 40, 2623(2001)
3.102 B. Hammer, J.K. Nørskov, Nature 376, 238(1995)
3.103 T. Zambelli, H. Tang, J. Lagoute, S. Gauthier, A. Gourdon, C. Joachim, Chem. Phys. Lett. 348, 1(2001)

3.104 A. Kühnle, T.R. Linderoth, B. Hammer, F. Besenbacher, Nature 415, 891(2002)
3.105 R. Otero, Y. Naitoh, F. Rosei, P. Jiang, P. Thostrup, A. Gourdon, E. Lægsgaard, I. Stensgaard, C. Joachim, F. Besenbacher, Angew. Chem. 43, 2092(2004)
3.106 B.C. Stipe, M.A. Rezaei, W. Ho, Science 280, 1732(1998)
3.107 N. Lorente, M. Persson, Phys. Rev. Lett. 85, 2997(2000)
3.108 J.C. Dunphy, M. Ross, S. Behler, D.F. Ogletree, M. Salmeron, P. Sautet, Phys. Rev. B57, R12705(1998)
3.109 S. Ichihara, J. Yoshinobu, H. Ogasawara, M. Nantoh, M. Kawai, K. Domen, J. Electron Spectroscopy Related Phenomena 88-91, 1003(1998)
3.110 L.R. Nassimbeni, Acc. Chem. Res. 36, 631(2003)
3.111 G.F. Swiegers, T.J. Malefetse, Chem. Rev. 100, 3483(2000)
3.112 A. Dmitriev, N. Lin, J. Weckesser, J.V. Barth, K. Kern, J. Phys. Chem. B106, 6907(2002)
3.113 S. Griessl, M. Lackinger, M. Edelwirth, M. Hietschold, W.M. Heckl, Single Mol. 3, 25(2002)
3.114 J.A. Theobald, N.S. Oxtoby, M.A. Phillips, N.R. Champness, P.H. Beton, Nature 424, 1029(2003)
3.115 J. Lu, Q.D. Zeng, C. Wang, Q.Y. Zheng, L.J. Wan, C.L. Bai, J. Mater. Chem. 12, 2856(2002)
3.116 J. Lu, S.B. Lei, Q.D. Zeng, S.Z. Kang, C. Wang, L.J. Wan, C.L. Bai, J. Phys. Chem. B108, 5161(2004)
3.117 P. Wu , Q.H. Fan, Q.D. Zeng, C. Wang, G.J. Deng, C.L. Bai, Chem. Phys. Chem 3, 633(2002)
3.118 S.D. Xu, Q.D. Zeng, J. Lu, C. Wang, L.J. Wan, C.L. Bai, Surf. Sci. 538, L451(2003)
3.119 Y.H. Liu, S.B. Lei, S.X. Yin, S.L. Xu, Q.Y Zheng, Q.D. Zeng, C. Wang, L.J. Wan, C.L. Bai, J. Phys. Chem. B106, 12569(2002)
3.120 S.R. Forrest, Chem. Rev. 97, 1793(1997)
3.121 F. Rosei, M. Schunack, P. Jiang, A. Gourdon, E. Lægsgaard, I. Stensgaard, C. Joachim, F. Besenbacher, Science 296, 328(2002)
3.122 B.G. Briner, M. Doering, H.-P. Rust, A.M. Bradshaw, Science 278, 257(1997)
3.123 T. Mitsui, M.K. Rose, E. Fomin, D.F. Ogletree, M. Salmeron, Science 297, 1850(2002)
3.124 M.C. Gerstenberg, F. Schreiber, T.Y.B. Leung, G. Bracco, S.R. Forrest, G. Scoles, Phys. Rev. B61, 7678(2000)
3.125 P. Samori, N. Severin, C.C. Simpson, K. Müllen, J.P. Rabe, J. Am. Chem. Soc. 124, 9454(2002)
3.126 B. Xu, S. Yin, C. Wang, X. Qui, Q. Zeng, C. Bai, J. Phys. Chem. B104, 10502(2000)
3.127 S.B. Lei, S.X. Yin, C. Wang, L.J. Wan, C.L. Bai, J. Phys. Chem. B108, 224(2004)

3.128 S. Hoeppener, J. Wonnemann, L.F. Chi, G. Erker, H. Fuchs, Chem. Phys. Chem 4, 490(2003)
3.129 S.B. Lei, C. Wang, L.J. Wan, C.L. Bai, Langmuir 19, 9759(2003)
3.130 S.B. Lei, C. Wang, L.J. Wan, C.L. Bai, J. Phys. Chem. B108, 1173(2004)
3.131 Y. Manassen, R.J. Hamers, J.E. Demuth, A.J. Castellano Jr., Phys. Rev. Lett. 62, 2531(1989)
3.132 Y. Manassen, J. Mag. Reson. 126, 133(1997)
3.133 Y. Manassen, I. Mukhopadhyay, N.R. Rao, Phys. Rev. B61, 16223(2000)
3.134 C. Durkan, M.E. Welland, Appl. Phys. Lett. 80, 458(2002)

Chapter 4

4.1   G.A. Somorjai, Introduction to Surface Science and Catalysis, Wiley, New York, 1994
4.2   G.L. Kellogg, Surf. Sci. Rep. 21, 1(1994)
4.3   R. Viswanathan, D.R. Burgess, P.C. Stair, E. Weitz, J. Vac. Sci. Technol. 20, 605(1982)
4.4   S.M. George, A.M. DeSantolo, R.B. Hall, Surf. Sci. 159, L425(1985)
4.5   E.G. Seebauer, L.D. Schmidt, Chem. Phys. Lett. 123, 129(1986)
4.6   R. Gomer, Rep. Prog. Phys. 53, 917(1990)
4.7   J. Crank, Mathematics of Diffusion, Oxford University Press, 1970
4.8   B.G. Briner, M. Doering, H.-P. Rust, A.M. Bradshaw, Science 278, 257(1997)
4.9   L. Bartels, G. Meyer, K.-H. Rieder, Surf. Sci. 432, L621(1999)
4.10 T. Komeda, Y. Kim, M. Kawai, Surf. Sci. 502-503, 12(2002)
4.11 T. Komeda, Y. Kim, M. Kawai, B.N.J. Persson, H. Ueba, Science 295, 2055(2002)
4.12 T. Mitsui, M.K. Rose, E. Fomin, D.F. Ogletree, M. Salmeron, Science 297, 1850(2002)
4.13 L. Bartels, F. Wang, D. Möller, E. Knoesel, T.F. Heinz, Science 305, 648(2004)
4.14 J.A. Stroscio, D.M. Eigler, Science 254, 1319(1991)
4.15 L.J. Whitman, J.A. Stroscio, R.A. Dragoset, R.J. Celotta, Science 251, 1206(1991)
4.16 Y.W. Mo, Phys. Rev. Lett. 71, 2923(1993)
4.17 Y.W. Mo, Science 261, 886(1993)
4.18 Y.W. Mo, Phys. Rev. Lett. 69, 3643(1992); Y.W. Mo, Phys. Rev. B48, 17233(1993)
4.19 Y.W. Mo, J. Kleiner, M.B. Webb, M.G. Lagally, Phys. Rev. Lett. 66, 1998(1991)
4.20 B.S. Swartzentruber, Phys. Rev. Lett. 76, 459(1996)
4.21 T.W. Fishlock, A. Oral, R.G. Egdell, J.B. Pethica, Nature 404, 743(2003)

4.22 U. Durig, O. Zuger, D.W. Poll, J. Microscopy 152, 259(1988)
4.23 J.A. Stroscio, D.M. Eigler, Science 254, 1319(1991)
4.24 D.M. Eigler, E.K. Schweizer, Nature 344, 524(1990)
4.25 J. Tersoff, D.R. Hamann, Phys. Rev. B31, 805(1985)
4.26 N.D. Lang, Phys. Rev. Lett. 56, 1164(1986)
4.27 N.D. Lang, Phys. Rev. Lett. 58, 45(1987)
4.28 D.M. Eigler, P.S. Weiss, E.K. Schweizer, N.D. Lang, Phys. Rev. Lett. 66, 1189(1991)
4.29 P.S. Weiss, D.M. Eigler, Phys. Rev. Lett. 69, 2240(1992)
4.30 D.M. Eigler, C.P. Lutz, W.E. Rudge, Nature 352, 600(1991)
4.31 I.W. Lyo, Ph. Avouris, Science 253, 173(1991)
4.32 H. Haberland, T. Kolar, T. Reiners, Phys. Rev. Lett. 63, 1219(1989)
4.33 K.S. Ralls, D.C. Ralph, R.A. Burnham, Phys. Rev. B40, 11561(1989)
4.34 N. Nilius, T.M. Wallis, W. Ho, Science 297, 1853(2002)
4.35 T.M. Wallis, N. Nilius, W. Ho, Phys. Rev. Lett. 89, 236802(2002)
4.36 G. Meyer, S. Zöphel, K.-H. Rieder, Phys. Rev. Lett. 77, 2113(1996)
4.37 L. Bartels, G. Meyer, R.-H. Rieder, Phys. Rev. Lett. 79, 697(1997)
4.38 L. Bartels, G. Meyer, K.-H. Rieder, D. Velic, E. Knoesel, A. Kotzel, M. Wolf, G. Ertl, Phys. Rev. Lett. 80, 2004(1998)
4.39 H.J. Lee, W. Ho, Science 286, 1719(1999)
4.40 M. Kageshima, H. Ogiso, H. Tokumoto, Surf. Sci. 517, L557(2002)
4.41 S. Renisch, R. Schuster, J. Wintterlin, G. Ertl, Phys. Rev. Lett. 82, 3839(1999)
4.42 G. Ertl, J. Mol. Catal. A 182-183, 5(2002)
4.43 J. Wintterlin, J. Trost, S. Renisch, R. Schuster, T. Zambelli, G. Ertl, Surf. Sci. 394, 159(1997)
4.44 J. Wintterlin, R. Schuster, G. Ertl, Phys. Rev. Lett. 77, 123(1996)
4.45 B.C. Stipe, M.A. Rezaei, W. Ho, S. Gao, M. Persson, B.I. Lundqvist, Phys. Rev. Lett. 78, 4410(1997)
4.46 S. Paavilainen, J.A. Nieminen, Surf. Sci. 521, 69(2002)
4.47 T. Zambelli, J.V. Barth, J. Wintterlin, G. Ertl, Nature 390, 495(1997)
4.48 G. Zheng, E.I. Altman, Surf. Sci. 462, 151(2000)
4.49 G. Zheng, E.I. Altman, Surf. Sci. 504, 253(2002)
4.50 P.W. Murray, F.M. Leibsle, Y. Li, Q. Guo, M. Bowker, G. Thornton, V.R. Dhanak, K.C. Prince, R. Rose, Phys. Rev. B47, 12976(1993)
4.51 A.M. Baró, G. Binnig, H. Rohrer, Ch. Gerber, E. Stoll, A. Baratoff, F. Salvan, Phys. Rev. Lett. 52, 1304(1984)
4.52 L. Ruan, F. Besenbacher, I. Stensgaard, E. Lægsgaard, Phys. Rev. Lett. 69, 3523(1992)
4.53 F.M. Chua, Y. Kuk, P.J. Silverman, Phys. Rev. Lett. 63 386(1989)
4.54 N. Hartmann, R.J. Madix, Surf. Sci. 488, 107(2001)
4.55 E. Bertel, Prog. Surf. Sci. 59, 207(1998)
4.56 K. Kern, H. Niehus, A. Schatz, P. Zeppenfeld, J. George, G. Comsa, Phys. Rev. Lett. 67, 855(1991)
4.57 D.J. Coulman, J. Wintterlin, R.J. Behm, G. Ertl, Phys. Rev. Lett. 64, 1761(1990)

4.58 F. Jensen, F. Besenbacher, E. Lægsgaard,, I. Stensgaard, Phys. Rev. B41, 10233(1990)
4.59 T. Mitsui, M.K. Rose, E. Fomin, D.F. Ogletree, M. Salmeron, Nature 422, 705(2003)
4.60 S. Horch, H.T. Lorensen, S. Helveg, E. Lægsgaard, I. Stensgaard, K.W. Jacobsen, J.K. Nørskov, F. Besenbacher, Nature 398, 134(1999)
4.61 T. Zambelli, J. Wintterlin, J. Trost, G. Ertl, Science 273, 1688(1996)
4.62 R. Hammers, Phys. Rev. Lett. 83, 3681(1999)
4.63 K.H. Hansen, Ž. Šljivančanin, B. Hammer, E. Lægsgaard, F. Besenbacher, I. Stensgaard, Surf. Sci. 496, 1(2002)
4.64 J.I. Pascual, N. Lorente, Z. Song, H. Conrad, H.-P. Rust, Nature 423, 525(2003)
4.65 J. Wintterlin, S. Völkening, T.V.W. Janssens, T. Zambelli, G. Ertl, Science 278, 1931(1997)
4.66 A. Alavi, P. Hu, T. Deutsch, P.L. Silvestrelli, J. Hutter, Phys. Rev. Lett. 80, 5862(1998)
4.67 F.M. Leibsle, P.W. Murray, S.M. Francis, G. Thornton, M. Bowker, Nature 363, 706(1993)
4.68 W.W. Crew, R.J. Madix, Surf. Sci. 319, L34(1994)
4.69 L.J. Lauhon, W. Ho, Surf. Sci. 451, 219(2000); J. Phys. Chem. A 104, 2463(2000)
4.70 Y. Kim, T. Komeda, M. Kawai, Surf. Sci. 502-503, 7(2002); ibid, Phys. Rev. Lett. 89, 12610(2002)
4.71 J. Gaudioso, H.J. Lee, W. Ho, J. Am. Chem. Soc. 121, 8479(1999)
4.72 A.D. Zhao, Q.X. Li, L. Chen, H.J. Xiang, W.H. Wang, S. Pan, B. Wang, X.D. Xiao, J.L. Yang, J.G. Hou, Q.S. Zhu, Science 309, 1542(2005)
4.73 S.-W. Hla, L. Bartels, G. Meyer, K.-H. Rieder, Phys. Rev. Lett. 85, 2777(2000)
4.74 S.-W. Hla, G. Meyer, K.-H. Rieder, Chem. Phys. Chem. 2, 361(2001); S.-W. Hla, K.-H. Rieder, Ann. Rev. Phys. Chem. 54, 307(2003)
4.75 N. Lin, A. Dmitriev, J. Weckesser, J.V. Barth, K. Kern, Angew. Chem. 41, 4779(2002)
4.76 S. Haq, F.M. Leibsle, Surf. Sci. 355, L345(1996)
4.77 F.M. Leibsle, S. Haq, B.G. Frederick, M. Bowker, N.V. Richarson, Surf. Sci. 343, L1175(1995)
4.78 M. Bowker, E. Rowbotham, F.M. Leibsle, S. Haq, Surf. Sci. 349, 97(1996)
4.79 T.E. Jones, C.J. Baddeley, Surf. Sci. 519, 237(2002)
4.80 T.L. Land, T. Michely, R.J. Behm, J.C. Hemminger, G. Comsa, J. Chem. Phys. 97, 6774(1992)
4.81 F.M. Leibsle, S.M. Francis, R. Davis, N. Xiang, S. Haq, M. Bowker, Phys. Rev. Lett. 72, 2569(1994)
4.82 X.C. Guo, R.J. Madrix, Acc. Chem. Res. 36, 471(2003)
4.83 B.J. McIntyre, M. Salmeron, G.A. Somorjai, Science 265, 141(1994)

4.84 L. Ruan, I. Stensgaard, E. Lægsgaard, F. Besenbacher, Surf. Sci. 314, L873(1994)
4.85 F.M. Leibsle, Surf. Sci. 311, 45(1994)
4.86 X.C. Guo, R.J. Madix, Surf. Sci. 496, 39(2002)
4.87 M. McEllistrem, M. Allgeier, J.J. Boland, Science 279, 544(1998)
4.88 E.J. Buehler, J.J. Boland, Science 290, 506(2000)
4.89 R.J. Hamers, Ph. Avouris, F. Bozso, Phys. Rev. Lett. 59, 2071(1987)
4.90 R. Wolkow, Ph. Avouris, Phys. Rev. Lett. 60, 1049(1988)
4.91 G. Dujardin, A.J. Mayne, F. Rose, Phys. Rev. Lett. 82, 3448(1999)
4.92 A.J. Mayne, F. Rose, G. Dujardin, Surf. Sci. 523, 157(2003)
4.93 R. Martel, Ph. Avouris, I.-W. Lyo, Science 272, 385(1996)
4.94 Ph. Avouris, I.-W. Lyo, F. Bozso, J. Vac. Sci. Technol. B9, 424(1991)
4.95 Ph. Avouris, D. Cahill, Ultramicroscopy 42-44, 838(1992)
4.96 G. Dujardin, R.E. Walkup, Ph. Avouris, Science 255, 1232(1992)
4.97 Y. Wang, M.J. Bronikowski, R.J. Hamers, Surf. Sci. 311, 64(1994)
4.98 Y. Liu, D.P. Masson, A.C. Kummel, Science 276, 1681(1997)
4.99 J.A. Jensen, C. Yan, A.C. Kummel, Science 267, 493(1995)
4.100 C. Yan, J.A. Jensen, A.C. Kummel, J. Chem. Phys. 102, 3381(1995)
4.101 J.A. Jensen, C. Yan, A.C. Kummel, Phys. Rev. Lett. 76, 1388(1996)
4.102 R.G. Egdell, F.H. Jones, J. Mater. Chem. 8, 469(1998)
4.103 H.-J. Freund, H. Kuhlenbeck, V. Staemmler, Rep. Prog. Phys. 59, 283(1996)
4.104 V.E. Henrich, P.A. Cox, The Surface Science of Metal Oxides, Cambridge University Press, Cambridge, 1994
4.105 U. Diebold, J.F. Anderson, K.O. Ng, D. Vanderbilt, Phys. Rev. Lett. 77, 1322(1996)
4.106 K.O. Ng, D. Vanderbilt, Phys. Rev. B56, 10544(1997)
4.107 H. Onishi, Y. Iwasawa, Surf. Sci. 313, L783(1994); H. Onishi, Y. Iwasawa, Phys. Rev. Lett. 77, 3851(1996)]
4.108 C.L. Pang, S.A. Haycock, H. Raza, P.W. Murray, G. Thornton, O. Gülseren, R. James, D.W. Bullett, Phys. Rev. B58, 1586(1998)
4.109 R.A. Bennett, P. Stone, N.J. Price, M. Bowker, Phys. Rev. Lett. 82, 3831(1999)
4.110 U. Diebold, Surf. Sci. Rep. 48, 53(2003)
4.111 N. Hartmann, J. Biener, R.J. Madix, Surf. Sci. 505, 81(2002)
4.112 R. Schaub, E. Wahlstöm, A. Rønnau, E. Lægsgaard, I. Stensgaard, F. Besenbacher, Science 299, 377(2003)
4.113 E. Wahlström, E.K. Vestergaard, R. Schaub, A. Rønnau, M. Vestergaard, E. Lægsgaard, I. Stensgaard, F. Besenbacher, Science 303, 511(2004)
4.114 U. Diebold, W. Hebenstreit, G. Leonardelli, M. Schmid, P. Varga, Phys. Rev. Lett. 81, 405(1998)
4.115 E.L.D. Hebenstreit, W. Hebenstreit, H. Geisler, C.A. Ventrice Jr., D.A. Hite, P.T. Sprunger, U. Diebold, Surf. Sci. 505, 336(2002)
4.116 I.M. Brookes, C.A. Muryn, G. Thornton, Phys. Rev. Lett. 87, 266103(2001)

4.117 J. Schnadt, J. Schiessling, J.N. O'Shea, S.M. Gray, L. Patthey, M.K.-J. Johansson, M. Shi, J. Krempaský, J. Åhlund, P.G. Karlsson, P. Persson, N. Mårtensson, P.A. Brühwiler, Surf. Sci. 540, 39(2003)
4.118 Y. Maeda, M. Okumura, M. Daté, S. Tsubota, M. Haruta, Surf. Sci. 514, 267(2002)
4.119 H. Over, Y.D. Kim, A.P. Seitsonen, S. Wendt, E. Lundgren, H. Schmid, P. Varga, A. Morgante, G. Ertl, Science 287, 1474(2000)
4.120 G. Ertl, J. Mol. Catal. A 182-183, 5(2002)
4.121 S.B. Basame, D. Habel-Rodriguez, D.L. Keller, Appl. Surf. Sci. 183, 62(2001)
4.122 R.L. Burwell, G.L. Haller, K.C. Taylor, J.F. Read, Adv. Catal. 20, 1(1969)
4.123 W. Weiss, W. Ranke, Prog. Surf. Sci. 70, 1(2002)
4.124 K.T. Rim, J.P. Fitts, T. Müller, K. Adib, N. Camillone III, R.M. Osgood, S.A. Joyce, G.W. Flynn, Surf. Sci. 541, 59(2003)
4.125 F.H. Jones, R. Rawlings, J.S. Foord, R.G. Egdell, J.B. Pethica, B.M.R. Wanklyn, S.C. Parker, P.M. Oliver, Surf. Sci. 359, 107(1996)
4.126 R.E. Tanner, P. Meethunkij, E.I. Altman, J. Phys. Chem. B104, 12315(2000)
4.127 M.A. Barteau, J.E. Lyons, I.K. Song, J. Catal. 216, 236(2003)
4.128 N. Magg, J.B. Giorgi, T. Schroeder, M. Bäumer, H.-J. Freund, J. Phys. Chem. B106, 8756(2002)
4.129 S. Surnev, M. Sock, G. Kresse, J.N. Andersen, M.G. Ramsey, F.P. Netzer, J. Phys. Chem. B107, 4777(2003)
4.130 C.I. Carlisle, D.A. King, M.-L. Bocquet, J. Cerdá, P. Sautet, Phys. Rev. Lett. 84, 3899(2000)
4.131 M.-L. Bocquet, P. Sautet, J. Cerda, C.I. Carlisle, M.J. Webb, D.A. King, J. Am. Chem. Soc. 125, 3119(2003)

Chapter 5

5.1 G. Binnig, C.F. Quate, C. Gerber, Phys. Rev. Lett. 56, 930(1986)
5.2 D. Sarid, Scanning Force Microscopy, Oxford University Press, Oxford, 1990
5.3 A. Noy, D.V. Vezenov, C.M. Lieber, Ann. Rev. Mater. Sci. 27, 381(1997)
5.4 H. Takano, J.R. Kenseth, S.-S. Wong, J.C. O'Brien, M.D. Porter, Chem. Rev. 99, 2845(1999)
5.5 J.E. Sader, Rev. Sci. Instrum. 74, 2438(2003)
5.6 J.E. Sader, R.C. Sader, Appl. Phys. Lett. 83, 3195(2003)
5.7 T.R. Albrecht, S. Akamine, T.E. Carver, C.F. Quate, J. Vac. Sci. Technol. A8, 3386(1990)
5.8 T.R. Albrecht, C.F. Quate, J. Vac. Sci. Technol. A6, 271(1988)
5.9 G. Meyer, N.M. Amer, Appl. Phys. Lett. 53, 1045(1988)
5.10 G. Meyer, N.M. Amer, Appl. Phys. Lett. 57, 2089(1990)

5.11 S. Alexander, L. Hellemans, O. Marti, J. Schneir, V. Elings, P.K. Hansma, M. Longmire, J. Gurley, J. Appl. Phys. 65, 164(1989)
5.12 Y. Martin, C.C. Williams, H.K. Wickramasinghe, J. Appl. Phys. 61, 4723(1987)
5.13 D. Sarid, D. Iams, V. Weissenberger, L.S. Bell, Opt. Lett. 13, 1057(1988)
5.14 G. Binnig, C.F.Quate, Ch.Gerber, Phys. Rev. Lett. 56, 930(1986)
5.15 H.J. Hug, Th. Jung, H.-J. Güntherodt, Rev. Sci. Instrum. 63, 3900(1992)
5.16 M.D. Kirk, T.R. Albrecht, C.F. Quate, Rev. Sci. Instrum. 59, 833(1988)
5.17 T. Göddenhenrich, U. Hartmann, M. Anders, C. Heiden, J. Vac. Sci. Technol. A8, 383(1990)
5.18 G. Neubauer, S.R. Cohen, G.M. McClelland, Rev. Sci. Instr. 90, 2296(1990)
5.19 T. Göddenhenrich, U. Hartmann, M. Anders, C. Heiden, J. Microscopy 152, 527(1988)
5.20 M. Anders, C. Heiden, J. Microscopy 152, 643(1988)
5.21 J. Tansock, C.C. Williams, Ultramicroscopy 42-44, 1464(1992)
5.22 M. Tortonese, R.C. Barrett, C.F. Quate, Appl. Phys. Lett. 62, 834(1993)
5.23 F.J. Giessbl, B.M. Trafas, Rev. Sci. Instrum. 65, 1923(1994)
5.24 C.W. Yuan, E. Batalla, M. Zacher, A.D. de Lozanne, M.D. Kirk, M. Tortonese, Appl. Phys. Lett. 65, 1308(1994)
5.25 R.E. Thomson, Rev. Sci. Instrum. 70, 3369(1999)
5.26 A. Schemmel, H.E. Gaub, Rev. Sci. Instrum. 70, 1313(1999)
5.27 J.L. Hutter, J. Bechhoefer, Rev. Sci. Instrum. 64, 1868(1993)
5.28 T.E. Schäffer, J.P. Cleveland, F. Ohnesorge, D.A. Walters, P.K. Hansma, J. Appl. Phys. 80, 3622(1996)
5.29 H.-J. Butt, M. Jaschke, Nanotechnology 6, 1(1995)
5.30 E.F. Florin, M. Rief, H. Lehmann, M. Ludwig, C. Dornmair, V.T. Moy, H.E. Gaub, Biosensors Bioelectronics 10, 895(1995)
5.31 J.P. Cleveland, S. Manne, D. Bocek, P.K. Hansma, Rev. Sci. Instrum. 64, 403(1993)
5.32 Y.Q. Li, N.J. Tao, J. Pan, A.A. Garcia, S.M. Lindsay, Langmuir 9, 637(1993)
5.33 N. Sasaki, M. Tsukada, Phys. Rev. B52, 8471(1995)
5.34 P.K. Hansma, V.B. Elings, O. Marti, C.E. Braker, Science 242, 209(1988)
5.35 L. Ruan, C. Bai, H. Wang, Z. Hu, M. Wan, J. Vac. Sci. Technol. B9, 1134(1991)
5.36 E. Meyer, R. Overney, R. Lüthi, D. Brodbeck, L. Howald, J. Frommer, H.-J. Güntherodt, O. Wolter, M. Fujihira, H. Takano, Y. Gotoh, Thin Solid Films 220, 132(1992)
5.37 K. Ekelund, E. Sparr, J. Engblom, H. Wennerström, S. Engström, Langmuir 15, 6946(1999)

5.38 K. Tanaka, A. Takahara, T. Kajiyama, Macromolecules 29, 3232(1996)
5.39 S. Manne, C.M. Zaremba, R. Giles, L. Huggins, D.A. Walters, A. Belcher, D.E. Morse, G.D. Stucky, J.M. Didymus, S. Mann, P.K. Hansma, Proc. R. Soc. Lond. B256, 17(1994)
5.40 M. Fritz, A.M. Belcher, M. Radmacher, D.A. Walters, P.K. Hansma, G.D. Stucky, D.E. Morse, S. Mann, Nature 371, 49(1994)
5.41 B.R. Heywood, S. Mann, Chem. Mater. 6, 311(1994)
5.42 J.B. Thompson, G.T. Paloczi, J.H. Kindt, M. Michenfelder, B.L. Smith, G. Stucky, D.E. Morse, P.K. Hansma, Biophysical J. 79, 3307(2000)
5.43 B.L. Smith, G.T. Paloczi, P.K. Hansma, R.P. Levine, J. Crystal Growth 211, 116(2000)
5.44 A.M. Belcher, X.H. Wu, R.J. Christensen, P.K. Hansma, G.D. Stucky, D.E. Morse, Nature 381, 56(1996)
5.45 B.L. Smith, T.E. Schäffer, M. Viani, J.B. Thompson, N.A. Frederick, J. Kindt, A. Belcher, G.D. Stucky, D.A. Morse, P.K. Hansma, Nature 399, 761(1999)
5.46 J.B. Thompson, J.H. Kindt, B. Drake, H.G. Hansma, D.E. Morse, P.K. Hansma, Nature 414, 773(2001)
5.47 J.K.H. Hörber, M.J. Miles, Science 302, 1002(2003)
5.48 P.C. Zhang, C.L. Bai, Y.M. Huang, H. Zhao, Y. Fang, N.X. Wang, Q. Li, Scanning Microscopy 9, 981(1995)
5.49 Y. Lyubchenko, L. Shlyakhtenko, R. Harrington, R. Oden, S. Lindsay, Proc. Natl. Acad. Sci. USA 90, 2137(1993)
5.50 S. Manne, P.K. Hansma, J. Massie, V.B. Elings, A.A. Gewirth, Science 251, 183(1991)
5.51 H.G. Hansma, I. Revenko, K. Kim, D.E. Laney, Nucl. Acids Res. 24, 713(1996)
5.52 Y. Lyubchenko, L. Shlyakhtenko, R. Harrington, R. Oden, S. Lindsay, Proc. Natl. Acad. Sci. USA 90, 2137(1993)
5.53 T. Thundat, D.P. Allison, R.J. Warmack, Nucl. Acids Res. 22, 4224(1994)
5.54 D.C. Schwartz, X.J. Li, L.I. Hernandez, P.R. Ramnarrain, E.J. Huff, Y.K. Wang, Science 262, 110(1993)
5.55 R.M. Zimmermann, E.C. Cox, Nucl. Acids Res. 22, 492(1994)
5.56 A. Bensimon, A. Simon, A. Chiffaudel, V. Croquette, F. Heslot, D. Bensimon, Science 265, 2096(1994)
5.57 D. Bensimon, A. Simon, V. Croquette, A. Bensimon, Phys. Rev. Lett. 74, 4754(1995)
5.58 X. Michalet, R. Ekong, F. Fougerousse, S. Rousseaux, C. Schurra, N. Hornigold, M.V. Slegtenhorst, J. Wolfe, S. Povey, J.S. Beckmann, A. Bensimon, Science 277, 1518(1997)
5.59 J.W. Li, C.L. Bai, C. Wang, C.F. Zhu, Z. Lin, Q. Li, E.H. Cao, Nucl. Acids Res. 26, 4785(1998)
5.60 C.R. Clemmer, T.P. Beebe Jr., Science 251, 640(1991)

5.61 H. Chang, A.J. Bard, Langmuir 7, 1143(1991)
5.62 D.P. Allison, P.S. Kerper, M.J. Doktycz, T. Thundat, P. Modrich, F.W. Larimer, D.K. Johnson, P.R. Hoyt, M.L. Mucenski, R.J. Warmack, Genomics 41, 379(1997)
5.63 D.P. Allison, P.S. Kerper, M.J. Doktycz, J.A. Spain, P. Modrich, F.W. Larimer, T. Thundat, R.J. Warmack, Proc. Natl. Acad. Sci. USA 93, 8826(1996)
5.64 H. Yokota, D.A. Nickerson, B.J. Trask, G. van den Engh, M. Hirst, Anal. Biochem. 264, 158(1998)
5.65 G.H. Seong, T. Niimi, Y. Yanagida, E. Kobatake, M. Aizawa, Anal. Chem. 72, 1288(2000)
5.66 H. Lin, D.O. Clegg, R. Lal, Biochemistry 38, 9956(1999)
5.67 I. Rousso, E. Khachatryan, Y. Gat, I. Brodsky, M. Ottolenghi, M. Sheves, Proc. Natl. Acad. Sci. USA 94, 7937(1997)
5.68 G. Meyer, N.M. Amer, Appl. Phys. Lett. 57, 2089(1990)
5.69 H. Hipp, H. Bielefelt, J. Colchero, O. Marti, J. Mlynek, Ultramicroscopy 42-44, 1498(1992)
5.70 K. Fukushima, K. Gohda, J. Phys. Chem. B103, 3582(1999)
5.71 S.-S. Wong, H. Takano, M.D. Porter, Anal. Chem. 70, 5209(1998)
5.72 Q. Zhong, D. Inniss, K. Kjoller, V.B. Elings, Surf. Sci. 290, L688(1993)
5.73 C.A.J. Putman, K.O. van der Werf, B.G. de Grooth, N.F. van Hulst, J. Greve, Appl. Phys. Lett. 64, 2454(1994)
5.74 J.P. Spatz, S. Seiko, M. Moller, R.G. Winkler, P. Reineker, O. Marti, Nanotechnology 6, 40(1995)
5.75 N.A. Burnham, O.P. Behrend, F. Oulevey, G. Gremand, P.-J. Gallo, D. Gourdon, E. Dupas, A.J. Kulik, H.M. Pollock, G.A.D. Briggs, Nanotechnology 8, 67(1997)
5.76 R. García, R. Pérez, Surf. Sci. Rep. 47, 197(2002)
5.77 R. García, J. Tamayo, M. Calleja, F. García, Appl. Phys. A66, S309(1998)
5.78 R.G. Winkler, J.P. Spatz, S. Sheiko, M. Möller, P. Reineker, O. Marti, Phys. Rev. B54, 8908(1996)
5.79 J. Tamayo, R. García, Appl. Phys. Lett. 71, 2394(1997)
5.80 R. Boisgard, D. Michel, J.P. Aimé, Surf. Sci. 401, 199(1998)
5.81 L. Nony, R. Boisgard, J.P. Aimé, J. Chem. Phys. 111, 1615(1999)
5.82 M.-H. Wangbo, G. Bar, R. Brandsch, Surf. Sci. 411, L794(1998)
5.83 L. Wang, Surf. Sci. 429, 178(1999)
5.84 J.P. Cleveland, B. Anczykowski, A.E. Schmid, V.B. Elings, Appl. Phys. Lett. 72, 2613(1998)
5.85 X. Chen, M.C. Davies, C.J. Roberts, S.J.B. Tendler, P.M. Williams, J. Davies, A.C. Dawkes, J.C. Edwards, Ultramicroscopy 75, 171(1998)
5.86 O.P. Behrend, L. Odoni, J.L. Loubet, N.A. Burnham, Appl. Phys. Lett. 75, 2552(1999)

5.87	G. Bar, M. Ganter, R. Brandsch, L. Delineau, M.-H. Wangbo, Langmuir 16, 5702(2000)
5.88	A. Knoll, R. Magerle, G. Krausch, Macromolecules 34, 4259(2001)
5.89	X. Li, Y. Han, L. An, Langmuir 18, 5293(2002)
5.90	M.S. Marcus, M.A. Eriksson, D.Y. Sasaki, R.W. Carpick, Ultramicrosopy 97, 145(2003)
5.91	W.H. Han, S.M. Lindsay, T.W. Jing, Appl. Phys. Lett. 69, 4111(1996)
5.92	W.H. Han, S.M. Lindsay, Appl. Phys. Lett. 72, 1656(1998)
5.93	A. Buguin, O. Du Roure, P. Silberzan, Appl. Phys. Lett. 78, 2982(2001)
5.94	T.R. Albrecht, P. Grütter, D. Horne, D. Rugar, J. Appl. Phys. 69, 668(1991)
5.95	S. Morita, R. Wiesendanger, E. Meyer, Non-contact Atomic Force Microscopy, Springer, Berlin Heidelberg New York, 2002
5.96	F.J. Giessibl, Rev. Mod. Phys. 75, 949(2003)
5.97	F.J. Giessibl, Science 267, 68(1995)
5.98	S. Kitamura, M. Iwatsuki, Jpn. J. Appl. Phys. 35, L145(1995)
5.99	F.J. Giessibl, S. Hembacher, H. Bielefeldt, J. Mannhart, Science 289, 422(2000)
5.100	Y. Sugawara, M. Ohta, H. Ueyama, S. Morita, Science 270, 5242(1995)
5.101	A.L. Shluger, A.I. Livshits, A.S. Foster, C.R.A. Catlow, J. Phys. Condens. Matter 11, R295(1999)
5.102	W. Allers, A. Schwarz, U.D. Schwarz, R. Wiesendanger, Europhys. Lett. 48, 276(1999)
5.103	S. Hembacher, F.J. Giessibl, J. Mannhart, C.F. Quate, Proc. Natl. Acad. Sci. USA 100, 12539(2003)
5.104	S. Hembacher, F.J. Giessibl, J. Mannhart, Science 305, 380(2004)
5.105	T. Uchihashi, T. Ishida, M. Komiyama, M. Ashino, Y. Sugawara, W. Mizutani, K. Yokoyama, S. Morita, H. Tokumoto, M. Ishikawa, Appl. Surf. Sci. 157, 244(2000)
5.106	A. Sasahara, H. Uetsuka, H. Onishi, Langmuir 19, 7474(2003)
5.107	Y. Martin, D. Rugar, H.K. Wickramasinghe, Appl. Phys. Lett. 52, 244(1988)
5.108	H.J. Mamin, D. Rugar, J.E. Stern, B.D. Terris, S.E. Lambert, Appl. Phys. Lett. 53, 1563(1988)
5.109	Y. Martin, C.C. Williams, H.K. Wickramasinghe, J. Appl. Phys. 61, 4723(1987)
5.110	U. Hartmann, Ann. Rev. Mater. Sci. 29, 53(1999)
5.111	P. Grutter, A. Wadas, E. Meyer, H. Heinzelmann, H.-R. Hiber, H.-J. Guntherodt, J. Vac. Sci. Technol. A8, 406(1990)
5.112	F. Tian, C. Wang, G.Y. Shang, N.X. Wang, C.L. Bai, J. Magn. Magn. Mat. 171, 135(1997)
5.113	M.R. Koblischka, U. Hartmann, Ultramicroscopy 97, 103(2003)
5.114	N. Yoshida, T. Arie, S. Akita, Y. Nakayama, Physica B323, 149(2002)
5.115	R. Sessoli, D. Gatteschi, M. Novak, Nature 365, 149(1993)

5.116 K. Wieghart, K. Pohl, I. Jibril, G. Huttner, Angew. Chem. 23, 77(1984)
5.117 H. Nishide, T. Ozawa, M. Miyasaka, E. Tsuchida, J. Am. Chem. Soc. 123, 5942(2001)
5.118 K.R. Dunbar, ed., Proceedings Seventh International Conference on Molecular-Based Magnets, San Antonio, TX, Polyhedron 20(11)(2001)
5.119 D. Rugar, R. Budakian, H.J. Mamin, B.W. Chui, Nature 430, 329(2004)
5.120 D. Ruiz-Molina, M. Mas-Torrent, J. Gómez, A.I. Balana, N. Domingo, J. Tejada, M.T. Martínez, C. Rovira, J. Veciana, Adv. Mater. 15, 42(2003)
5.121 J.N. Israelachivili, Intermolecular and Surface Forces, Academic Press, New York 1985
5.122 G.U. Lee, L.A. Chrisey, R.J. Colton, Science 266, 771(1994)
5.123 V.T. Moy, E.L. Florin, H.E. Gaub, Science 265, 257(1994); V.T. Moy, V.T., E.L. Florin, H.E. Gaub, Science 266, 5183(1994)
5.124 E.L. Florin, V.T. Moy, H.E. Gaub, Science 264, 415(1994)
5.125 U. Dammer, O. Popescu, P. Wagner, D. Anselmetti, H.J. Güntherodt, G.N. Misevic, Science 267, 5201(1995)
5.126 M. Rief, F. Oesterhelt, B. Heymann, H.E. Gaub, Science 275, 1295(1997)
5.127 R.M. Overney, D.P. Leta, C.F. Pictroski, M.H. Rafailovich, Y. Liu, J. Quinn, J. Sokolov, A. Eisenberg, G. Overney, Phy. Rev. Lett. 76, 1272(1996)
5.128 O.H. Willemsen, M.M.E. Snel, K.O. van der Werf, B.G. de Grooth, J. Greve, P. Hinderdorfer, H.J. Gruber, H. Schindler, Y. van Kooyk, C.G. Figdor, Biophys. J. 75, 2220(1998)
5.129 K.L. Johnson, K. Kendall, A.D. Roberts, Proc. R. Soc. Lond. A324, 301(1971)
5.130 B.V. Derjaguin, V.M. Muller, Y.P. Toporov, J. Colloid Interface Sci. 53, 314(1975)
5.131 A. Noy, C.D. Frisbie, L.F. Rozsnyai, M.S. Wrighton, C.M. Lieber, J. Am. Chem. Soc.117, 7943(1995)
5.132 S.K. Sinniah, A.B. Steel, C.J. Miller, J.E. Reutt-Robey, J. Am. Chem. Soc. 118, 8925(1996)
5.133 S.S. Wong, J.D. Harper, P.T. Lansbury Jr., C.M. Lieber, J. Am. Chem. Soc. 129, 603(1998)
5.134 S.S. Wong, A.T. Woolley, T.W. Odom, J.-L. Huang, P. Kim, D.V. Vezenov, C.M. Lieber, Appl. Phys. Lett. 73, 3465(1998)
5.135 H. Hafner, C.L. Cheung, C.M. Lieber, Nature 398, 761(1999)
5.136 H. Nishijima, S. Kamo, S. Akita, Y. Nakayama, K.I. Hohmura, S.H. Yoshimura, K. Takeyasu, Appl. Phys. Lett. 74, 4061(1999)
5.137 C.L. Cheung, J.H. Hafner, T.W. Odom, K. Kim, C.M. Lieber, Appl. Phys. Lett. 76, 3136(2000)

5.138 R. Stevens, C. Nguyen, A. Cassell, L. Delzeit, M. Meyyappan, J. Han, Appl. Phys. Lett. 77, 3453(2000)
5.139 J.H. Hafner, C.L. Cheung, T.H. Oosterkamp, C.M. Lieber, J. Phys. Chem. B105, 743(2001)
5.140 S.S. Wong, E. Joselevich, A.T. Woolley, C.L. Cheung, C.M. Lieber, Nature 394, 52(1998)
5.141 S.S. Wong, A.T. Woolley, E. Joselevich, C.L. Cheung, C.M. Lieber, J. Am. Chem. Soc. 120, 8557(1998)
5.142 S.S. Wong, A.T. Woolley, E. Joselevich, C.M. Lieber, Chem. Phys. Lett. 306, 219(1999)

Chapter 6

6.1 E. Evans, Ann. Rev. Biophys. Biomol. Struct. 30, 105(2001)
6.2 E. Evans, K. Richie, R. Merkel, Biophys. J. 68, 2580(1995)
6.3 D.A. Simson, F. Ziemann, M. Strigl, R. Merkel, Biophys. J. 74, 2080(1998)
6.4 E. Evans, R. Skalak, Mechanics and Thermodynamics of Biomembranes, CRC Press, Boca Raton, FL, 1980
6.5 F. Ziemann, J. Rädler, E. Sackmann, Biophys. J. 66, 2210(1994)
6.6 J.N. Israelachvili, Surf. Sci. Rep. 14, 109(1992)
6.7 H. Yoshizawa, P. McGuiggan, J. Israelachvili, Science 259, 1305(1993)
6.8 H. Yoshizawa, Y.-L. Chen, J. Israelachvili, J. Phys. Chem. 97, 4128(1993)
6.9 A. Ashkin, Phys. Rev. Lett. 24, 156(1970)
6.10 A. Ashkin, J. Biophys. 61, 569(1991)
6.11 R. Gussgard, T. Lindmo, I. Brevik, J. Opt. Soc. Am. B9, 1922(1992)
6.12 J.P. Barton, D.R. Alexander, S.A. Schaub, J. Appl. Phys. 66, 4594–4601(1989)
6.13 H. Furukawa, I. Yamaguchi, Opt. Lett. 23, 216(1998)
6.14 P.C. Ke, M. Gu, Applied Optics 38, 160(1999)
6.15 A. Ashkin, J.M. Dziedzic, Science 235, 1517(1987)
6.16 S. Chu, Science 253, 861(1991)
6.17 S.C. Kuo, M.P. Sheetz, Science 260, 232(1993)
6.18 K. Svoboda, C.F. Schmidt, B.J. Schnapp, S.M. Block, Nature 365, 721(1993)
6.19 T. Perkins, D.E. Smith, S. Chu, Science 264, 819(1994)
6.20 A. Ashkin, J.M. Dziedzic, J.E. Bjorkholm, S. Chu, Opt. Lett. 11, 288(1986)
6.21 K. Svoboda, S.M. Block, Opt. Lett. 19, 930(1994)
6.22 J.H. Hoh, J.P. Cleveland, C.B. Prater, J.-P. Revel, P.K. Hansma, J. Am. Chem. Soc. 114, 4917(1992)
6.23 K.L. Johnson, K. Kendall, A.D. Roberts, Proc. R. Soc. Lond. A324, 301(1971)

6.24 E.W. van der Vegte, G. Hadziioannou, Langmuir 13, 4357(1997)
6.25 J.M. Williams, T. Han, T.P. Beebe Jr., Langmuir 12, 1291(1996)
6.26 L.A. Wenzler, G.L. Moyes, G.N. Raiker, R.L. Hansen, J.M. Harris, T.P. Beebe Jr., Anal. Chem. 67, 2855(1997)
6.27 Y.-S. Lo, N.D. Huefuer, W.S. Chan, F. Stevens, T.P. Beebe Jr., Langmuir 15, 1373(1999); F. Stevens, Y.-S. Lo, J.M. Harris, T.P. Beebe Jr., Langmuir 15, 207(1999)
6.28 Z.Q. Wei, C. Wang, C.F. Zhu, C.Q. Zhou, B. Xu, C.L. Bai, Surf. Sci. 459, 401(2000)
6.29 T. Boland, B.D. Ratner, Proc. Natl. Acad. Sci. USA 92, 5297(1995)
6.30 G.U. Lee, L.A. Chrisey, R.J. Colton, Science 266, 771(1994)
6.31 U. Bockelmann, B. Essevaz-Roulet, F. Heslot, Phys. Rev. Lett. 79, 4489(1997)
6.32 M.A. Lantz, H.J. Hug, R. Hoffmann, P.J.A. van Schendel, P. Kappenberger, S. Martin, A. Baratoff, H.-J. Güntherodt, Science 291, 2580(2001)
6.33 M. Grandbois, M. Beyer, M. Rief, H. Clausen-Schaumann, H.E. Gaub, Science 283, 1727(1999)
6.34 D. Kruger, H. Fuchs, R. Rousseau, D. Marx, M. Parrinello, Phys. Rev. Lett. 89, 186402(2002)
6.35 H. Skulason, C.D. Frisbie, J. Am. Chem. Soc. 122, 9750(2000)
6.36 H. Skulason, C.D. Frisbie, J. Am. Chem. Soc. 124, 15125(2002)
6.37 Y. Leng, S. Jiang, J. Am. Chem. Soc. 124, 11764(2002)
6.38 D.L. Patrick, J.F. Flanagan IV, P. Kohl, R.M. Lynden-Bell, J. Am. Chem. Soc. 125, 6761(2003)
6.39 M. Rief, H. Grubmüller, Chem. Phys. Chem 3, 255(2002)
6.40 P. Hinterdorfer, W. Baumgartner, H.J. Gruber, K. Schilcher, H. Schindler, Proc. Natl. Acad. Sci. USA 93, 3477(1996)
6.41 A. Raab, W. Han, D. Badt, S.J. Smith-Gill, S.M. Lindsay, H. Schindler, P. Hinterdorfer, Nat. Biotechnol. 17, 902(1999)
6.42 U. Dammer, M. Hegner, D. Anselmetti, P. Wagner, M. Dreier, W. Huber, H.-J. Guntherodt, Biophys. J. 70, 2437(1996)
6.43 E.-L Florin, V.T. Moy, H.E. Gaub, Science 264, 415(1994)
6.44 R. Ros, F. Schwesinger, D. Anselmetti, M. Kubon, R. Schäfer, A. Plückthun, L. Tiefenauer, Proc. Natl. Acad. Sci. USA 95, 7402(1998)
6.45 U. Dammer, O. Popescu, P. Wagner, D. Anselmetti, H.-J. Guntherodt, G.N. Misevic, Science 267, 1173(1995)
6.46 S.S. Wong, E. Joselevich, A.T. Woolley, C.L. Cheung, C.M. Lieber, Nature 394, 52(1998)
6.47 H. Grubmüller, B. Heymann, P. Tavan, Science 271, 997(1996)
6.48 V.T. Moy, E.-L. Florin, H.E. Gaub, Science 266, 257(1994)
6.49 O.H. Willemsen, M.M.E. Snel, K.O. van der Werf, B.G. de Grooth, J. Greve, P. Hinterdorfer, H.J. Gruber, H. Schindler, Y. van Kooyk, C.G. Figdor, Biophys. J. 75, 2220(1998)

6.50 H. Schönherr, M.W.J. Beulen, J. Bügler, J. Huskens, F.C.J.M. van Veggel, D. Reinhoudt, G.J. Vancso, J. Am. Chem. Soc. 122, 4963(2000)
6.51 W. Zhang, S.X. Cui, Y. Fu, X. Zhang, J. Phys. Chem. B106, 12705(2002)
6.52 D.J. Müller, W. Baumeister, A. Engel, Proc. Natl. Acad. Sci. USA 96, 13170(1999)
6.53 P. Cluzel, A. Lebrun, C. Heller, R. Lavery, J.-L. Viovy, D. Chatenay, F. Caron, Science 271, 792(1996)
6.54 S.B. Smith, Y. Cui, C. Bustamente, Science 271, 795(1996)
6.55 D. Bensimon, A.J. Simon, V. Croquette, A. Bensimon, Phys. Rev. Lett. 74, 4754(1995)
6.56 M. Rief, H. Clausen-Schaumann, H.E. Gaub, Nat. Struct. Biol. 6, 346(1999)
6.57 G.V. Shivashankar, A. Libchaber, Appl. Phys. Lett. 71, 3727(1997)
6.58 J.-C. Meiners, S.R. Quake, Phys. Rev. Lett. 84, 5014(2000)
6.59 C. Bustamente, S.B. Smith, J. Liphardt, D. Smith, Curr. Opin. Struct. Biol. 10, 279(2000)
6.60 M.W. Konrad, J.I. Bolonick, J. Am. Chem. Soc. 118, 10989(1996)
6.61 B. Maier, D. Bensimon, V. Croquette, Proc. Natl. Acad. Sci. USA 97, 12002(2000)
6.62 Y. Zhang, H. Zhou, Z.-C. Ou-Yang, Biophys. J. 81, 1133(2001)
6.63 A. Montanari, M. Mézard, Phys. Rev. Lett. 86, 2178(2001)
6.64 M.-N. Dessinges, B. Maier, Y. Zhang, M. Peliti, D. Bensimon, V. Croquette, Phys. Rev. Lett. 89, 248102(2002)
6.65 G.J.L. Wuite, S.B. Smith, M. Young, D. Keller, C. Bustamante, Nature 404, 103(2000)
6.66 Y. Murayama, Y. Sakamaki, M. Sano, Phys. Rev. Lett. 90, 18102(2003)
6.67 T.R. Strick, V. Croquette, D. Bensimon, Nature 404, 901(2000)
6.68 T. Strick, J. Allemand, D. Bensimon, A. Bensimon, V. Croquette, Science 271, 1835(1996)
6.69 A.F. Oberhauser, P.E. Marszalek, H.P. Erickson, J.M. Fernandez, Nature 393, 181(1998)
6.70 M. Carrion-Vazquez, A.F. Oberhauser, S.B. Fowler, P.E. Marszalek, S.E. Broedel, J. Clarke, J.M. Fernandez, Proc. Natl. Acad. Sci. USA 96, 3694(1999)
6.71 L. Tskhovrebova, J. Trinick, J.A. Sleep, R.M. Simmons, Nature 387, 308(1997)
6.72 M. Rief, M. Gautel, F. Oesterhelt, J.M. Fernandez, H.E. Gaub, Science 276, 1109(1997)
6.73 G. Yang, C. Cecconi, W.A. Baase, I.R. Vetter, W.A. Breyer, J.A. Haack, B.W. Matthews, F.W. Dahlquist, C. Bustamante, Proc. Natl. Acad. Sci. USA 97, 139(2000)
6.74 F. Oesterhelt, D. Oesterhelt, M. Pfeiffer, A. Engel, H.E. Gaub, D.J. Müller, Science 288, 143(2000)

6.75 M.S.Z. Kellermayer, S.B. Smith, H.L. Granzier, C. Bustamante, Science 276, 1112(1997)
6.76 A. Ptak, S. Takeda, C. Nakamura, J. Miyake, M. Kageshima, S.P. Jarvis, H. Tokumoto, J. Appl. Phys. 90, 3095(2001)
6.77 S.J. Koch, M.D. Wang, Phys. Rev. Lett. 91, 28103(2003)
6.78 A. Samorí, Chem. Eur. J. 6, 4249(2000)
6.79 E. Evans, K. Ritchie, Biophys. J. 76, 2439(1999)
6.80 C. Bustamante, C. Rivetti, D.J. Keller, Curr. Opin. Struct. Biol. 7, 709(1997)
6.81 M. Rief, F. Oesterhelt, B. Heymann, H.E. Gaub, Science 275, 1295(1997)
6.82 H.B. Li, M. Rief, F. Oesterhelt, H.E. Gaub, X. Zhang, J.C. Shen, Chem. Phys. Lett. 305, 197(1999)
6.83 Q. Xu, S. Zou, W. Zhang, X. Zhang, Macromol. Rapid Commun. 22, 1163(2001)
6.84 M. Rief, J.M. Fernandez, H.E. Gaub, Phys. Rev. Lett. 81, 4764(1998)
6.85 T. Hugel, N.B. Holland, A. Cattani, L. Moroder, M. Seitz, H.E. Gaub, Science 296, 1103(2002)
6.86 T. Hugel, M. Grosholz, H. Clusen-Schaumann, A. Pfau, H. Gaub, M. Seitz, Macromolecules 34, 1039(2001)
6.87 F. Oesterhelt, M. Rief, H.E. Gaub, New. J. Phys. 1, 6(1999)
6.88 B. Heymann, H. Grubmüller, Chem. Phys. Lett. 307, 425(1999)
6.89 P.E. Marszalek, Y.P. Pang, H.B. Li, J.E. Yazal, A.F. Oberhauser, J.M. Fernandez, Proc. Natl. Acad. Sci. USA 96, 7894(1999)
6.90 E. Evans, K. Ritchie, Biophys. J. 72, 1541(1997)
6.91 B. Heymann, H. Grubmüller, Phys. Rev. Lett. 84, 6126(2000)
6.92 E. Galligan, C.J. Roberts, M.C. Davies, S.J.B. Tendler, P.M. Williams, J. Chem. Phys. 3208(2002)
6.93 M. Rief, H. Grubmüller, Chem. Phys. Chem 3, 255(2002)
6.94 R. Merkel, P. Nassoy, A. Leung, K. Ritchie, E. Evans, Nature 397, 50(1999)
6.95 T. Strunz, K. Oroszlan, R. Schäfer, H.-J. Güntherodt, Proc. Natl. Acad. Sci. USA 96, 11277(1999)
6.96 W. Grange, T. Strunz, I. Schumakovitch, H.-J. Güntherodt, M. Hegner, Single Mol. 2, 75(2001)
6.97 R. de Paris, T. Strunz, K. Oroszlan, H.J. Güntherodt, M. Hegner, Single Mol. 1, 285(2000)
6.98 P.M. Williams, A. Moore, M.M. Stevens, S. Allen, M.C. Davies, C.J. Roberts, S.J.B. Tendler, J. Chem. Soc., Perk. Trans. 2, 5(2000)
6.99 E. Evans, D. Berk, A. Leung, Biophys. J. 59, 838(1991)
6.100 D.A. Simson, M. Strigl, M. Hohenadl, R. Merkel, Phys. Rev. Lett. 83, 652(1999)
6.101 J. Fritz, A.G. Katopodis, F. Kolbinger, D. Anselmetti, Proc. Natl. Acad. Sci. USA 95, 12283(1998)

6.102 F. Schwesinger, R. Ros, T. Strunz, D. Anselmetti, H.-J. Güntherodt, A. Honegger, L. Jermutus, L. Tiefenauer, A. Plückthun, PNAS 97, 9972(2000)
6.103 G.U. Lee, D.A. Kidwell, R.J. Colton, Langmuir 10, 354(1994)
6.104 S.M. Block, L.S.B. Goldstein, B.J. Schnapp, Nature 348, 348(1990)
6.105 A. Ashkin, K. Schütze, J.M. Dziedzic, U. Euteneuer, M. Schliwa, Nature 348, 346(1990)
6.106 S.C. Kuo, M.P. Sheetz, Science 260, 232(1993)
6.107 K. Svoboda, C.F. Schmidt, B.J. Schnapp, S.M. Block, Nature 365, 721(1993)

Chapter 7

7.1 (a) F.L. Carter, ed., Molecular Electronic Designs I, Marcel Dekker, New York, 1987; (b) F.L. Carter, ed., Molecular Electronic Designs II, Marcel Dekker, New York, 1987
7.2 H.M. McConnell, J. Chem. Phys. 35, 508(1961)
7.3 M.A. Fox, Acc. Chem. Res. 32, 201(1999)
7.4 M.N. Paddon-Row, Acc. Chem. Res. 27, 18(1994)
7.5 M.A. Ratner, J. Jortner, pp 5-72 in Molecular Electronics, M.A. Ratner, J. Jortner, eds., Blackwell, Oxford, UK, 1997
7.6 R. Landauer, IBM J. Res. Dev. 1, 223(1957)
7.7 A. Nitzan, Annu. Rev. Phys. Chem. 52, 681(2001)
7.8 Y. Xue, S. Datta, M.A. Ratner, J. Chem. Phys. 115, 4292(2001)
7.9 B. Larade, J. Taylor, H. Guo, Phys. Rev. B64, 075420(2001)
7.10 D.A. Adams, L. Brus, C.E.D. Chidsey, S. Creager, C. Creutz, C.R. Kagan, P.V. Kamat, M. Lieberman, S. Lindsay, R.A. Marcus, R.M. Metzger, M.E. Michel-Beyerle, J.R. Miller, M.D. Newton, D.R. Rolison, O. Sankey, K.S. Schanze, J. Yardley, X.Y. Zhu, J. Phys. Chem. B107, 6668(2003)
7.11 C.P. Hsu, R.A. Marcus, J. Chem. Phys. 106, 584(1997)
7.12 X.D. Cui, X. Zarate, J. Tomfohr, A. Primak, A.L. Moore, T.A. Moore, D. Gust, G. Harris, O.F. Sankey, S.M. Lindsay, Nanotechnology 13, 5(2002)
7.13 B. Mann, H. Kuhn, J. Appl. Phys. 42, 4389(1971)
7.14 S. Roth, One-Dimensional Metals: Physics and Materials Science, Wiley New York, 1995
7.15 C. Joachim, J.K. Gimzewski, A. Aviram, Nature 408, 541(2000)
7.16 T. Kato, Science 295, 2414(2002)
7.17 S. Fernandez-Lopez, H.S. Kim, E.C. Choi, M. Delgado, J.R. Granja, A. Khasanov, K. Kraehenbuehl, G. Long, D.A. Weinberger, K.M. Wilcoxen, M. R. Ghadiri, Nature 412, 452(2001)
7.18 O. Ikkala, G. ten Brinke, Science 295, 2407(2002)
7.19 G.W. Gokel, O. Murillo, Acc. Chem. Res. 29, 425(1996)
7.20 G.R. Newkome, C.N. Moorefield, F. Vögtl, Dendritic Molecules. Concepts, Synthesis, Perspectives, VCH, Weinhem, 1996

7.21 S.M. Grayson, J.M.J. Fréchet, Chem. Rev. 101, 3819(2001)
7.22 A.W. Bosman, H.W. Janssen, E.W. Meijer, Chem. Rev. 99, 1665(1999)
7.23 M.A. Reed, C. Zhou, C.J. Muller, T.P. Burgin, J.M. Tour, Science 278, 252(1997)
7.24 C. Kergueris, J.P. Bourgoin, S. Palacin, D. Esteve, C. Urbina, M. Magoga, C. Joachim, Phys. Rev. B59, 12505(1999)
7.25 (a) M.A. Rampi, O.J.A. Schueller, G.M. Whitesides, Appl. Phys. Lett. 72, 1781(1998); (b) R. Haag, M.A. Rampi, R.E. Holmin, G.M. Whitesides, J. Am. Chem. Soc. 121, 7895(1999); (c) R.E. Holmlin, R. Haag, M.L. Chabinyc, R.F. Ismagilov, A.E. Cohen, A. Terfort, M.A. Rampi, G.M. Whitesides, J. Am. Chem. Soc. 123, 5075(2001); (d) M.A. Rampi, G.M. Whitesides, Chem. Phys. 281, 373(2002)
7.26 (a) K. Slowinski, K.U. Slowinski, M. Majda, J. Phys. Chem. B103, 8544(1999); (b) K. Slowinski, H.K.Y. Fong, M. Majda, J. Am. Chem. Soc. 121, 7257(1999)
7.27 J.K.N. Mbindyo, T.E. Mallouk, J.B. Mattzela, I. Kratochvilova, B. Razavi, T.N. Jackson, T.S. Mayer, J. Am. Chem. Soc. 124, 4020(2002)
7.28 C. Zhou, M.R. Deshpande, M.A. Reed, L. Jones, J.M. Tour, Appl. Phys. Lett. 71, 611(1997)
7.29 (a) X.L. Fan, C. Wang, D.L. Yang, L.J. Wan, C.L. Bai, Chem. Phys. Lett. 361, 465(2002); (b) Y.H. Liu, X.L. Fan, D.L. Yang, C. Wang, L.J. Wan, C.L. Bai, Langmuir 20, 855(2004)
7.30 L.A. Bumm, J.J. Arnold, M.T. Cygan, T.D. Dunbar, T.P. Burgin, L. Jones, D.L. Allara, J.M. Tour, P.S. Weiss, Science 271, 1705(1996)
7.31 L.A. Bumm, J.J. Arnold, T.D. Dunbar, D.L. Allara, P.S. Weiss, J. Phys. Chem. B103, 8122(1999)
7.32 Y.Q. Xue, S. Datta, S.H. Hong, R. Reifenberger, J.I. Henderson, C.P. Kubiak, Phys. Rev. B59, R7852(1999)
7.33 S. Datta, W.D. Tian, S.H. Hong, R. Reifenberger, J.I. Henderson, C.P. Kubiak, Phys. Rev. Lett. 79, 2530(1997)
7.34 C.G. Zeng, H.Q. Wang, B. Wang, J.L. Yang, J.G. Hou, Appl. Phys. Lett. 77, 3595(2000)
7.35 K. Takanashi, S. Mitani, J. Chiba, H. Fujimori, J. Appl. Phys. 87, 6331(2000)
7.36 X.D. Cui, A. Primak, X. Zarate, J. Tomfohr, O.F. Sankey, A.L. Moore, T.A. Moore, D. Gust, G. Harris, S.M. Lindsay, Science 294, 571(2001)
7.37 F.R.F. Fan, J.P. Yang, L.T. Cai, D.W. Price, Jr., S.M. Dirk, D.V. Kosynkin, Y.X. Yao, A.M. Rawlett, J.M. Tour, A.J. Bard, J. Am. Chem. Soc. 124, 5550(2002)
7.38 D.J. Wold, R. Haag, M.A. Rampi, C.D. Frisbie, J. Phys. Chem. B106, 2813(2002)

7.39	J.G. Kushmerick, D.B. Holt, S.K. Pollack, M.A. Ratner, J.C. Yang, T.L. Schull, J. Naciri, M.H. Moore, R. Shashidhar, J. Am. Chem. Soc. 124, 10654(2002)
7.40	M. Dorogi, J. Gomez, R. Osifchin, R.P. Andres, R. Reifenberger, Phys. Rev. B52, 9071(1995)
7.41	H. Park, A.K.L. Lim, A.P. Alivisatos, J. Park, P.L. McEuen, Appl. Phys. Lett. 75, 301(1999)
7.42	K.A. Son, H.I. Kim, J.E. Houston, Phys. Rev. Lett. 86, 5357(2001)
7.43	D. Di Ventra, S.T. Pantelides, N.D. Lang, J. Chem. Phys. 76, 3448(2000)
7.44	C.K. Chiang, C.R. Fincher, Y.W. Park, A.J. Heeger, H. Shirakawa, E.J. Louis, S.C. Gau, A.G. MacDiarmid, Phys. Rev. Lett. 39, 1098(1977)
7.45	T.A. Skotheim, J.R. Reynolds, R.L. Elsenbaumer, eds., Handbook of Conducting Polymers, 2nd edn., Marcel Dekker, New York, 1997
7.46	R.P. Andres, T. Bein, M. Dorogi, S. Feng, J.I. Henderson, C.P. Kubiak, W. Mahoney, R.G. Osifchin, R. Reifenberger, Science 272, 1323(1996)
7.47	S.N. Yaliraki, M. Kemp, M.A. Ratner, J. Am. Chem. Soc. 121, 3428(1999)
7.48	J.M. Seminario, C.E. de la Cruz, P.A. Derosa, J. Am. Chem. Soc. 123, 5616(2001)
7.49	R.E. Martin, U. Gubler, C. Boudon, V. Gramlich, C. Bosshard, J.P. Gisselbrecht, P. Gunter, M. Gruss, F. Diederich, Chem. Eur. J. 3, 1505(1997)
7.50	R.M. Martin, T. Mader, F. Diederich, Angew. Chem. 38, 817(1999)
7.51	J.M. Tour, Acc. Chem. Res. 33, 791(2000)
7.52	S. Creager, C.J. Yu, C. Bamdad, S. O'Connor, T. MacLean, E. Lam, Y. Chong, G.T. Olsen, J.Y. Luo, M. Gozin, J.F. Kayyem, J. Am. Chem. Soc. 121, 1059(1999)
7.53	S.P. Dudek, H.D. Sikes, C.E.D. Chidsey, J. Am. Chem. Soc. 123, 8033(2001)
7.54	A. Aviram, M.A. Ratner, Chem. Phys. Lett. 29, 277(1974)
7.55	R.M. Metzger, B. Chen, U. Hopfner, M.V. Lakshmikantham, T.V. Hughes, H. Sakurai, J.W. Baldwin, C. Hosch, M.P. Cava, L. Brehmer, G.J. Ashwell, J. Am. Chem. Soc. 119, 10455(1997)
7.56	R.M. Metzger, Acc. Chem. Res. 32, 950(1999)
7.57	R.M. Metzger, Chem. Rev. 103, 3803(2003)
7.58	G.J. Ashwell, R. Hamilton, L.R.H. High, J. Mater. Chem. 13, 1501(2003)
7.59	N. Okazaki, J.R. Sambles, M.J. Jory, C.J. Ashwell, Appl. Phys. Lett. 81, 2300(2002)
7.60	D. Evans, R. Wampler, J. Phys. Chem. B103, 4666(1999)
7.61	M. Di Ventra, S.T. Pantelides, N.D. Lang, Phys. Rev. Lett. 84, 979(2000)

7.62 G.K. Ramachandran, T.J. Hopson, A.M. Rawlett, L.A. Nagahara, A. Primak, S.M. Lindsay, Science 300, 1413(2003)
7.63 G.J. Ashwell, D.S. Gandolfo, J. Mater. Chem. 11, 246(2001)
7.64 G.J. Ashwell, D.S. Gandolfo, J. Mater. Chem. 12, 411(2002)
7.65 G.J. Ashwell, D.S. Gandolfo, R. Hamilton, J. Mater. Chem. 12, 416(2002)
7.66 G.J. Ashwell, M.A. Amiri, J. Mater. Chem. 12, 2181(2002)
7.67 R. Hamilton, G.A. Ashwell, R.M. Metzger, J. Phys. Chem. B106, 12158(2002)
7.68 Y.J. Zhang, Y.S. Li, Q.S. Liu, J. Jin, B.Q. Ding, Y.L. Song, L. Jiang, X.G. Du, Y.Y. Zhao, T.J. Li, Synth. Met. 128, 43(2002)
7.69 S.Q. Zhou, Y.Q. Liu, Y. Xu, W.P. Hu, D.B. Zhu, X.H. Qiu, C. Wang, C.L. Bai, Chem. Phys. Lett. 297, 77(1998)
7.70 R.M. Metzger, Chem. Rev. 103, 3803(2003)
7.71 C. Krzeminski, G. Allan, C. Delerue, D. Vuillaume, R.M. Metzger, Phys. Rev. B64, 085405(2001)
7.72 P.E. Kornilovitch, A.M. Bratkovsky, R.S. Williams, Phys. Rev. B66, 165436(2002)
7.73 S. Lenfant, C. Krzeminski, C. Delerue, G. Allan, D. Vuillaume, Nano Letters 3, 741(2003)
7.74 M. Pomerantz, A. Aviram, R.A. McCorkle, L. Li, A.G. Schrott, Science 255, 1115(1992)
7.75 (a) J. Chen, M.A. Reed, A.M. Rawlett, J.M. Tour, Science 286, 1550(1999); (b) J. Chen, W. Wang, M.A. Reed, A.M. Rawlett, D.W. Price, J.M. Tour, Appl. Phys. Lett. 77, 1224(2000)
7.76 J.M. Seminario, A.G. Zacarias, P.A. Derosa, J. Chem. Phys. 116, 1671(2002)
7.77 (a) H.J. Gao, K. Sohlberg, Z.Q. Xue, H.Y. Chen, S.M. Hou, L.P. Ma, X.W. Fang, S.J. Pang, S.J. Pennycook, Phys. Rev. Lett. 84, 1780(2000); (b) D.X. Shi, Y.L. Song, D.B. Zhu, H.X. Zhang, S.S. Xie, S.J. Pang, H.J. Gao, Adv. Mater. 13, 1103(2001); (c) D.X. Shi, Y.L. Song, H.X. Zhang, P. Jiang, S.T. He, S.S. Xie, S.J. Pang, H.J. Gao, App. Phys. Lett. 77, 3203(2000)
7.78 S. Ranganathan, I. Steidel, F. Anariba, R.L. McCreery, Nano Lett. 1, 491(2001)
7.79 Z.J. Donhauser, B.A. Mantooth, K.F. Kelly, L.A. Bumm, J.D. Monnell, J.J. Stapleton, D.W. Price Jr., A.M. Rawlett, D.A. Allara, J.M. Tour, P.S. Weiss, Science 292, 2303(2001)
7.80 A.O. Solak, S. Ranganathan, T. Itoh, R.L. McCreery, Electrochem. Solid State Lett. 5, E43(2002)
7.81 J.M. Seminario, C.E. de la Cruz, P.A. Derosa, J. Am. Chem. Soc. 123, 5616(2001)
7.82 A. Nitzan, M.A. Ratner, Science 300, 1384(2003)
7.83 G.V. Nazin, X.H. Qiu, W. Ho, Science 302, 77(2003)
7.84 Y.H. Liu, X.L. Fan, D.L. Yang, C. Wang, L.J. Wan, C.L. Bai, Chem. Phys. Lett. 380, 767(2003)

7.85 (a) A. Yazdani, D.M. Eigler, N.D. Lang, Science 272, 1921(1996); (b) N.D. Lang, Phys. Rev. B55, 4113(1997)
7.86 L. Olesen, E. Lægsgaard, I. Stensgaard, F. Besenbacher, J. Schiøtz, P. Stoltze, K.W. Jacobsen, J.K. Nørkov, Phys. Rev. Lett. 72, 2251(1994)
7.87 J.I. Pascual, J. Méndez, J. Gómez-Herrero, A.M. Baró, N. Garcia, U. Landman, W.D. Luedtke, E.N. Bogachek, H.P. Cheng, Science 267, 1793(1995)
7.88 M. Brandbyge, J. Schiøtz, M.R. Sørensen, P. Stoltze, K.W. Jacobsen, J.K. Nórkov, L. Olesen, E. Lægsgaard, I. Stensgaard, F. Besenbacher, Phys. Rev. B52, 8499(1995)
7.89 N. Agraït, G. Rubio, S. Vieira, Phys. Rev. Lett. 74, 3995(1995)
7.90 J.L. Costa-Kramer, N. Garcia, P. Gacia-Mochales, P.A. Serena, M.I. Marques, A. Correia, Phys. Rev. B55, 5416(1997)
7.91 M. Brandbyge, K.W. Jacobsen, J.K. Norskov, Phys. Rev. B55, 2637(1997)
7.92 H. Mehrez, S. Ciraci, A. Buldum, I.P. Batra, Phys. Rev. B55, 1981(1997)
7.93 H.R. Zeller, I. Giaever, Phys. Rev. 181, 789(1969)
7.94 T.A. Fulton, G.J. Dolan, Phys. Rev. Lett. 59, 109(1987)
7.95 J.B. Barner, S.T. Ruggiero, Phys. Rev. Lett. 59, 807(1987)
7.96 S. Chen, R.S. Ingram, M.J. Hostetler, J.J. Pietron, R.W. Murray, T.G. Schaaff, J.T. Khoury, M.M. Alvarez, R.L. Whetten, Science 280, 2098(1998)
7.97 B.Q. Xu, N.J. Tao, Science 301, 1221(2003)
7.98 E. Sheer, N. Agraït, J.C. Cuevas, A.L. Yeyati, B. Ludoph, A. Martín-Rodero, G.R. Bollinger, J.M. Van Ruitenbeek, C. Urbina, Nature 394, 154(1998)
7.99 W. Liang, M.P. Shores, M. Bockrath, J.R. Long, H. Park, Nature 417, 725(2002)
7.100 J. Park, A.N. Pasupathy, J.I. Goldsmith, C. Chang, Y. Yaish, J.R. Petta, M. Rinkoski, J.P. Sethna, H.D. Abruña, P.L. McEuen, D.A. Ralph, Nature 417, 722(2002)
7.101 R.H.M. Smit, Y. Noat, C. Untiedt, N.D. Lang, M.C. van Hemert, J.M. van Ruitenbeek, Nature 419, 906(2002)
7.102 H. Park, J. Park, A.K.L. Lim, E.H. Anderson, A.P. Alivisatos, P. McEuen, Nature 407, 57(2000)
7.103 (a) C. Joachim, J.K. Gimzewski, R.R. Schlittler, C. Chavy, Phys. Rev. Lett. 74, 2102(1995); (b) C. Joachim, J.K. Gimzewski, H. Tang, Phys. Rev. B58, 16407(1998)
7.104 J. Taylor, H. Guo, J. Wang, Phys. Rev. B63, 121104(2001)
7.105 C.G. Zeng, H.Q. Wang, B. Wang, J.L. Yang, J.G. Hou, App. Phys. Lett. 77, 3595(2000)
7.106 Z.J. Donhauser, B.A. Mantooth, K.F. Kelly, L.A. Bumm, J.D. Monnell, J.J. Stapleton, D.W. Price Jr., A.M. Rawlett, D.A. Allara, J.M. Tour, P.S. Weiss, Science 292, 2303(2001)
7.107 P.M. Ajayan, Chem. Rev. 99, 1787(1999)

284 References

7.108 Ph. Avouris, Acc. Chem. Res. 35, 1026(2002)
7.109 D.L. Carroll, X. Blase, J.-C. Charlier, S. Curran, Ph. Redlich, P.M. Ajayan, S. Roth, M. Ruhle, Phys. Rev. Lett. 81, 2332(1998)
7.110 A.M. Rao, P.C. Eklund, S. Bandow, A. Thess, R.E. Smalley, Nature 388, 257(1997)
7.111 R.S. Lee, H.J. Kim, J.E. Fischer, A. Thess, R.E. Smalley, Nature 388, 255(1997)
7.112 C.W. Zhou, J. King, W. Yenilmez, H.J. Dai, Science 290, 1552(2000)
7.113 J. Kong, N.R. Franklin, C.W. Zhou, M.G. Chapline, S. Peng, K. Cho, H.J. Dai, Science 287, 622(2000)
7.114 Y. Cui, Q. Wei, H. Park, C.M. Lieber, Science 293, 1289(2001)
7.115 C.J. Murphy, M.R. Arkin, Y. Jenkins, N.D. Ghatlia, S.H. Bossmann, N.J. Turro, J.K. Barton, Science 262, 1025(1993)
7.116 A. Harriman, Angew. Chem. 38, 945(1998)
7.117 G.B. Schuster, Acc. Chem. Res. 33, 253(2000)
7.118 B. Giese, Acc. Chem. Res. 33, 631(2000)
7.119 M. Bixon, J. Jortner, J. Am. Chem. Soc. 123, 12556(2001)
7.120 U. Diederichsen, Angew. Chem. 36, 2317(1997)
7.121 M.W. Grinstaff, Angew. Chem. 38, 3629(1999)
7.122 K.-H. Yoo, D.H. Ha, J.-O. Lee, J.W. Park, J. Kim, J.J. Kim, H.-Y. Lee, T. Kawai, H.Y. Choi, Phys. Rev. Lett. 87, 198102(2001)
7.123 P. Tran, B. Alavi, G. Gruner, Phys. Rev. Lett. 85, 1564(2000)
7.124 E.M. Conwell, S.V. Rakhmanova, Proc. Natl. Acad. Sci. USA 97, 4556(2000)
7.125 Z. Hermon, S. Caspi, E. Ben-Jacob, Europhys. Lett. 43, 482(1998)
7.126 D.N. Beratan, S. Priyadarshy, S.M. Risser, Chem. Biol. 4, 3(1997)
7.127 P.J. de Pablo, F. Moreno-Herrero, J. Colchero, J.G. Herrero, P. Herrero, A.M. Baró, P. Ordejón, J.M. Soler, E. Artacho, Phys. Rev. Lett. 85, 4992(2000)
7.128 (a) J.M. Tour, M. Kozaki, J.M. Seminario, J. Am. Chem. Soc. 120, 8486(1998); (b) J.M. Tour, Acc. Chem. Res. 33, 791(2000)
7.129 C.P. Collier, E.W. Wong, M. Belohradsky, F.M. Raymo, J.F. Stoddart, P.J. Kuekes, R.S. Williams, J.R. Heath, Science 285, 391(1999)
7.130 C.P. Collier, G. Mattresteig, E.W. Wong, Y. Luo, K. Beverly, J. Sampaio, F.M. Raymo, J.F. Stoddart, J.R. Heath, Science 289, 1172(2000)
7.131 C.L. Brown, U. Jonas, J.A. Preece, H. Ringsdorf, M. Seitz, J.F. Stoddart, Langmuir 16, 1924(2000)
7.132 C.P. Collier, J.O. Jeppesen, Y. Luo, J. Perkins, E.W. Wong, J.R. Heath, J.F. Stoddart, J. Am. Chem. Soc. 123, 12632(2001)
7.133 A.R. Pease, J.O. Jeppesen, J.F. Stoddart, Y.Luo, C.P. Collier, J.R. Heath, Acc. Chem. Res. 34, 433(2001)
7.134 Y. Luo, C.P. Collier, J.O. Jeppesen, K.A. Nielsen, E. Delonno, G. Ho, J. Perkins, H.R. Tseng, T. Yamamoto, J.F. Stoddart, J.R. Hcath, Chem. Phys. Chem 3, 519(2002)
7.135 F.L. Carter, Physica D10, 175(1984)

7.136 S. Ami, M. Hliwa, C. Joachim, Chem. Phys. Lett. 367, 662(2003)
7.137 S. Ami, C. Joachim, Phys. Rev. B65, 155419(2002)
7.138 A. Bachtold, P. Hadley, T. Nakanishi, C. Dekker, Science 294, 1317(2001)
7.139 V. Derycke, R. Martel, J. Appenzeller, Ph. Avouris, Nano Lett. 1, 453(2001)
7.140 Y. Huang, X. Duan, Y. Cui, L.J. Lauhon, K.-H. Kim, C.M. Lieber, Science 294, 1313(2001)

Chapter 8

8.1 J.R. Lakowicz, Principles of Fluorescence Spectroscopy, Plenum Press, New York, 1983
8.2 R. Chang, Basic Principles of Spectroscopy, McGraw-Hill, New York, 1971
8.3 S. Nie, R.N. Zare, Annu. Rev. Biophys. Biomol. Struct. 26, 567(1997)
8.4 W.E. Moerner, J. Chem. Phys. 117, 10925(2002)
8.5 X.S. Xie, J.K. Trautman, Annu. Rev. Phys. Chem. 49, 441(1998)
8.6 R. Hanbury-Brown, R.Q. Twiss, Nature 177, 27(1956)
8.7 H.Z. Cummins, E.R. Pike, eds., Photon Correlation and Light Beating Spectroscopy, Plenum Press, New York, 1974; R. Rigler, J. Widengren, U. Mets, p. 13 in Fluorescence Spectroscopy, E. Wolfbeis, ed., Springer, Berlin Heidelberg New York, 1992
8.8 M. Orrit, Single Mol. 3, 255(2002)
8.9 T. Basché, W.E. Moerner, M. Orrit, H. Talon, Phys. Rev. Lett. 69, 1516(1992)
8.10 W.P. Ambrose, P.M. Goodwin, J. Enderlein, D.J. Semin, J.C. Martin, R.A. Keller, Chem. Phys. Lett. 269, 365(1997)
8.11 L. Fleury, J.-M. Segura, G. Zumofen, B. Hecht, U.P. Wild, Phys. Rev. Lett. 84, 1148(2000)
8.12 M. Wu, P.G. Goodwin, W.P. Ambrose, R.A. Keller, J. Phys. Chem. 100, 17406(1996)
8.13 W.P. Ambrose, T. Basché, W.E. Moerner, J. Chem. Phys. 95, 7150(1991)
8.14 X. Michalet, A.N. Kapanidis, T. Laurence, F. Pinaud, S. Doose, M. Pflughoefft, S. Weiss, Annu. Rev. Biophys. Biomol. Struct. 32, 161(2003)
8.15 T. Plakhotnik, E.A. Donley, U.P. Wild, Ann. Rev. Phys. Chem. 48, 181(1997)
8.16 J. Azoulay, A. Débarre, R. Jaffiol, P. Tchénio, Single Mol. 2, 241(2001)
8.17 Th. Basché, W.E. Moerner, M. Orrit, U.P. Wild, eds., Single Molecule Optical Detection, Imaging and Spectroscopy, VCH, Weinheim, 1997
8.18 M. Orrit, J. Chem. Phys. 117, 10938(2002)
8.19 Th. Tamarat, A. Maali, B. Lounis, M. Orrit, J. Phys. Chem. A104, 1(2000)

8.20 R. Kühnemuth, C.A.M. Seidel, Single. Mol. 2, 251(2001)
8.21 E.K. Hill, A.J. de Mello, Analyst 125, 1033(2000)
8.22 R.M. Dickson, D.J. Norris, Y.L. Tzeng, W.E. Moerner, Science 274, 966(1996)
8.23 M.J. Levene, J. Korlach, S.W. Turner, M. Foquet, H.G. Craighead, W.W. Webb, Science 299, 682(2003)
8.24 W.E. Moerner, L. Kador, Phys. Rev. Lett. 62, 2535(1989)
8.25 M. Orrit, J. Bernard, Phys. Rev. Lett. 65, 2716(1990)
8.26 L. Kador, D.E. Horne, W.E. Moerner, J. Phys. Chem. 94, 1237(1990)
8.27 A. Kulzer, S. Kummer, R. Matzke, C. Bräuchle, Th. Basché, Nature 387, 688(1997)
8.28 L. Fleury, Ph. Tamarat, B. Lounis, J. Bernard, M. Orrit, Chem. Phys. Lett. 236, 87(1995)
8.29 M. Pirotta, A. Renn, M.H.V. Werts, U.P. Wild, Chem. Phys. Lett. 250, 576(1996)
8.30 I.I. Ibram, R.A. Auerbach, R.R. Birge, B.E. Kohler, J.M. Stevenson, J. Chem. Phys. 63, 2473(1975)
8.31 É.V. Shpol'skii, A.A. Il'ina, L.A. Klimova, Dokl. Akad. Nauka SSSR 87, 935(1952)
8.32 E.J. Bowen, B. Brocklehurst, J. Chem. Soc. 4, 4320(1955)
8.33 W.G. van Dorp, M. Soma, J.A. Kooter, J.H. van der Waals, Mod. Phys. 28, 1551(1974)
8.34 A.M. Merle, W.M. Pitts, M.A. El-Sayed, Chem. Phys. Lett. 54, 211(1978)
8.35 A. Bloeß, Y. Durand, M. Matsushita, J. Schmidt, E.J.J. Groenen, Chem. Phys. Lett. 344, 55(2001)
8.36 A. Müller, W. Richter, L. Kador, Chem. Phys. Lett. 241, 547(1995)
8.37 R. Brown, J. Wrachtrup, M. Orrit, J. Bernard, C. von Borczyskowski, J. Chem. Phys. 100, 7182(1994)
8.38 S. Kummer, Th. Basché, C. Bräuchle, Chem. Phys. Lett. 229, 309(1994)
8.39 S. Mais, J. Tittel, Th. Basché, C. Bräuchle, W. Göhde, H. Fuchs, G. Müllen, J. Phys. Chem. A101, 8435(1997)
8.40 A.-M. Boiron, B. Lounis, M. Orrit, J. Chem. Phys. 105, 3969(1996)
8.41 A.M. van Oijen, M. Ketelaars, J. Köhler, T.J. Aartsma, J. Schmidt, Science 285, 400(1999)
8.42 M.A. Bopp, Y. Jia, L. Li, R.J. Cogdell, R.M. Hochstrasser, Proc. Natl. Acad. Sci. USA 94, 10630(1997)
8.43 A. Lounis, F. Jelezko, M. Orrit, Phys. Rev. Lett. 78, 3673(1997)
8.44 F. Jelezko, B. Lounis, M. Orrit, J. Chem. Phys. 107, 1692(1997)
8.45 Ch. Brunel, B. Lounis, Ph. Tamarat, M. Orrit, Phys. Rev. Lett. 81, 2679(1998)
8.46 Ph. Tamarat, B. Lounis, J. Bernard, M. Orrit, S. Kummer, R. Kettner, S. Mais, Th. Basché, Phys. Rev. Lett. 75, 1514(1995)
8.47 U.P. Wild, F. Güttler, M. Pirotta, A. Renn, Chem. Phys. Lett. 193, 451(1992)

8.48 M. Orrit, J. Bernard, A. Zumbusch, R.I. Personov, Chem. Phys. Lett. 196, 595(1992)
8.49 D. Walser, T. Plakhotnik, A. Renn, U.P. Wild, Chem. Phys. Lett. 270, 16(1997)
8.50 T. Plakhotnik, D. Walser, M. Pirotta, A. Renn, U.P. Wild, Science 271, 1703(1996)
8.51 T. Plakhotnik, D. Walser, A. Renn, U.P. Wild, Phys. Rev. Lett. 77, 5365(1996)
8.52 E.J. Sánchez, L. Novotny, G.R. Holtom, X.S. Xie, J. Phys. A101, 7019(1997)
8.53 J. Mertz, C. Xu, W.W. Webb, Opt. Lett. 20, 2532(1995)]
8.54 J. Köhler, J.A.J.M. Disselhorst, M.C.J.M. Donckers, E.J.J. Groenen, J. Schmidt, W.E. Moerner, Nature 363, 242(1993)
8.55 J. Wrachtrup, C. von Borczyskowski, J. Bernard, M. Orrit, R. Brown, Nature 363, 244(1993)
8.56 J. Wrachtrup, C. von Borczyskowski, J. Bernard, M. Orrit, R. Brown, Phys. Rev. Lett. 71, 3565(1993)
8.57 R. Brown, J. Wrachtrup, M. Orrit, J. Bernard, C. von Borczyskowski, J. Chem. Phys. 100, 7182(1994)
8.58 A.C.J. Brouwer, E.J.J. Groenen, J. Schmidt, Phys. Rev. Lett. 80, 3944(1998)
8.59 A.J. van Strien, J. Schmidt, Chem. Phys. Lett. 70, 513(1980)
8.60 J. Köhler, A.C.J. Brouwer, E.J.J. Groenen, J. Schmidt, Science 268, 1457(1995)
8.61 J. Köhler, A.C.J. Brouwer, E.J.J. Groenen, J. Schmidt, Chem. Phys. Lett. 228, 47(1994)
8.62 A. Gruber, M. Vogel, J. Wrachtrup, C. von Borczyskowski, Chem. Phys. Lett. 242, 465(1995)
8.63 J. Köhler, A.C.J. Brouwer, E.J.J. Groenen, J. Schmidt, J. Am. Chem. Soc. 120, 2212(1998)
8.64 P.M. Goodwin, W.P. Ambrose, R.A. Keller, Acc. Chem. Res. 29, 607(1996)
8.65 W.P. Ambrose, P.M. Goodwin, J.H. Jett, A. Van Orden, J.H. Werner, R.A. Keller, Chem. Rev. 99, 2929(1999)
8.66 M. Sauer, K.H. Drexhage, C. Zander, J. Wolfrum, Chem. Phys. Lett. 254, 223(1996)
8.67 M.M. Collinson, R.M. Wightman, Science 268, 1883(1995)
8.68 X.H. Fang, W.H. Tan, Anal. Chem. 71, 3101(1999)
8.69 S. Nie, D.T. Chiu, R.N. Zare, Science 266, 1018(1994)
8.70 M.A. Summers, M.R. Robinson, G.C. Bazan, S.K. Buratto, Chem. Phys. Lett. 364, 542(2002)
8.71 L. Edman, U. Mets, R. Rigler, Proc. Natl. Acad. Sci. USA 93, 6710(1996)
8.72 S. Wennmalm, L. Edman, R. Rigler, Proc. Natl. Acad. Sci. USA 94, 10641(1997)

8.73 Y. Jia, A. Sytnik, L. Li, S. Vladimirov, B.S. Cooperman, R.M. Hochstrasser, Proc. Natl. Acad. Sci. USA 94, 7932(1997)
8.74 C. Eggeling, J.R. Fries, L. Brand, R. Günther, C.A.M. Seidel, Proc. Natl. Acad. Sci. USA 95, 1556(1998)
8.75 D.W. Pierce, N. Hom-Booher, R.D. Vale, Nature 388, 338(1997)
8.76 G.S. Harms, L. Cognet, P.H.M. Lommerse, G.A. Blab, T. Schmidt, Biophys. J. 80, 2396(2001)
8.77 W. Trabesinger, G.J. Schütz, H.J. Gruber, H. Schindler, Th. Schmidt, Anal. Chem. 71, 279(1999)
8.78 G.J. Schütz, W.Trabesinger, Th. Schmidt, Biophys. J. 74, 2223(1998)
8.79 C.D. Talley, R.C. Dunn, J. Phys. Chem. B103, 10214(1999)
8.80 T. Funatsu, Y. Harada, M. Tokunaga, K. Saito, T. Yanagida, Nature 374, 555(1995)
8.81 H.P. Lu, L. Xu, X.S. Xie, Science 282, 1877(1998)
8.82 Q. Xue, E.S. Yeung, Nature 373, 881(1995)
8.83 D.B. Craig, E.A. Arriaga, J.C.Y. Wong, H. Lu, N.J. Dovichi, J. Am. Chem. Soc. 118, 5245(1996)
8.84 I. Sase, H. Miyata, S. Ishiwata, K. Kinosita Jr., Proc. Natl. Acad. Sci. USA 94, 5646(1997)
8.85 H. Sosa, E.J. Peterman, W.E. Moerner, L.S.B. Goldstein, Nat. Struct. Biol. 8, 540(2001)
8.86 H. Noji, R. Yasuda, M. Yoshida, K. Kinosita Jr., Nature 386, 299(1997)
8.87 J. Gelles, B.J. Schnapp, M.P. Sheetz, Nature 331, 450(1988)
8.88 Y. Sako, S. Minoghchi, T. Yanagida, Nat. Cell Biol. 2, 168(2000)
8.89 T. Funatsu, Y. Harada, M. Tokunaga, K. Saito, T. Yanagida, Nature 374, 555(1995)
8.90 C.W. Wilkerson, Jr., P.M. Goodwin, W.P. Ambrose, J.C. Martin, R.A. Keller, Appl. Phys. Lett. 62, 2030(1993)
8.91 T.D. Lacoste, X.P. Michalet, F. Pinaud, D.S. Chemla, A.P. Alivisatos, S. Weiss, Proc. Natl. Acad. Sci. USA. 97, 9461(2000)
8.92 T.A. Byassee, W.C.W. Chan, S.M. Nie, Anal. Chem. 72, 5606(2000)
8.93 K. Fushimi, A.S. Verkman, J. Cell Biol. 112, 719(1991)
8.94 K. Luby-Phelps, S. Mujumdar, R.B. Mujumdar, L.A. Ernst, W. Galbraith, A.S. Waggoner, Biophys. J. 65, 236(1993)
8.95 M. Goulian, S.M. Simon, Biophys. J. 79, 2188(2000)
8.96 M. Ueda, Y. Sako, T. Tanaka, P. Devreotes, T. Yanagida, Science 294, 864(2001)
8.97 Y. Sako, S. Minoghchi, T. Yanagida, Nat. Cell Biol. 2, 168(2000)
8.98 L. Oddershede, J.K. Dreyer, S. Grego, S. Brown, K. Berg-Sørensen, Biophys. J. 83, 3152(2002)
8.99 R. Iino, I. Koyama, A. Kusumi, Biophys. J. 80, 2667(2001)
8.100 W.E. Moerner, J. Chem. Phys. 117, 10925(2002)
8.101 G. Seisenberger, M.U. Ried, T. Endreß, H. Büning, M. Hallek, C. Bräuchle, Science 294, 1929(2001)
8.102 T. Ha, Methods 25, 78(2001)

References

8.103 S. Weiss, Science 283, 1676(1999)
8.104 S. Weiss, Nat. Struct. Biol. 7, 724(2000)
8.105 A. Schuler, E.A. Lipman, W.A. Eaton, Nature 419, 743(2002)
8.106 T. Ha, A.Y. Ting, J. Liang, W.B. Caldwell, A.A. Deniz, D.S. Chemla, P.G. Schultz, S. Weiss, Proc. Natl. Acad. Sci. USA 96, 893(1999)
8.107 B.W. Van Der Meer, G. Coker, III, S.S.-Y. Chen, Resonance Energy Transfer, Theory and Data, VHC, New York, 1994
8.108 A.A. Deniz, T.A. Laurence, G.S. Beligere, M. Dahan, A.B. Martin, D.S. Chemla, P.E. Dawson, P.G. Schultz, S. Weiss, Proc. Natl. Acad. Sci. USA 97, 5179(2000)
8.109 S. Xie, Single Mol. 2, 229(2001)
8.110 J.R. Grunwell, J.L. Glass, T.D. Lacoste, A.A. Deniz, D.S. Chemla, P.G. Schutz, J. Am. Chem. Soc. 123, 4295(2001)
8.111 G. Yao, X.H. Fang, H. Yokota, T. Yanagida, W.H. Tan, Chem. Eur. J. 9, 5686(2003)
8.112 T. Ha, X. Zhuang, H.D. Kim, J.W. Orr, J.R. Williamson, S. Chu, Proc. Natl. Acad. Sci. USA 96, 9077(1999)
8.113 A.A. Deniz, M. Dahan, J.R. Grunwell, T. Ha, A.E. Faulhaber, D.S. Chemla, S. Weiss, P.G. Schultz, Proc. Natl. Acad. Sci. USA 96, 3670(1999)
8.114 S. Brasselet, E.J.G. Peterman, A. Miyawaki, W.E. Moerner, J. Phys. Chem. B104, 3676(2000)
8.115 A. Miyawaki, J. Llopis, R. Heim, J.M. McCaffrey, J.A. Adams, M. Ikura, R.Y. Tsien, Nature 388, 882(1997)
8.116 S. Brasselet, W.E. Moerner, Single Mol. 1, 15(2000)
8.117 G.J. Schütz, W. Trabesinger, T. Schmidt, Biophys. J. 74, 2223(1998)
8.118 M.F. García-Parajó, J.-A. Veerman, R. Bouwhuis, R. Vallée, N.F. van Hulst, Chem. Phys. Chem 2, 347(2001)
8.119 J.J. Macklin, J.K. Trautman, T.D. Harris, L.E. Brus, Science 272, 255(1996)
8.120 W. Lukosz, R.E. Kunz, Opt. Commun. 20, 195(1977)
8.121 L. Novotny, Appl. Phys. Lett. 69, 3806(1996)
8.122 H. Gersen, M.F. García-Parajó, L. Novotny, J.A. Veerman, L. Kuipers, N.F. van Hulst, Phys. Rev. Lett. 85, 5312(2000)
8.123 A. Kramer, W. Trabesinger, B. Hecht, U.P. Wild, Appl. Phys. Lett. 80, 1652(2002)
8.124 J.Y. Ye, M. Ishikawa, O. Yogi, T. Okada, Y. Maruyama, Chem. Phys. Lett. 288, 885(1998)
8.125 L.A. Deschenes, D.A. Vanden Bout, Science 292, 255(2001)
8.126 R. Kettner, T. Tittel, Th. Basché, C. Bräuchle, J. Phys. Chem. 98, 6671(1994)
8.127 A. Zumbusch, L. Fleury, R. Brown, J. Bernard, M. Orrit, Phys. Rev. Lett. 70, 3584(1993)
8.128 D.A. Vanden Bout, W.T. Yip, D.H. Hu, D.K. Fu, T.M. Swager, P.F. Barbara, Science 277, 1074(1997)

8.129 A. Köhn, J. Hofkens, U.-M. Wiesler, M. Cotlet, M. van der Auweraer, K. Müllen, F.C. De Schryver, Chem. Eur. J. 7, 4126(2001)
8.130 F. Jäckel, S. De Feyter, J. Hofkens, F. Köhn, F.C. De Schryver, C. Ego, A. Grimsdale, K. Müllen, Chem. Phys. Lett. 362, 534(2002)
8.131 P. Tchénio, A.B. Myers, W.E. Moerner, J. Phys. Chem. 97, 2491(1993)
8.132 P. Tchénio, A.B. Myers, W.E. Moerner, Chem. Phys. Lett. 213, 325(1993)
8.133 R.M. Dickson, A.B. Cubitt, R.Y. Tsien, W.E. Moerner, Nature 388, 355(1997)
8.134 X.H. Xu, E.S. Yeung, Science 275, 1106(1997)
8.135 X.H. Xu, E.S. Yeung, Science 281, 1650(1998)
8.136 Th. Schmidt, G.J. Schütz, W. Baumgartner, H.J. Gruber, H. Schindler, Proc. Natl. Acad. Sci. USA 93, 2926(1996)
8.137 G.J. Schütz, H. Schindler, Th. Schmidt, Biophys. J. 73, 1073(1997)
8.138 Th. Schmidt, G.J. Schütz, W. Baumgartner, H.J. Gruber, H. Schindler, J. Phys. Chem. 99, 17662(1995)
8.139 Th. Schmidt, G.J. Schütz, H.J. Gruber, H. Schindler, Anal. Chem. 68, 4397(1996)
8.140 M. Ishikawa, O. Yogi, J.Y. Ye, T. Yasuda, Anal. Chem. 70, 5198(1998)
8.141 R.K. Bauer, P. Mayo, W.R. Ware, K.C. Wu, J. Phys. Chem. 86, 3781(1982)
8.142 Y. Liang, A.M.P. Goncalves, J. Phys. Chem. 89, 3290(1985)
8.143 M. Lee, J. Kim, J. Tang, R.M. Hochstrasser, Chem. Phys. Lett. 359, 412(2002)
8.144 M. Jaroniec, R. Madey, Physical Adsorption on Heterogeneous Solids, Elsevier, Amsterdam, 1988
8.145 H.P. Lu, X.S. Xie, J. Phys. Chem. B101, 2753(1997)
8.146 L.A. Peyser, A.E. Vinson, A.P. Bartko, R.M. Dickson, Science 291, 103(2001)
8.147 T.H. Lee, J.I. Gonzalez, R.M. Dickson, Proc. Natl. Acad. Sci. USA 99, 10272(2002)
8.148 B.L. Lounis, H.A. Bechtel, D. Gerion, P. Alivosatos, W.E. Moerner, Chem. Phys. Lett. 329, 399(2000)
8.149 M. Nirmal, L. Brus, Acc. Chem. Res. 32, 407(1999)
8.150 A. Gruber, A. Dräbenstedt, C. Tietz, L. Fleury, J. Wrachtrup, C. von Borczyskowski, Science 276, 2012(1997)
8.151 B. Lounis, W.E. Moerner, Nature 407, 491(2000)
8.152 Ch. Brunel, B. Lounis, Ph. Tamaratm, M. Orrit, Phys. Rev. Lett. 83, 2722(1999)
8.153 B. Kurtsiefer, S. Mayer, P. Zarda, H. Weinfurter, Phys. Rev. Lett. 85, 290(2000)
8.154 A. Beveratos, R. Brouri, T. Gacoin, J.-P. Poizat, P. Grangier, Phys. Rev. A64, R61802(2001)

8.155 R. Berndt, R. Gaisch, J.K. Gimzewski, B. Reihl, R.R. Schlittler, W.D. Schneider, M. Tschudy, Science 262, 1425(1993)
8.156 K. Sakamoto, K. Meguro, R. Arafune, M. Satoh, Y. Uehara, S. Ushioda, Surf. Sci. 502, 149(2002)
8.157 X.H. Qiu, G.V. Nazin, W. Ho, Science 299, 542(2003)
8.158 D. Fujita, T. Ohgi, W.L. Deng, H. Dejo, T. Okamoto, S. Yokoyama, K. Kamikado, S. Mashiko, Surf. Sci. 454, 1021(2000)
8.159 M. Doi, S. Edwards, The Theory of Polymer Dynamics, Clarendon, Oxford, 1986
8.160 P.G. de Gennes, J. Chem. Phys. 55, 572(1971)
8.161 P.G. de Gennes, Macromolecules 9, 587(1976)
8.162 M. Doi, S. Edwards, J. Chem. Soc. Faraday Trans. 2, 74, 1789(1978); ibid, 74, 1802(1978)
8.163 D.E. Smith, T.T. Perkins, S. Chu, Phys. Rev. Lett. 75, 4146(1995)
8.164 D. Lumma, S. Keller, T. Vilgis, J.O. Rädler, Phys. Rev. Lett. 90, 218301(2003)
8.165 W.D. Volkmuth, T. Duke, M.C. Wu, R.H. Austin, A. Szabo, Phys. Rev. Lett. 72, 2117(1994)
8.166 D. Ertas, Phys. Rev. Lett. 80, 1548(1998)
8.167 T.A.J. Duke, R.H. Austin, Phys. Rev. Lett. 80, 1552(1998)
8.168 B.B. Haab, R.A. Mathies, Anal. Chem. 67, 3253(1995)
8.169 T. Anazawa, H. Matsunaga, E.S. Yeung, Anal. Chem. 74, 5033(2002)
8.170 E.M. Sevick, D.R.M. Williams, Phys. Rev. Lett. 76, 2595(1996)
8.171 O.B. Bakajin, T.A.J. Duke, C.F. Chou, S.S. Chan, R.H. Austin, E.C. Cox, Phys. Rev. Lett. 80, 2737(1998)
8.172 D.E. Smith, H.P. Babcock, S. Chu, Science 283, 1724(1999)
8.173 T.T. Perkins, D.E. Smith, S. Chu, Science 276, 2016(1997)
8.174 D.E. Smith, S. Chu, Science 281, 1335(1998)
8.175 D. Wirtz, Phys. Rev. Lett. 75, 2436(1995)
8.176 E. Neher, B. Sakmann, Nature 260, 799(1976)
8.177 O. Orwar, K. Jardemark, I. Jacobson, A. Moscho, H.A. Fisherman, Science 272, 1779(1996)
8.178 B. Sakmann, E. Neher, eds., Single-Channel Recording, 2nd edn., Plenum, New York, 1995
8.179 A. Meller, D. Branton, Electrophoresis 23, 2583(2002)
8.180 A. Meller, L. Nivon, D. Branton, Phys. Rev. Lett. 86, 3435(2001)
8.181 L.Q. Gu, O. Braha, S. Conlan, S. Cheley, H. Bayley, Nature 398, 686(1999)
8.182 D. Branton, J. Golovchenko, Nature 398, 660(1999)
8.183 S. Howorka, L. Movileanu, O. Braha, H. Bayley, Proc. Natl. Acad. Sci. USA 98, 12996(2001)
8.184 A.F. Sauer-Budge, J.A. Nyamwanda, D.A. Lubensky, D. Branton, Phys. Rev. Lett. 90, 238101(2003)
8.185 J. Han, S.W. Turner, H.G. Craighead, Phys. Rev. Lett. 83, 1688(1999)

8.186 M. Cabodi, S.W.P. Turner, H.G. Craighead, Anal. Chem. 74, 5169(2002)
8.187 S.W.P. Turner, M. Cabodi, H.G. Craighead, Phys. Rev. Lett. 88, 128103(2002)
8.188 A. Maier, J.O. Rädler, Phys. Rev. Lett. 82, 1911(1999)
8.189 N. Pernodet, V. Samuilov, K. Shih, J. Sokolov, M.H. Rafailovich, D. Gersappe, B. Chu, Phys. Rev. Lett. 85, 5651(2000)
8.190 D.A. Erie, G. Yang, H.C. Schultz, C. Bustanmante, Science 266, 1562(1994)
8.191 W. Han, S.M. Lindsay, M. Dlakic, R.E. Harrington, Nature 386, 563(1997)
8.192 C. Ma, V.A. Bloomfield, Biophys. J. 67, 1678(1994)
8.193 V.A. Bloomfield, Biopolymers 31, 1471(1991)
8.194 J.-L. Sikorav, J. Pelta, F. Livolant, Biophys. J. 67, 1387(1994)
8.195 I. Flink, D.E. Pettijhon, Nature 253, 62(1975)
8.196 U.K. Laemmli, Proc. Natl. Acad. Sci. USA 72, 4288(1975)
8.197 L.C. Gosule, J.A. Schellman, Nature 259, 333(1976)
8.198 H.V. Hud, K.H. Downing, R. Balhorn, Biochem. Biophys. Res. Commun. 193, 1347(1993)
8.199 J. Widom, R.L. Baldwin, J. Mol. Biol. 144, 431(1980)
8.200 M. Haynes, R.A. Garret, W.B. Gratzer, Biochemistry 9, 4410(1970)
8.201 M. Ueda, K. Yoshikawa, Phys. Rev. Lett. 77, 2133(1996)
8.202 S.H. Kang, M.R. Shortreed, E.S. Yeung, Anal. Chem. 73, 1091(2001)
8.203 D.K. Chattoraj, L.C. Gosule, J.A. Schellman, J. Mol. Biol. 121, 327(1978)
8.204 R.W. Wilson, V.A. Bloomfield, Biochemistry 18, 2192(1979)
8.205 L.C. Gosule, J.A. Schellman, J. Mol. Biol. 121, 311(1978)
8.206 K.A. Marx, G.C. Ruben, J. Biomol. Struct. Dyn. 1, 1109(1984)
8.207 S.A. Allison, J.C. Herr, J.M. Schurr, Biopolymers 20, 469(1981)
8.208 A.G. Arscott, A.-Z. Li, V.A. Bloomfield, Biopolymers 30, 619(1990)
8.209 G.S. Manning, Biophys. Chem. 7, 95(1977)
8.210 G.S. Manning, Biopolymers 19, 37(1980)
8.211 G.S. Manning, Cell Biophys. 7, 57(1985)
8.212 K.A. Marx, T.C. Reynolds, Proc. Natl Acad. Sci. USA 79, 6484(1982)
8.213 K.A. Marx, G.C. Ruben, Nucl. Acids Res. 11, 1839(1983)
8.214 N.V. Hud, K.H. Downing, R. Balhorn, Proc. Natl Acad. Sci. USA 92, 3581(1995)
8.215 H. Noguchi, S. Saito, S. Kidoaki, K. Yoshikawa, Chem. Phys. Lett. 261, 527(1996)
8.216 D.D. Dunlap, A. Maggi, M.R. Soria, L. Monaco, Nucl. Acids Res. 25, 3095(1997)
8.217 M. Becker, R. Misselwitz, H. Damaschun, G. Damaschun, D. Zirwer, Nucl. Acids Res. 7, 1297(1979)
8.218 J.A. Schellman, N. Parthasarathy, J. Mol. Biol. 175, 313(1984)

8.219 J. Pelta, F. Livolant, J.-L. Sikorav, J. Biol. Chem. 271, 5656(1996)
8.220 Z. Lin, C. Wang, X.Z. Feng, M.Z. Liu, J.W. Li, C.L. Bai, Nucl. Acids Res. 26, 3228(1998)

Chapter 9

9.1 E. Abbe, Archiv f. Mikroskop. Anat. 9, 413(1873)
9.2 K. Lieberman, S. Harush, A. Lewis, R. Kopelman, Science 247, 59(1990)
9.3 E. Betzig, J.K. Troutman, T.D. Harris, J.S. Weiner, R.L. Kostelak, Science 251, 1468(1991); E. Betzig, R.J. Chichester, Science 262, 1422(1993)
9.4 W. Tan, Z.-Y. Shi, S. Smith, D. Birnbaum, R. Kopelman, Science 258, 778(1992); R. Kopeltman, W. Tan, Science 262, 1382(1993)
9.5 J.-J. Greffet, R. Carninati, Prog. Surf. Sci. 56, 133(1997)
9.6 C. Girard, A. Dereux, Rep. Prog. Phys. 59, 657(1996); C. Girard, C. Joachim, S. Gauthier, Rep. Prog. Phys. 63, 893(2000)
9.7 W.P. Ambrose, P.M. Goodwin, J.C. Martin, R.A. Keller, Phys. Rev. Lett. 72, 160(1994)
9.8 A. Lewis, S. Lieberman, S. Haroush, V. Habib, R. Kopelman, M. Isaacson, Light microscopy beyond the limits of diffraction and to the limits of single molecule resolution, in Optical Microscopy for Biology, ed. by B. Herman, K. Jacobson (Wiley-Liss, New York 1990) pp.615-639
9.9 E. Betzig, J.K. Troutman, Science 257, 189 (1992)
9.10 R. Kopelman, W. Tan, pp. 227-254, in Spectroscopic and Microscopic Imaging of the Chemical State, M.D. Morris ed., Dekker, New York, 1993
9.11 G.W. White, Introduction to Microscopy, Butterworths, London, 1966
9.12 B. Herman, K. Jacobson, Optical Microscopy for Biology, Wiley-Liss, New York, 1989
9.13 M.D. Morris, Spectroscopic and Microscopic Imaging of the Chemical State, Dekker, New York, 1993
9.14 L.M. Loew, ed., Spectroscopic Membrane Probes, CRS Press, Boca Raton, 1988
9.15 K. Lieberman, A. Lewis, Appl. Phys. Lett. 62, 1335(1993)
9.16 M. Garcia-Parajo, E. Cambril, Y. Chen, Appl. Phys. Lett. 65, 1498(1994)
9.17 J.K. Trautman, E. Betzig, J.S. Weiner, D.J. DiGiovani, T.D. Harris, F. Hellman, E.M. Gyorgy, J. Appl. Phys. 71, 465(1992)
9.18 X.S. Xie, Acc. Chem. Res. 29, 598(1996)
9.19 A. Lewis, M. Isaacson, A. Muray, A. Harootunian, Ultramicroscopy 13, 227(1984)
9.20 D.W. Pohl, W. Denk, M. Lanz, Appl. Phys. Lett. 44, 651(1984)

9.21 W. Tan, Z.-Y. Shi, R. Kopelman, Anal. Chem. 64, 2985(1992)
9.22 Th. Lacosta, Th. Huser, R. Prioli, H. Heinzelmann, Ultramicroscopy 71, 333(1998)
9.23 M. Muranishi, K. Sato, S. Hosaka, A. Kikukawa, T. Shintani, K. Ito, Jpn. J. Appl. Phys. 2, 36, L942(1997)
9.24 F. Baida, D. Courjon, G. Tribillon, p. 71, in Near Field Optics, D.W. Pohl, D. Courjon eds., NATO ASI Series, Kluwer, Dordrecht, 1993
9.25 J.A. Veerman, A.M. Otter, L. Kuipers, N.F. van Hulst, Appl. Phys. Lett. 72, 3115(1998)
9.26 J.A. Veerman, A.M. Otter, L. Kuipers, N.F. van Hulst, Appl. Phys. Lett. 72, 3115(1998)
9.27 H. Zhou, A. Midha, G. Mills, S. Thoms, S.K. Murad, J.M.R. Weaver, J. Vac. Sci. Technol. B16, 54(1998)
9.28 H. Zhou, A. Midha, G. Mills, L. Donaldson, J.M.R. Weaver, Appl. Phys. Lett. 75, 1824(1999)
9.29 G. Schürmann, P.F. Indermühle, U. Staufer, N.F. de Rooij, Surf. Interface Anal. 27, 299(1999)
9.30 C. Mihalcea, W. Scholz, S. Werner, S. Münster, E. Oesterschulze, R. Kassing, Appl. Phys. Lett. 68, 3531(1996)
9.31 H. Muramatsu, N. Chiba, K. Homma, K. Nakajima, T. Ataka, S. Ohta, A. Kusumi, M. Fujihira, Appl. Phys. Lett. 66, 3567(1996)
9.32 Nanonic Imaging Ltd.,
9.33 R.C. Dunn, Chem. Rev. 99, 2891(1999)
9.34 I. Rousso, E. Khachatryan, Y. Gat, I. Brodsky, M. Ottolenghi, M. Sheves, A. Lewis, Proc. Natl. Acad. Sci. USA 94, 7937(1997)
9.35 R. Eckert, J.M. Freyland, H. Gersen, H. Heinzelmann, G. Schürmann, W. Noell, U. Staufer, N.F. de Rooij, Appl. Phys. Lett. 77, 3695(2000)
9.36 B. Gross Levi, Phys. Today 52, 18(1999)
9.37 E. Betzig, D.L. Finn, J.S. Weiner, Appl. Phys. Lett. 60, 2484(1992)
9.38 R. Toledo-Crow, P.C. Yang, Y. Chen, M. Vaez-Iravani, Appl. Phys. Lett. 60, 2957(1992)
9.39 A. Shchemelinin, M. Rudman, K. Lieberman, A. Lewis, Rev. Sci. Instrum. 64, 3538(1993)
9.40 K. Karrai, R.D. Grober, Appl. Phys. Lett. 66, 1842(1995)
9.41 J. Barenz, O. Hollricher, O. Marti, Rev. Sci. Instrum. 67, 1912(1996)
9.42 O. Hollricher, R. Brunner, O. Marti, Ultramicroscopy 71, 143(1998)
9.43 D. Birnbaum, S. Kook, R. Kopelman, J. Phys. Chem. 97, 3091 (1993)
9.44 J.K. Leong, C.C. Williams, Appl. Phys. Lett. 66, 1432(1995)
9.45 J.W.P. Hsu, M. Lee, B.S. Weaver, Rev. Sci. Instrum. 66, 3177(1995)
9.46 P.F. Barbara, D.M. Adams, D.B. O'Connor, Annu. Rev. Mater. Sci. 29, 433(1999)
9.47 F.H. Lei, G.Y. Shang, M. Troyon, M. Spajer, H. Morjani, J.F. Angiboust, M. Manfait, Appl. Phys. Lett. 79, 2489(2001)
9.48 M.A. Bopp, A.J. Meixner, G. Tarrach, I. Zschokke-Gränacher, L. Novotny, Chem. Phys. Lett. 263, 721(1996)
9.49 C.E. Talley, M.A. Lee, R.C. Dunn, Appl. Phys. Lett. 72, 2954(1998)

9.50 J. Michaelis, C. Hettich, J. Mlynek, V. Sandoghdar, Nature 405, 325(2000)
9.51 G.C. des Francs, C. Girard, J. Chem. Phys. 117, 4659(2002)
9.52 H.F. Hess, E. Betzig, T.D. Harris, L.N. Pfeiffer, K.W. West, Science 264, 1740(1994)
9.53 J.K. Trautman, J.J. Macklin, L.E. Brus, E. Betzig, Nature 369, 40(1994)
9.54 X.S. Xie, R.C. Dunn, Science 265, 361(1994)
9.55 H.P. Lu, X.S. Xie, Nature 385, 143(1997)
9.56 R.C. Dunn, G.A. Holtom, L. Mets, X.S. Xie, J. Phys. Chem. 98, 3094(1994)
9.57 D.A. Higgins, J. Kerimo, D.A. Vanden Bout, P.F. Barbara, J. Am. Chem. Soc. 118, 4049(1996)
9.58 R.X. Bian, R.C. Dunn, X.S. Xie, P.T. Leung, Phys. Rev. Lett. 75, 4772(1995)
9.59 W.P. Ambrose, P.M. Goodwin, J.C. Martin, R.A. Keller, Science 265, 364(1994)
9.60 W.E. Moerner, T. Plakhotnik, T. Irngartinger, U.P. Wild, D. Pohl, B. Hecht, Phys. Rev. Lett. 73, 2764(1994)
9.61 D.J. Norris, M. Kuwata-Gonokami, W.E. Moerner, Appl. Phys. Lett. 71, 297(1997)
9.62 T. Ha, Th. Enderle, D.F. Ogletree, D.S. Chemla, P.R. Selvin, S. Weiss, Proc. Natl. Acad. Sci. USA 93, 6264(1996)
9.63 G.T. Shubeita, S.K. Sekatskii, M. Chergui, G. Dietler, Appl. Phys. Lett. 74, 3453(1999)
9.64 G.T. Shubeita, S.K. Sekatskii, G. Dietler, V.S. Letokhov, Appl. Phys. Lett. 80, 2625(2002)
9.65 W.R. Seitz, CRC Crit. Rev. Anal. Chem. 19, 135(1988)
9.66 W. Tan, Z.-Y Shi, B.A. Thorsrud, C. Harris, R. Kopelman, SPIE 2068, 10(1993)
9.67 R. Kopelman, A. Lewis, K. Lieberman, W. Tan, J. Lumin. 48, 871(1991)
9.68 R. Kopelman, in Physical and Chemical Mechanisms in Molecular Radiation Biology, W.A. Glass, M. Varma eds., Plenum, New York, 1991, Basic Life Sci. 58, 475(1991)

Chapter 10

10.1 D.A. Long, Raman Spectroscopy, McGraw-Hill, New York, 1977
10.2 M. Fleischmann, P.J. Hendra, A.J. McQuillan, Chem. Phys. Lett. 26, 163(1974)
10.3 D.L. Jeanmaire, R.P. Van Duyne, J. Electroanal. Chem. 84, 1(1977)
10.4 M.G. Albrecht, J.A. Creighton, J. Am. Chem. Soc. 99, 5215(1977)
10.5 M. Moskovits, Rev. Mod. Phys. 57, 783(1985)
10.6 H. Metiu, P. Das, Annu. Rev. Phys. Chem. 35, 507(1984)

10.7	K. Kneipp, H. Kneipp, I. Itzkan, R.R. Dasari, M.S. Feld, Chem. Rev. 99, 2957(1999)
10.8	K. Kneipp, H. Kneipp, F. Seifeit, I. Itzkan, R.R. Dasari, M.S. Feld, Phys. Rev. Lett. 76, 2444(1996)
10.9	A.M. Michaels, M. Nirmal, L.E. Brus, J. Am. Chem. Soc. 121, 9932(1999)
10.10	D.A. Weitz, S. Garoff, J.I. Gersten, A. Nitzan, J. Chem. Phys. 78, 5324(1983)
10.11	A. Campion, Annu. Rev. Phys. Chem. 36, 549(1985)
10.12	H. Xu, J. Aizpurua, M. Kall, P. Apell, Phys. Rev. E62, 4318(2000)
10.13	F.J. García-Vidal, J.B. Pendry, Phys. Rev. Lett. 77, 1163(1996)
10.14	A. Campion, P. Kambhampati, Chem. Soc. Rev. 27, 241(1998)
10.15	J.R. Lombardi, R.L. Birke, T. Lu, J. Xu, J. Chem. Phys. 84, 4174(1986)
10.16	R.J. Hinde, M.J. Sepaniak, R.N. Compton, J. Nordling, N. Lavrik, Chem. Phys. Lett. 339, 167(2001)
10.17	K. Ujihara, J. Appl. Phys. 43, 2376(1972)
10.18	E.J. Zeman, G.C. Schatz, J. Phys. Chem. 91, 634(1987)
10.19	S. Nie, S.R. Emory, Science 275, 1102(1997)
10.20	K.L. Kelly, E. Coronado, L.L. Zhao, G.C. Schatz, J. Phys. Chem. B107, 668(2003)
10.21	C.L. Haynes, R.P. Van Duyne, J. Phys. Chem. B107, 7426(2003)
10.22	C.L. Haynes, A.D. McFarland, L.L. Zhao, R.P. Van Duyne, G.C. Schatz, L. Gunnarsson, J. Prikulis, B. Kasemo, M. Käll, J. Phys. Chem. B107, 7337(2003)
10.23	K. Kneipp, Y. Wang, H. Kneipp, L.T. Perelman, I. Itzkan, R.R. Dasari, M.S. Feld, Phys. Rev. Lett. 78, 1667(1997)
10.24	W.E. Doering, S. Nie, J. Phys. Chem. B106, 311(2002)
10.25	A.M. Michaels, M. Nirmal, L.E. Brus, J. Am. Chem. Soc. 121, 9932(1999)
10.26	A.M. Michaels, J. Jiang, L. Brus, J. Phys. Chem. B104, 11965(2000)
10.27	K.A. Bosnick, J. Jiang, L.E. Brus, J. Phys. Chem. B106, 8096(2002)
10.28	J. Jiang, K.A. Bosnick, M. Maillard, L.E. Brus, J. Phys. Chem. B107, 9964(2003)
10.29	K. Kneipp, H. Kneipp, I. Itzkan, R.R. Dasari, M.S. Feld, Chem. Phys. 247, 155(1999)
10.30	L.D. Ziegler, J. Raman Spectrosc. 21, 769(1990)
10.31	K. Kneipp, H. Kneipp, G. Deinum, I. Itzkan, R.R. Dasari, M.S. Feld, Appl. Spectrosc. 52, 175(1998)
10.32	A. Weiss, G. Haran, J. Phys. Chem. B105, 12348(2001)
10.33	C.J.L. Constantino, T. Lemma, P.A. Antunes, R. Aroca, Anal. Chem. 73, 3674(2001)
10.34	R.C. Maher, L.F. Cohen, P. Etchegoin, Chem. Phys. Lett. 352, 378(2002)
10.35	Y.D. Suh, G.Y. Schenter, L. Zhu, H.P. Lu, Ultramicroscopy 97, 89(2003)

10.36 M. Ishikawa, Y. Maruyama, J.-Y. Ye, M. Futamata, J. Biol. Phys. 28, 573(2002)
10.37 G. Xu, E.J. Bjerneld, M. Käll, L. Börjesson, Phys. Rev. Lett. 83, 4357(1999)
10.38 E.J. Bjerneld, Z. Földes-Papp, M. Käll, R. Rigler, J. Phys. Chem. B106, 1213(2002)
10.39 E.J. Bjerneld, P. Johansson, M. Käll, Single Mol. 3, 239(2000)
10.40 B. Pettinger, G. Picardi, R. Schuster, G. Ertl, Single Mol. 5, 285(2002)
10.41 B. Pettinger, B. Ren, G. Picardi, R. Schuster, G. Ertl, Phys. Rev. Lett. 92, 096101(2004)
10.42 M.S. Anderson, Appl. Phys. Lett. 76, 3130(2000)
10.43 R.M. Stöckle, Y.D. Suh, V. Deckert, R. Zenobi, Chem. Phys. Lett. 318, 131(2000)
10.44 D. Zeisel, B. Dutoit, V. Deckert, T. Roth, R. Zenobi, Anal. Chem. 69, 749(1997)
10.45 V. Deckert, D. Zeisel, R. Zenobi, Anal. Chem. 70, 2646(1998)
10.46 D. Zeisel, V. Deckert, R. Zenobi, T. Vo-Dinh, Chem. Phys. Lett. 283, 381(1998)
10.47 C.L. Jahncke, M.A. Paesler, H.D. Hallen, Appl. Phys. Lett. 67, 2483(1995)
10.48 D.A. Smith, S. Webster, M. Ayad, S.D. Evans, D. Fogherty, D. Batchelder, Ultramicroscopy 61, 247(1995)
10.49 M.S. Dresselhaus, A. Jorio, A.G.S. Filho, G. Dresselhaus, R. Saito, Physica B323, 15(2002)
10.50 Z. Yu, L. Brus, J. Phys. Chem. B105, 1123(2001)
10.51 M.S. Dresselhaus, P.C. Eklund, Adv. Phys. 49, 705(2000)
10.52 A. Hartschuh, H.N. Pedrosa, L. Novotny, T.D. Krauss, Science 301, 1354(2003)

Index

0–0 transition   192
1,3-butadiene(C_4H_6)   86
1,3,5-benzenetricarboxylic acid   57
1,3,5-triazine-2,4,6-triamine   58
3-aminopropyltriethoxysilane   115
4, 4', 7, 7'-tetrachlorothioindigo   117

absorption spectrum   189
abstraction reaction   96
acetylene   55, 86
acidified graphite surface   168
acoustically coupled excitation   120
AC-Stark effect   194
activation energy   70, 72, 84
active light source   226
added-row model   99
adenine   123
adhesion force   26
adsorbate-induced restructuring   37
adsorbate-induced surface reconstruction   80
adsorption configuration   53
amino acid   63
anisotropic diffusion barrier   61
antibunching effect   184
antigen–antibody interaction   139
anti-Stokes line   241
apertureless approach   228
atomic switch   74
atomistic rupturing process   138
atom–jellium model   31
attenuation factor   159, 165
Autler-Townes splitting   194
autocorrelation function   141, 190, 234
avidin   140

barrier height   6, 25, 34, 36
benzene   42, 86
benzoic acid   102

bichromophoric molecule   209
bimolecular array   53
biomembrane force probe   131, 155
biotin   140
bistable conductance behavior   170
bistable element   74
bond-rupture force   134
break junction   164
Brownian motion   201
bunching effect   91
burst-integrated fluorescence lifetime scheme   198

cantenane   179
cantilever   107
cantilever-shaped probe   228
capillary force   114, 136
Carbon nanotube   176
carrier transport kinetic   159
cellular autofluorescence   200
center atom   95
charge exchange interaction   247
charge transfer   101
chemical enhancement effect   244
chemical identification   66
chemically decorated probe   129
chemically specific tip   37
chemifluorescence reaction   197
chemisorbed   44
C_{60} heptomers   58
C–H stretch   88
Clostridium histolyticum collagenases   117
CO   37
coherent transformation   144
coil-to-helix transition   153
collagen I   117
collection efficiency   186

collection mode   225
collisional quenching   185
conditional probability density   211
confocal excitation   196
confocal microscopy   188
conformational dynamic   197
constant-current mode   18
constant-height mode   19
contact mode   112
coordinatively unsaturated site   103
copper phthalocyanine   60
copper(II) phthalocyanine   31
corner atom   95
Coulomb blockade effect   173
cross-correlation function   145
cross-wire junction   165
Cu-tetra[3,5 di-t-butylphenyl]porphyrin   48
cyclodextrin   141

dangling bond   46
dark state   184
decaborane   96
decomposition reaction   91
deflection detection   109
Dehydrogenation reaction   86
density function theory calculation   84
dephasing effect   194
deprotonation   90
desorption force   142
detection volume   185
deuterium isotope effect   56
diatomic molecule   36
dielectric effect   245
dielectric interface   207
diffraction limit   223
diffusion barrier   45, 79, 83, 103
diffusion coefficient   70
diffusion trajectory   210
discrete dipole approximation   245
disilane   96
dissociation efficiency   84
dissociation rate   88
dissociative adsorption   79, 101
DNA condensation   147
donor–spacer–acceptor model   167
dressed molecule   193
driving frequency   125
dynamic force microscope   140
dynamic force spectrum   156

elastic regime   144
electrical breakdown voltage   166
electrodynamic trap   196
electromigration   76
electron beam lithography   227
electron charging effect   166
electron-scattering quantum chemistry   42
electron-spin resonance effect   66
electrophoretic flow   199
electrophoretic mobility   214
elongational flow   215
enantioselective hydrogenation   91
enhancement factor   242
entropic regime   144
enzymatic process   204
ethene   105
ethylene   56, 91
ethylidyne   92
evanescent field   189, 224, 230
evanescent field excitation   201
exchange correlation potential   25

fatty acid   114
feedback parameter   119
field evaporation   75
field-induced diffusion   72
flavin adenine dinucleotide   198
flavinoid   200
fluorescence correlation spectroscopy   187
fluorescence polarization   210
fluorescence resonance energy transfer   202, 235
fluorescence spectra   232
fluorine-substituted carboxylate   123
focused ion beam   227
force-distance curves   120
force sensor   110
force spectrum   128
formaldehyde   92
fractioning effect   218
Frank-Condon weighted density   161
free oscillation   120
freely jointed chain   146
frequency-modulation Stark double modulation   190
frequency-modulation ultrasound double modulation   190

friction force microscope   117
Förster's equation   202

g-factor   68
glass transition temperature   208
green fluorescence protein   198
guest–host interaction   57, 141

hairpin structure   146
Hanbury-Brown-Twiss correlator   184
hemoglobin   251
heptadecanethiol   117
heterocyclic molecule   51
heterogeneous adsorption site   61
heterogeneous organic structure   61
hexadecanethiol   117
hexa-tert-butyl-decacyclene   54
hopping rate   45
horseradish peroxidase   253
hydrodynamic focusing effect   196
hydrodynamic screening effect   214
hydrogen bonding   57
hydrogen-filled dimer   93
hyperfine interaction   195

illumination mode   224
iminobiotin   140
immiscible blend   114
immunoglobin   148
induced dipole moment   35
inelastic electron tunneling
    spectroscopy   56, 77
inelastic tunneling process   8
infrared-type spectra   9
intensity trajectory   197
interdigitated alkyl part   47
intermediate state   14
intracellular diffusion   201
intramolecular logic gate   180
intramolecular vibrational mode   79
inverted epifluorescence
    microscope   199
iodobenzene   88
iodochloride   96
ionization potential   33
iron oxide film   103
isometric force   158

Johnson-Kendall-Roberts (JKR) theory
    134

kinesin   158
Kondo effect   174

Landauer mechanism   160
Lander molecule   54
Larmor angular processing frequency
    67
laser-induced diffusion   71
lateral diffusion   65
light–matter interaction   233
light-shift effect   193
local density of state   18
logic function   179
luminescent center   212
lysozyme   140

magnetic bead method   133
magnetic excitation   120
magnetic force microscope   123
magnetic resonance effect   195
magnetic resonance force microscopy
    127
magnetic storage media   126
manipulation-assisted reaction   89
matrix molecule   64
membrane protein   149
mercury column junction   165
metal-clad configuration   189
metal ligand complex   89
metal phthalocyanine   46
methanol   92
methylacetoacetate   91
microcontact printing   117
microtubule   158
Mie regime   132
mode-selective excitation   85
modified long wavelength
    approximation   245
molecular beacon   206
molecular device   179
molecular exciton microscopy   239
molecular magnet   126
molecular orbital   31
multiphoton absorption   184
multiple-atom tip   112
myosin   198

n-alkanethiolate   117
nanopore experiment   169

naphthalene   54
near-field-induced Raman scattering   256
near-field optic   224
near-field optical chemical sensor   237
near-field scanning optical microscopy   187, 224
near-field scanning optical spectroscopy   232
near-field spectroscopy   230
negative differential resistance   169, 174
non-contact AFM   122
non-linear optical effect   194
non-linear Raman effect   248
non-radiative rate   185
non-specific interaction   135
normalized tunneling current   23
numerical aperture   188

oligo(phenylene ethynylene)   165
oligo(phenylene vinylene)   165
O–O stretch mode   42
optical potential well   132
optical tweezers   131, 144
osmotically pressurized capsule   131
overlayer structure   42
overstreching transition   144
oxidative dehydration   105
oxidative reaction   79
oxygen vacancy   99

passive optical probe   226
peak broadening   13
pentacene   189
permanent dipole moment   11, 35
persistence length   144
perylene tetra-carboxylic-di-imide   58
phase imaging   120
phase separation   114
phosphorescent   183
photobleaching   184
photo-induced motion   117
photonic local density of state   231
photooxidation process   184
physisorbed   44
pivoting movement   120
point contact   25
Poisson statistics method   134
polarizability   34
polaron hopping mechanism   178
poly(ethylene-glycol)   154
polysaccharide   128, 152
porphyrin   48
position-sensitive detector   109
power spectrum   110
pre-exponential factor   159
preferential adsorption   52, 54
probe–surface interaction   127
probe wave function   18, 20
proteolysis process   117
pulling rate   148
pump-probe method   193
pyridine   86
pyridinecarboxylic acid   102

Q-factor   125

R,R-tartaric acid   91
radial breathing mode   256
radiative lifetime   207
radiative rate   185
radio-frequency   67
Raman spectroscopy   241
Raman-type spectra   9
random-walk behavior   230
random walk model   211
Rayleigh regime   132
reactive ion etching   227
recoiling process   217
rectification behavior   48
rectification effect   167
reduced-row model   99
reduction agent   185
replication activity   146
resonance frequency   111
resonant excitation   249
resonant inelastic electron tunneling process   79
resonant tunneling effect   14
rest atom   95
reverse imaging   123
reversible transfer   76
rhodamine   197
rotaxane   179
Rouse dynamic   218
rupture length   128

sample wave function   17

saturated hydrocarbon 34
scanner 15, 107
scanning confocal Raman
    microscope 250
scanning tunneling microscopy 15
secondary atomic defect 95
selective adsorption behavior 62
self-assembled monolayer 117
self-avoiding interaction 146
self-promoted autocatalytic
    process 93
semi-active light source 226
separation effect 217
Shpol'skii matrix 191
Si radical 67
signal-to-noise ratio 186
single atom tip 112
single-bond force 134
singlet state 183
site-blocking effect 92
Smoluchowski effect 45
spectral diffusion 208
spectral fluctuation 234, 244, 250
spool-like model 220
spring constant 110
static molecular dipole 13
static Stark effect 235
static tunneling model 5
stearic acetate 165
steric effect 54
sticking coefficient 93
stick-slip motion 137
Stokes line 241
strand-exchange reaction 117
stray magnetic field 125
streptavidin 140
submolecular feature 47
supercoiled DNA 148
superstructure 45
surface force apparatus 131

tapping mode 118
terrylene 190
tether molecule 151
thermal activation 72
thermal desorption 96
thermal dissociation 79
thermal noise spectrum 111

thymine 123
time-correlated single photon
    counting 187
tip-enhanced-Raman-scattering 253
tip-induced dissociation effect 45
tip-induced fluorescence 213
tip-induced movement 72
titanium dioxide surface 99
titin 148
toroidal structure 219
torsion force constant 117
total internal excitation 196
total-internal-reflection 185
$trans$-2-butene(C_4H_8) 86
transition Hamiltonian method 7, 17
transition law 10
transition matrix element 7
transmembrane channeling
    process 216
trapping effect 60
tridodecyl amine 34, 64
trimesic acid 57, 89
triplet state 183
tuning fork 229
tunneling-current-stimulated
    migration 72
tunneling spectroscopy 16
two-level system 208

under-potential-deposited
    monolayer 115
untilted dimer 93
unzipping rate 217

vacancy dynamic 82
vanadyl phthalocyanine 47
vibrational characteristic 11
vibrational dipole moment 13
vibrational excitation 84
vibrational heating 75

warm-like-chain model 146
weighted-quadratic-sum filter 196
Wentzel-Kramers-Brillouin 6

xenon 75
xenon-Cu(II) etioporphyrin I 51

zero-phonon line 192